Structural Systems

Henry J. Cowan
Head of the Department of Architectural Science
University of Sydney, Sydney, N.S.W.

Forrest Wilson
Professor of Architecture
Catholic University of America, Washington, D.C.

VNR VAN NOSTRAND REINHOLD COMPANY
New York Cincinnati Toronto London Melbourne

Copyright © 1981 by Van Nostrand Reinhold Company
Library of Congress Catalog Card Number 80-10698
ISBN 0-442-21713-7 (paper)
ISBN 0-442-21714-5 (cloth)

All rights reserved. No part of this work covered by the copyright hereon may be reproduced or used in any form or by any means—graphic, electronic, or mechanical, including photocopying, recording, taping, or information storage and retrieval systems—without written permission of the publisher.

Printed in the United States of America

Designed by Rose Delia Vasquez

Published by Van Nostrand Reinhold Company
135 West 50th Street
New York, NY 10020

Van Nostrand Reinhold Limited
1410 Birchmount Road
Scarborough, Ontario M1P 2E7, Canada

Van Nostrand Reinhold Australia Pty. Ltd.
17 Queen Street
Mitcham, Victoria 3132, Australia

Van Nostrand Reinhold Company Limited
Molly Millars Lane
Wokingham, Berkshire, England

16 15 14 13 12 11 10 9 8 7 6 5 4 3 2 1

Library of Congress Cataloging in Publication Data
Cowan, Henry J
  Structural systems.

  Bibliography: p.
  Includes index.
  1. Structures, Theory of. 2. Buildings.
3. Architecture. I. Wilson, Forrest, 1918-
joint author. II. Title.
TA645.C68   624.1'7   80-10698
ISBN 0-442-21714-5
ISBN 0-442-21713-7 pbk.

To
THISBE
and her Godfather

Contents

Preface 7
Foreword for Instructors 9

PART 1. PRINCIPLES 10

Chapter 1. Introduction 11
1.1 The Scope of This Book 12
1.2 An Historical Note 12
1.3 Why Do We Need Safety Factors? 15
1.4 The Influence of Structure on the Design of the Building 16
1.5 Preview of the Following Chapters 19

Chapter 2. Loads, Forces, and Equilibrium 20
2.1 Loads 21
2.2 Static Loads 21
2.3 Dynamic Loads 23
2.4 Fire Loads 26
2.5 The Effect of Temperature, Moisture, and Foundation Settlement 27
2.6 The Forces Produced by the Loads, and the Conditions of Equilibrium 28
2.7 How Traditional Structures Fail, and How Steel and Reinforced Concrete Structures Fail 30

Chapter 3. Structural Materials and Safety Factors 31
3.1 Elastic Deformation 32
3.2 Ductility 33
3.3 Brittleness 33
3.4 The Cure for Brittleness 35
3.5 Which is the Right Structural Material? 36
3.6 The Choice of the Structural Material for Domestic Buildings 40
3.7 The Choice of the Structural Material for Small and Medium-Sized Commercial and Industrial Buildings 42
3.8 The Choice of the Structural Material for Tall Buildings 42
3.9 The Choice of the Structural Material for Long-Span Buildings 43
3.10 Safety Factors 43
3.11 Serviceability 46

Chapter 4. The Problem of Span 48
4.1 A Brief History of Span 49
4.2 The Concept of Moment 51
4.3 Composition and Resolution of Forces 55
4.4 The Conditions of Equilibrium 56
4.5 Bending Moments and Shear Forces: Why Do Structural Engineers Talk So Much About Them? 57
4.6 A Bevy of Beams 58
4.7 Curved Structures and Trusses Use Less Material for Horizontal Spans 65
4.8 A Note on Vertical Spans 69

PART 2. PRELIMINARY STRUCTURAL DESIGN 70

Chapter 5. Structural Members 71
5.1 Structural Members and Structural Assemblies 72
5.2 Tension Members 73
5.3 Compression Members and Buckling 75
5.4 Lintels and Beams: Simple in Appearance but Complex in Theory 76
5.5 Reinforced Concrete: Concrete with Flexural Strength 79
5.6 Prestressed Concrete: Concrete Without Cracks 83
5.7 Jointing of Structural Members 86
5.8 Summary of Results for Ties, Columns, Beams, and Slabs 88
5.9 A Note on All Problems 89
5.10 Problems for Chapter 5 89
5.11 Problems on Steel and Timber Beams 90
5.12 Problems on Reinforced Concrete Slabs and Beams 95
5.13 Problems on Ties and Columns 98

Chapter 6. Some Secondary Problems in the Design of Beams 101
6.1 Shear 102
6.2 Torsion 104
6.3 Stress Trajectories 105
6.4 Stress Concentrations 106
6.5 Experimental Stress Analysis 106
6.6 Problems for Chapter 6 107

Chapter 7. Structural Assemblies 110
7.1 Pinned and Rigid Joints 111
7.2 Statically Determinate and Statically Indeterminate Structures 112
7.3 Roof Trusses: An Ancient Structural System Still Widely Used 114
7.4 Parallel-Chord Trusses with Pin Joints: Beams with Big Holes or an Assembly of Tension and Compression Members 116
7.5 Parallel-Chord Trusses with Rigid Joints, and Shear Walls 117
7.6 Portal Frames: Simple Frames with Three Pins and More Complicated Rigid Frames 118
7.7 Arches 123
7.8 Problems for Chapter 7 129
7.9 Problems on Trusses 129
7.10 Problems on Portal Frames and Arches 131

Chapter 8. Multistory Buildings and Tall Buildings 137
8.1 A Brief Historical Note 138
8.2 The Design of Low-Rise Steel Frames for Vertical Loads 138
8.3 The Design of Low-Rise Reinforced Concrete Frames for Vertical Loads 139
8.4 The Effect of Horizontal Loads on Simple Frames 139
8.5 The Vertical Structure in High-Rise and Low-Rise Buildings 142
8.6 The Floor Structure 147
8.7 The Rigid Frame Theory 147
8.8 Energy Dissipation Systems for Tall Buildings 149
8.9 Back to the Loadbearing Wall 149
8.10 Factors Limiting the Height of Buildings 152
8.11 Problems for Chapter 8 154

Chapter 9. Foundations 164
9.1 Foundation Materials 165
9.2 Soil Pressure 166
9.3 Footings for Walls and Single Columns 167
9.4 Combined Column Footings, Raft Foundations, and Piled Foundations 169
9.5 Retaining Walls and Basement Walls 172
9.6 Stability of Tall Buildings 173
9.7 Foundations for Long-Span Buildings 175
9.8 Problems for Chapter 9 176

Chapter 10. Curved Structures and Long-Span Buildings 181
10.1 The Simple Theory of Vaults and Domes 182
10.2 Problems in the Construction of Masonry Vaults and Domes 183
10.3 Reinforced Concrete Shell Domes and Cylindrical Vaults 187
10.4 Schwedler Domes, Geodesic Domes, Lattice Vaults, and Lamella Roofs 192
10.5 The Geometry of Curved Surfaces and the General Theory of Shells 196
10.6 Hypars and other Nontraditional Surfaces 198
10.7 Folded-Plate Roofs 202
10.8 Side-Lit and Top-Lit Roof Structures 207
10.9 Prestressed Shells 209
10.10 Suspension Structures: Why We Do Not Use Them More Often 211
10.11 Pneumatic Structures: Span Without Limit 214
10.12 Planar Space Frames: Three-Dimensional Trusses 216
10.13 A Postscript 217
10.14 Problems for Chapter 10 218

Chapter 11. The Structure and the Environment 231
11.1 The Cost of the Structural System and Its Interaction with the Environmental System Prior to the Nineteenth Century 232
11.2 The Cost of the Structure is No Longer the Largest Part of the Cost of a Building 233
11.3 The Energy Crisis 234
11.4 Solar Energy 234
11.5 Thermal Inertia 235
11.6 Thermal Insulation 237
11.7 Sunshading and Natural Lighting 237
11.8 The Structure and Sound Insulation 238
11.9 The Structure and the Building Services 239
11.10 Coordination of Structural and Environmental Design 239

Suggestions for Further Reading 241

Glossary 244

Notation 249

Units of Measurement 251

Tables 253

Index 254

Preface

The spans of buildings are much smaller than those of bridges, and therefore few structural systems in architecture are actually impossible to build; but some are unnecessarily complicated and expensive. This book offers a guide to preliminary structural design and a "shopping list" of standard solutions. It endeavors to describe structural forms and explain their mechanics. This is done with a minimum of mathematics; the reader needs to know only elementary arithmetic and a little trigonometry.

The introduction of computer-based methods, such as the matrix-displacement method and the finite-element method, has greatly increased the scope of structural design. This book briefly explains their basis without going into details; but it concentrates on the behavior of structures, which has not altered, although they are designed differently. It is still necessary to transfer the loads acting on the structure to the ground without overstressing the structural material in the process.

We have included a number of problems, so that readers may acquire a skill in determining the *approximate* size of the principal structural members, to assure themselves that their proposed structural system is realistic and reasonably economical. These are supplementary to the descriptive text. The book is complete without them, and omitting them causes no loss of continuity. However, we advise readers to attempt the problems, because they give realism to the presentation.

All dimensions and stresses are given both in metric SI units and in the customary American (formerly British) units. We also solve the problems in both systems of measurement, using roman type for metric units and italics for the American units. A reader using metric units should omit the lines in italics, and vice versa.

The historical sequence is briefly traced throughout the book. Traditional methods, such as loadbearing walls, masonry arches, and masonry domes, are discussed because they are still used at least in some parts of the world.

Concrete and steel domes are described in detail; at the present time the three longest-spanning architectural structures are domes. Other types of shell, which have waned in popularity since the 1960s, are treated in less detail. New structural types, such as cable roofs and air-supported membranes, are taking their place.

Rectangular frames remain the bread-and-butter of architectural structures. This book therefore gives due place to them; not merely to the superstructure but also to the foundations. Although tall buildings have not become much taller since the 1930s, their design has become more sophisticated and economical. The new methods are explained, together with the simpler methods that remain valid for smaller buildings.

We hope that the strong emphasis on graphical presentation will help the reader understand a subject that many people find abstract and difficult.

We are indebted to Mrs. Hilda Mioche for typing the manuscript; to Mr. H. Milton of the National Bureau of Standards, Mr. R. O. Disque of the American Institute of Steel Construction, Mr. H. A. Krentz of the Canadian Institute of Steel Construction, and to Mr. R. F. DeGrace of the Canadian Wood Council for advice on American metrication; and to the following for reading the whole or part of the text and commenting critically on it: Dr. R. M. Aynsley, Dr. K. Dunstan, Mr. D. Epstein, Prof. J. S. Gero, Prof. H. Harrison, Dr. Valerie Havyatt, Mr. P. Healy, Mr. W. Julian, Mr. A. Milston, Mr. O. P. Phillips, Prof. P. R. Smith, Mr. E. R. Taylor, Dr. P. Towson, and Mr. A. Wargon.

Foreword for Instructors

As mentioned in the Preface, this book can be used as a descriptive text without the problems; it can be used, as it has been written, with the problems, which give approximate design solutions; or the subject can be supplemented by more rigorous examples from standard textbooks on architectural structures. Any book listed in Section 12.3 of the Suggestions for Further Reading can be used; we recommend Cowan, *Architectural Structures,* or Salvadori and Levy, *Structural Design in Architecture*.

A formal course on architectural structures, whether it takes the form of lectures, tutorials, or seminars, is greatly assisted by visual demonstrations. We have not mentioned any in the text, because we wanted to keep it as short as possible and because the type of demonstration depends on the facilities available to the instructor. However, a wide range of visual aids, ranging from home-made models to sophisticated demonstrations, are described in Ref. 6.3, *Building Science Laboratory Manual* (which includes demonstrations for lectures and seminars).

We have assumed no prior knowledge of mechanics, and forces and moments are briefly explained in Chapters 3 and 4. However, these are difficult concepts, and students who are able to draw triangles of forces and take moments with ease will derive more benefit from this book. We therefore recommend that students be encouraged to take an introductory course in statics or mechanics beforehand. This is particularly desirable if the problems are to be included in the course. The problems use only simple arithmetic, but they assume that the student is capable of resolving forces and taking moments.

We have used metric units first, with customary American units in brackets, because metric units will be introduced in the United States during the lifetime of this book. In every other respect the units have been treated as equal, and the book can be used with American units.

Some of the problems require the use of standard steel sections. We have employed American standard sections in all problems using American units, and their metric equivalents in accordance with Canadian standards in the metric problems. Tables of these standard steel sections are given at the end of this book. Instructors in architecture schools elsewhere can easily convert to their national standard sections. American and Canadian standard sections have also been used in problems on timber and reinforced concrete.

Unfortunately, the various national standards institutes failed to grasp the opportunity to create a single international system of measurements, and we had to make a choice between MPa and N/mm^2 as units of stress. We have chosen the former, because it has been adopted by the United States, Canada, Australia, New Zealand, and South Africa, and we apologize to British schools for any inconvenience this choice may cause.

PART 1 PRINCIPLES

Chapter 1 Introduction

I am but a gatherer and disposer of other men's stuff. (Sir Henry Wotton, in his Preface to The Elements of Architecture, *first published in 1624.)*

This chapter is a preview of the subject matter of the subsequent chapters.

1.1 The Scope of This Book

Structural systems can be explained without numerical calculations, and this is one of the objects of this book. You can therefore omit the problems at the ends of chapters altogether if you do not wish to use arithmetic.

For many buildings, however, approximate structural sizes can be determined by quite elementary and brief calculations, and the problems show how these solutions are obtained. More elaborate calculations are needed for precise structural sizes, and there are many books which describe how they are performed (for example, Refs. 1.1, 1.2, and 1.3).

We will first consider the loads a building must support, the forces they cause, and the conditions necessary to ensure that the building does not collapse or deform too much under the action of these forces. Next we will look at the materials used in structures and their properties. We are then ready to examine the problems posed by span and by height and how they can be solved by individual structural members and by structural assemblies. Finally, we will discuss the interaction of structural and environmental design.

1.2 An Historical Note

Prior to the nineteenth century the sizes of structural members were determined by empirical rules. For example, one such rule states that for a particular purpose and material the depth of a beam must be at least one tenth of its span. These rules, based on long practical experience, were (and still are) quite satisfactory for buildings that have small spans and a small height. Theory is needed to determine the structural sizes for buildings with great height or long spans, and these became possible only after structural mechanics had been perfected.

Thus the longest-spanning building prior to the mid-nineteenth century was the Pantheon in Rome (Fig. 1.2.1), which was built as early as 123 A.D.; it has a span of 44 m (143 ft). The tallest "building" prior to the mid-nineteenth century was the Great Pyramid at Gizeh in Egypt (Fig. 1.2.2), built about 2580 B.C.; it has a height of 147 m (481 ft). A few medieval spires were slightly taller, but these collapsed long ago.*

---

*If you do not consider the Great Pyramid to be a building, the record belongs to Strasbourg Cathedral in France, whose spire, completed in 1439, has a height of 142 m (465 ft).

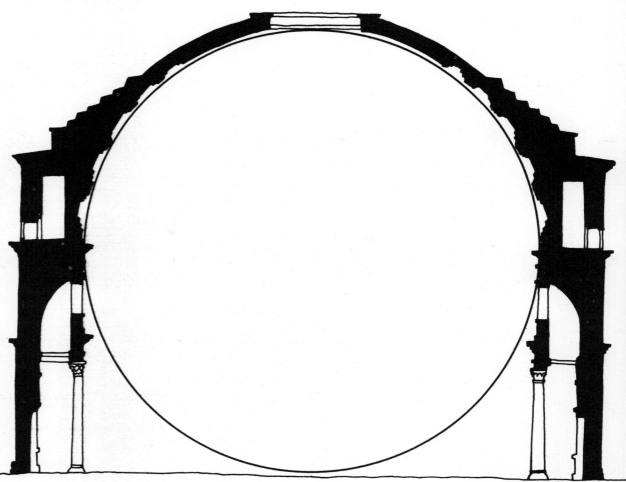

1.2.1. The Pantheon, built in Rome in 123 A.D., with a span of 44 m (143 ft); it was the longest-spanning building until the mid-nineteenth century.

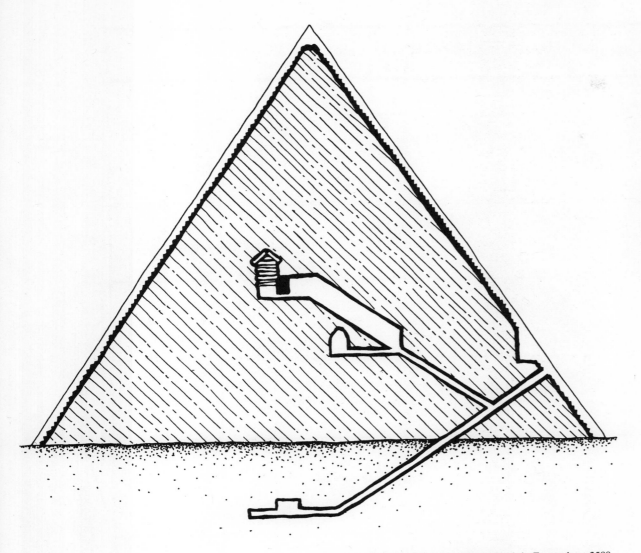

1.2.2. The Great Pyramid, built at Gizeh in Egypt about 2580 B.C., with a height of 147 m (481 ft); it was the tallest building until the mid-nineteenth century.

At the time of writing the longest span in a building is 207 m (680 ft), and the tallest building is 442 m (1450 ft) high (Fig. 2.3.6).

Theory is important not merely because it enables us to build to great heights or over large spans. Buildings designed in accordance with the theory of structures use far less material than those designed in similar materials by empirical rules; they are consequently much cheaper.

The domes of imperial Rome were built predominantly with concrete. Most of the long-spanning structures of the Middle Ages and the Renaissance were built from natural stone or brick. All these materials have strength similar to that of modern concrete. We can therefore compare concrete and masonry structures built over the past two thousand years and this is done in Table 1.1.

Table 1.1. Reduction in the amount of material used in concrete and masonry domes.

Comparative sizes	Year of completion A.D.	Name and construction of dome	Span in meters (feet)	Average thickness of dome or combined thickness of double dome in meters (inches)	Ratio of span to thickness
	123	Pantheon, Rome. A solid concrete dome, faced with masonry (Section 10.2).	44 (143)	4 (156)	11
	1434	S. Maria del Fiore, Florence. A double dome of brick and stone (Section 10.2).	42 (138)	2 (80)	21
	1710	St. Paul's Cathedral, London. A brick dome, surmounted by a brick cone (Section 7.7).	33 (109)	0.9 (36)	36
	1927	Planetarium, Jena, East Germany. The first reinforced concrete dome designed by the membrane theory (Section 10.3).	25 (82)	0.060 (2⅜)	420
	1953	Schwarzwaldhalle, Karlsruhe, West Germany. The first prestressed concrete saddle shell (Section 10.9).	73 (239)	0.058 (2¼)	1250
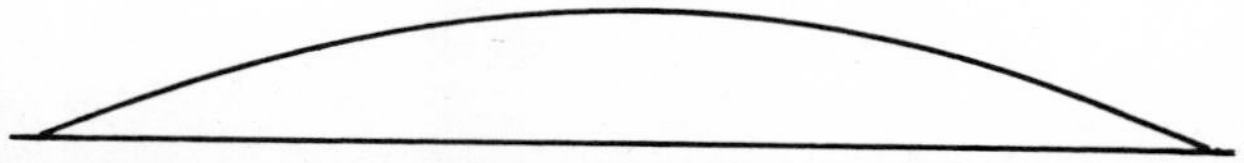	1958	C N I T Exhibition Hall, Paris. A double, reinforced concrete shell (Section 10.3).	206 (676)	0.120 (4¾)	1700

A hen's egg is also comparable to a concrete dome. Its shell is shaped like a dome, and the lime from which it is formed behaves like a weak concrete. An egg is much stronger than is generally thought. If carefully loaded, it can support 50 to 80 kg (110 to 170 lb) (Fig. 1.2.3). The average ratio of its diameter (that is, span) to the thickness of its shell is about 100.

As Table 1.1 shows, domes constructed by traditional methods never became as thin, comparatively, as a hen's egg, although with experience they became more economical. The dome of St. Paul's (eighteenth century) is thinner than that of S. Maria del Fiore (fifteenth century), and this in turn is thinner than that of the Pantheon (second century). The economy of modern concrete shells has also improved as the theory has become more accurate; all are, comparatively, thinner than a hen's egg.

There has been a similar reduction in the thickness of the walls of tall buildings. Surviving medieval towers are generally about 2 m (6 ft) thick at the base. A fraction of this thickness suffices for the much taller buildings of the twentieth century (Section 8.9).

Several books describe the structures of ancient Egypt, ancient Rome, and medieval and Renaissance Europe (for example, Refs. 1.4, 1.5, and 1.6). There was an enormous disparity of income between those who commissioned the great buildings at that time and those who did the actual work. This disparity has been greatly reduced since the nineteenth century and particularly since World War II, but the higher wages have also raised the cost of building and therefore the cost of the structure. During the nineteenth century it was still possible to revive the methods of construction used by the Gothic master builders, but today the cost would be prohibitive. The use of the theory of structures in designing large buildings has become an economic necessity.

1.3 Why Do We Need Safety Factors?

Building regulations have been used for several thousand years, but most early codes were primitive. The oldest code known to us is written on a stone column now in the Louvre (Fig. 1.3.1). It was promulgated by Hammurabi, king of Babylon, about 3800 years ago. One of its clauses states (Fig. 1.3.1):

If a builder has built a house for a man and his work is not strong, and if the house he has built falls and kills the householder, the builder shall be slain.

1.2.3. The shell of a hen's egg can support 50 to 80 kg (110 to 170 lb). It is, comparatively speaking, thinner than any classical masonry dome and thicker than any modern reinforced concrete shell.

If a builder has built a house . . . if the house he has built falls and kills the householder, the builder shall be slain.

1.3.1. The Code of Hammurabi, king of Babylon, promulgated about 3800 years ago.

This type of code, if strictly enforced, is likely to result not merely in structural safety but in excessively heavy and unadventurous structures, since the builder will make absolutely certain that there will be no structural collapse.

Most of the structures of the ancient world used a great amount of material and were by modern standards oversized. This applies even to ancient Rome, which created the greatest structures prior to the nineteenth century (Ref. 1.4).

In medieval Europe, on the other hand, the collapse of a big roof structure was not uncommon, as we know from the chronicles and from the evidence of the buildings themselves (Ref. 1.4). The Roman Empire had been split up into many smaller kingdoms, bishoprics, and city states, and slavery had been abolished. The labor available for building was therefore limited, and material was used sparingly. Furthermore, the strong religious feelings of the time and a lack of knowledge of structural mechanics encouraged an unwarranted belief in divine providence for the structure of cathedrals. Some Gothic stone columns and flying buttresses surviving from the Middle Ages are so slender that modern building codes would not permit their construction even in reinforced concrete, which is much stronger.

Modern structural design codes are intended to prevent failure while encouraging economy of material. They specify a margin of safety to allow for minor flaws in the materials (Section 3.10), inaccuracies in the setting out of the dimensions, and other unavoidable errors. These margins are small for aircraft structures, which are regularly inspected and have structural members replaced after a certain period of service. Buildings have a longer life than airplanes, and saving weight is not as important as saving money; their structural members are designed to last for the lifetime of the building, which may vary from twenty years for a temporary house to hundreds of years for an important public edifice. This requirement necessitates a higher margin of safety, but the margin has been progressively reduced because of advances in knowledge since theoretically based structural design was first introduced in the nineteenth century. Present-day building codes require structural safety factors which vary from 1½ (that is, a 50% excess margin) to 2½ (a 150% excess margin), depending on the material and the method of construction.

Structural mechanics enables us to combine the adventurous spirit of the Gothic cathedrals with the safety of Hammurabi's temples. It has been one of the success stories of modern technology. Structural failures are very rare, and a comparison between structures erected anywhere in the world two hundred years ago and today clearly shows the amount of material saved by theoretically based design (see Table 1.1 again).

1.4 The Influence of Structure on the Design of the Building

Structure has little significance for small spans and small heights. It needs to be considered in the design of buildings of medium height and with medium spans. It dominates the design when spans are very long or heights very great.

Any craftsman can build a canopy spanning 2 m (6 ft) or a platform 2 m high by using common sense (Fig. 1.4.1). There are no real structural problems in the design of buildings on the domestic scale, although the structural members have to be sized. Structure is always important when the building is more than seven stories high or has spans exceeding 15 m (50 ft).

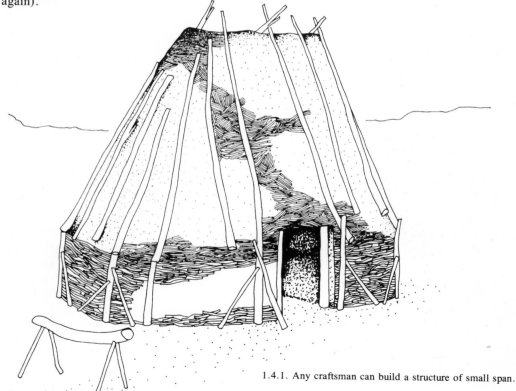

1.4.1. Any craftsman can build a structure of small span.

A hundred years ago both height and span were still limited by technical considerations. The tallest building was Rouen Cathedral in France, whose spire, completed in 1876, was 148 m (485 ft) high (Ref. 1.7). The longest interior span, inside St. Pancras Station in London, built in 1866, was 74 m (244 ft). At present the tallest building is the Sears Tower in Chicago, which is 442 m or 1450 ft tall (Fig. 1.4.2.a), and the Louisiana Superdome in New Orleans has the longest interior span, 207 m or 680 ft (Fig. 1.4.2.b).

There are no architectural structures that are actually impossible to build, although a building 300 stories high or a column-free auditorium holding a quarter of a million people would present many problems and would be very expensive. However, the question must be asked whether these buildings would serve any useful purpose; for example, would a quarter of a million people in a single auditorium enjoy the spectacle more than if they stayed home and watched it on their own color television sets?

At present the size of structures is limited mainly by transportation problems. A very tall building (whether it contained offices or apartments) or a large auditorium would attract large numbers of people; greater difficulties would be posed by moving many of them simultaneously to and from the building and by parking the automobiles (in which some of them would travel) than by the design of the structure.

While structure is no longer a limiting factor, it is a very important part of the design of a building. The structure has a decisive influence on the plan. The location of the vertical supports (columns, loadbearing walls, service core) determines the layout of the floors through which they pass. The structure has almost as great a significance for the elevation. The size and spacing of the columns influence the design of the facade, and the columns themselves are often the most visually prominent feature in the appearance of the building.

Basic structural decisions are of such importance

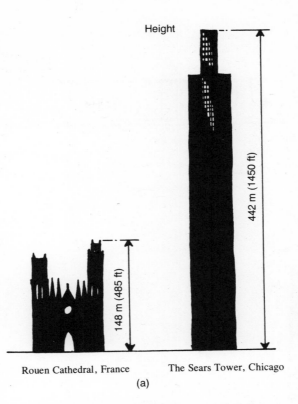

1.4.2. (a) The tallest buildings 100 years ago and today.
(b) The longest interior spans 100 years ago and today.

that they should be made at an early stage in the design. They should never be left to the structural consultant after the rest of the building has been designed.

Some people consider that a correctly designed structure is essential for a beautiful building, and some hold that a correctly designed structure automatically produces beauty. Such a structure is lighter, and to that extent may be more elegant, than one that uses material less efficiently. Connoisseurs with a knowledge of structural systems recognize a good solution and presumably derive satisfaction from the skill evident in its design. However, beauty does not automatically flow from a correct structure. This is evident if we look at some long-span bridges, whose appearance is determined by their structure to a greater extent than for buildings (Fig. 1.4.3).

Structures by great engineers, such as Thomas Telford and Gustave Eiffel (Fig. 1.4.4) in the nineteenth century and Eduardo Torroja and Pier Luigi Nervi in the twentieth, are much admired, but this is evidence of the esthetic sense rather than the mechanical genius of their designers. Correct structural design does not guarantee beauty; its value is rather that it saves a great deal of material, and this is just as important.

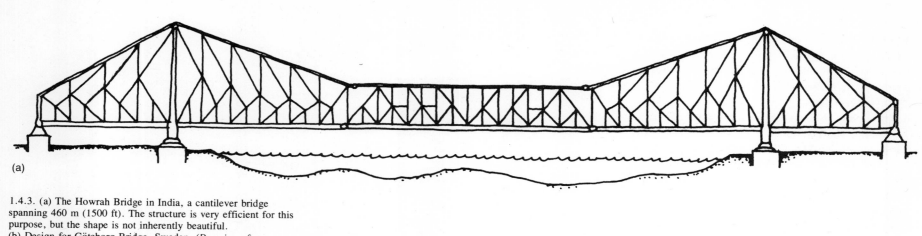

1.4.3. (a) The Howrah Bridge in India, a cantilever bridge spanning 460 m (1500 ft). The structure is very efficient for this purpose, but the shape is not inherently beautiful.
(b) Design for Göteborg Bridge, Sweden. (*Drawing after Ricardo Morandi.*)

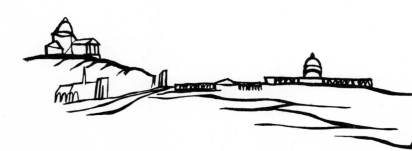

1.4.4. The Eiffel Tower, built by Gustave Eiffel for the International Exhibition of 1889; its height is 322 m (1056 ft.) and it has become a landmark of Paris. (*Drawing after Le Corbusier.*)

1.5 Preview of the Following Chapters

Chapter 2 describes the loads which a building must carry, the forces imposed by these loads, and the reactions of the building to these loads. These must be in equilibrium; otherwise, the building will collapse.

Chapter 3 sets out the criteria for structural design. The building must be safe; that is, there must be a sufficient margin between the *ultimate* loads that would cause it to collapse and the *service loads* actually imposed on it. It must also be serviceable; that is, the *service loads* must neither cause cracks in the structure nor cause it to deflect so much that cracks might occur in brittle finishes. The properties of the various structural materials and their relative cost and suitability are then discussed.

Chapter 4 deals with span, the basic problem of structural design. All buildings have a horizontal span in relation to vertical loads, and tall buildings have a vertical span in relation to horizontal loads (wind and earthquakes). Design, both of long-span roofs and of tall buildings, is concerned with finding a structural system that resists the bending moments due to these loads as efficiently as possible.

Chapter 5 examines in turn each of the principal structural members (tension members, columns, beams, and slabs) in each of the principal structural materials (timber, structural steel, reinforced concrete, and prestressed concrete). Loadbearing walls are covered in Chapter 8.

Chapter 6 contains some more advanced problems in the design of structural members, namely, shear, torsion, the maximum stresses at any point in a structure, and stress concentrations. This chapter could be omitted on a first reading.

Chapter 7 is the first of three chapters on assemblies of structural members. It covers single-story structures of limited span, such as roof trusses, trusses with parallel chords (both with and without diagonal members), portal frames, and arches.

Chapter 8 considers in some detail multistory buildings of medium height, both for vertical loading and for horizontal loading due to wind and earthquakes. The design of tall buildings is described in more general terms, and the structural systems in current use are briefly explained.

Chapter 9 briefly examines the foundations suitable for various buildings.

Chapter 10 deals with concrete and masonry domes (both ancient and modern), geodesic domes, cylindrical shells, hyperbolic paraboloid (hypar) shells, prestressed shells, shells with north lights or lanterns, lamella roofs, folded plate roofs, suspension structures, pneumatic membrane structures, and space frames. Special consideration is given to the suitability of the various curved roofs to long spans.

Chapter 11 is a postscript examining the interrelation between structural and environmental design and the extent to which the structure can be used to improve the interior environment of the building.

References

1.1 ARCHITECTURAL STRUCTURES by Henry J. Cowan. Elsevier, New York, 1976. 448 pp.

1.2 SIMPLIFIED ENGINEERING FOR ARCHITECTS AND BUILDERS by Harry Parker. Wiley, New York, 1975. 411 pp.

1.3 STRUCTURAL MECHANICS by W. Morgan and D.T. Williams. Pitman, London, 1958. 429 pp.

1.4 THE MASTERBUILDERS by Henry J. Cowan. Wiley, New York, 1977. 299 pp.

1.5 A HISTORY OF CIVIL ENGINEERING by Hans Straub. Leonard Hill, London, 1960. 258 pp.

1.6 THE ANCIENT ENGINEERS by L. Sprague de Camp. Doubleday, New York, 1963. 408 pp.

1.7 THE GUINNESS BOOK OF STRUCTURES by John H. Stephens. Guinness Superlatives, London, 1976. 288 pp.

Chapter 2 Loads, Forces, and Equilibrium

"What is the use of a book," thought Alice,
"without pictures?" (Lewis Carroll)

We consider the loads a building must support and the forces and reactions produced by those loads. Under normal conditions the forces and the reactions are in equilibrium. When a part of the building fails, the equilibrium is disturbed.

2.1 Loads

The loads a building must support may be divided into three groups:
1. Stationary (or static) loads due to the weight of the building, the weight of the contents, and the weight of the people within the building.
2. Moving (or dynamic) loads caused by wind and earthquakes.
3. Equivalent loads caused by changes in the temperature, by changes in the moisture content of the building materials, and by the settlement of the foundations.

These types of loads can be further subdivided, as described in the following sections.

2.2 Static Loads

Static loads are of two kinds: those which always act on the structure and those which may or may not act on the structure. The first are called *dead loads*, and the second are called *live loads*.

The dead load consists of the weight of the part of the building that is carried by the structural member under consideration, and it includes the weight of that structural member. Once the building has been designed, it is a simple matter to calculate the volumes of its component parts and multiply each by the unit weight of the material from which it is to be built. The unit weights of common building materials are given in tables that are included in building codes or standard specifications.

In practice this computation is not quite so simple, because at the design stage the sizes of many parts of the building have not yet been determined. We do not know the dimensions of the structural member we are designing, because the determination of its dimensions is the object of the design. However, since most buildings are of a standard type, we can guess the dimensions of their structural members from past experience for the purpose of determining the dead load. It is wise to guess a little too high if in doubt, because we then err on the safe side. As the weight of the structural member is often only a small part of the total load it carries, a small error is not significant. If the building is of an unusual type, our guess may be greatly in error, and in that case we must repeat the calculation until the dimensions assumed are sufficiently close to the dimensions calculated.

Partitions that are permanently fixed are included in the dead load, but movable partitions are part of the live load. The live load also includes furniture and equipment in the building and the people who live or work in it (Fig. 2.2.1). These people move around, and we must consider where they are likely to congregate and cause heavy load concentrations. This may happen in department stores during a sale, and in any building in a corridor which is used for rush hour traffic. All buildings have special routes for evacuation during a fire, and these may carry heavy traffic if a fire occurs, because of the crowding of people trying to leave the building (Figs. 2.2.2 and 2.2.3).

2.2.1(a). Dead loads and live loads.

It would be too laborious to weigh all the furniture and fittings in a building and to estimate for each building the number of people who may congregate in every part of it. Building codes specify loads to be assumed for each type of occupancy, for example, normal office buildings, library reading rooms, library stacks, law courts, hospital wards, hospital operating theatres, hotel bedrooms, and restaurants. These loads are based on surveys during which the furniture and equipment in representative buildings was weighed and the people counted (Ref. 2.1).

The live loads specified in building codes are of two kinds: an equivalent uniformly distributed load, which is specified in kilopascals (kPa) or in pounds per square foot (psf) and a concentrated load in kilonewtons or pounds* (Fig. 2.2.4). The concentrated load allows for heavy loads, such as tall bookshelves filled with books or heavy computer units.

---

*In American (the old British) units, mass and weight are numerically the same. Mass is most conveniently determined with a balance. If a man steps on a balance and registers 150 lb (pounds), then that is his mass. The force exerted on the man by the earth's gravitational attraction is expressed in pounds-force, and 1 lbf (pound-force) is the force that produces in a mass of 1 lb an acceleration g, where g is the acceleration of a body falling freely as a result of the earth's gravitational attraction. The value of g varies slightly in different locations; it has an average value of 32.2 ft/sec^2. Thus the man's weight is also 150 lbf. Most books on architecture use the term "pound" and its abbreviation "lb" both for "pound" and for "pound-force."

In the old metric units, used at present in most European countries, a mass of 1 kg (kilogram) has a weight of 1 kgf (kilogram-force). In the SI (Système International d'Unités) metric units, which are now in use in Great Britain, Australia, and Canada and are being introduced in the United States, mass and weight are not numerically equal. The newton (N) is the unit of force, and this is defined as the force that produces in a mass of 1 kg an acceleration of 1 m/sec^2. The average value of g in metric units is 9.807 m/sec^2, and therefore the weight of 1 kg is 9.807 N. If the man registers a mass of 70 kg on the balance, his weight is 686.5 N. Note that

$$1 \text{ kPa (kilopascal)} = 1 \text{ kN/m}^2 \text{ (kilonewton per square meter)}$$
$$1 \text{ kN} = 1000 \text{ N (newton)}$$

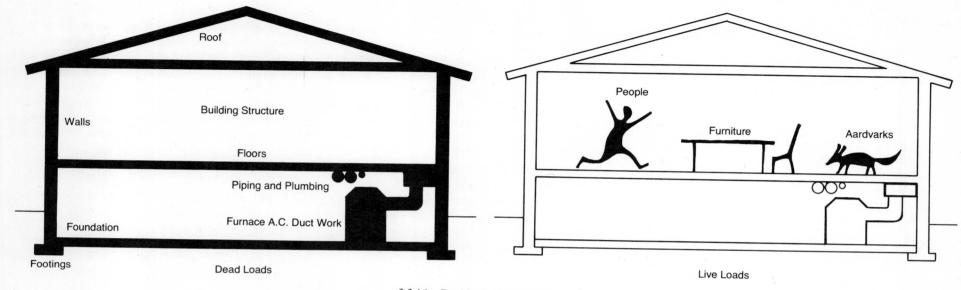

2.2.1(b). Dead loads and live loads.

2.2.2. Normal live load caused by people.

2.2.3. Live load caused by crowding in a building on fire.

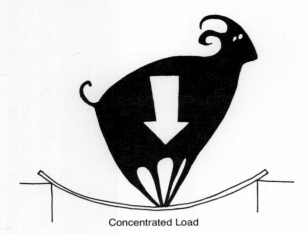

In cold climates snow may impose heavy live loads on roof structures. In tropical and subtropical regions no allowance for snow is required except in mountainous country.

Snow loads may be greatly increased by the effect of wind, which can cause snow to accumulate on low roofs adjacent to higher roofs, and behind roof monitors. In areas where high wind and heavy snow may be expected, it is best to avoid roof configurations that offer shelter for snow drifts.

Solar radiation can reduce snow loads by melting even when the air temperature does not rise above 0°C (32°F), if suitable drainage is provided. On the other hand, melting of the snow followed by subsequent freezing can produce an ice barrier at the edge of a sloping roof, which increases the depth of snow retained. This danger is increased by unsuitable heating of the roof space, that is, by heating the underside of the roof above the house but omitting to heat the lowest portion of the slope (Fig. 2.2.5).

2.3 Dynamic Loads

Wind loads (Fig. 2.3.1) are important even for low buildings if they are not sheltered by surrounding buildings, because the horizontal wind pressure can push a building over if it is not adequately *braced* (see the Glossary for a definition). When thin walls (for example, of galvanized steel sheeting) are used, it is necessary to provide diagonal wind bracing for the structural frame beneath the sheeting (Fig. 2.3.2).

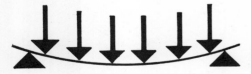

2.2.4. Concentrated load and uniformly distributed load.

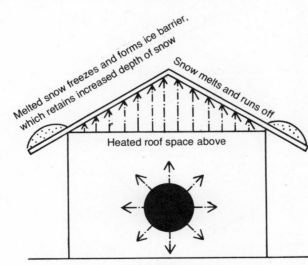

2.2.5. Melting of snow followed by subsequent freezing can produce an ice barrier. This is encouraged by heating the roof space without heating the lowest portion of the roof.

2.3.1. Wind load.

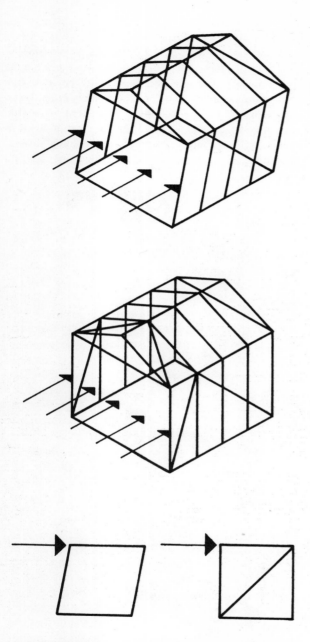

Substantial walls of brick or concrete within the structural frame perform the same function.

Wind produces both pressure and suction. The wind presses on the windward side of a building, but suction occurs on the other three walls, and also on a *flat roof* (Fig. 2.3.3). The wind produces suction not merely on the leeward side of a *sloping roof* (Fig. 2.3.4), but also on the windward side if the angle of the slope is less than 30°. In mountainous country, roofs of houses are sometimes weighed down with heavy stones to stop them from being lifted up by the high winds that occur at exposed positions.

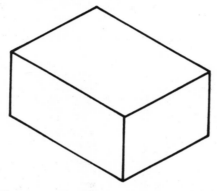

2.3.3. Wind pressure and suction on a building with a flat roof.

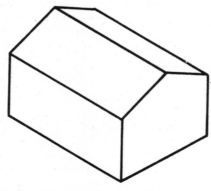

2.3.4. Wind pressure and suction on a building with a sloping roof. The windward roof surface is subject to pressure (instead of suction) when the angle of the roof slope becomes greater than 30°.

2.3.2. Diagonal bracing for a single-story steel-framed building with light cladding.

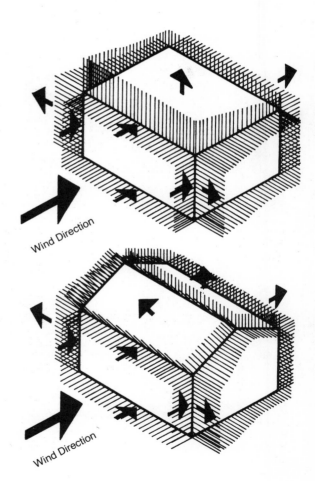

Wind loads are always important for tall buildings, which form a vertical *cantilever* (see Glossary) resisting the horizontal wind pressure on one side (the windward side) and horizontal suction on the other (leeward) side. The building behaves like a horizontal cantilevered beam resisting a vertical load (Fig. 2.3.5); for high-rise buildings the span of the cantilever is much greater than any horizontal span in a building (Fig. 2.3.6).

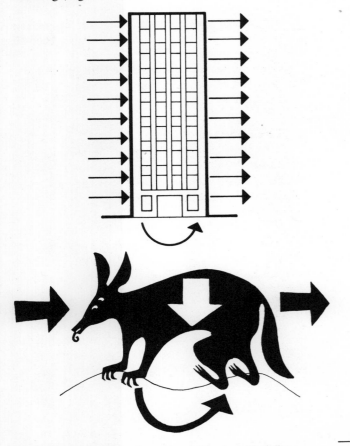

2.3.5. Wind pressure on the windward side of a tall building, plus suction on the leeward side, produces a bending moment, and the building behaves like a giant cantilever. The problem is precisely the same as for a horizontal cantilever carrying a vertical load.

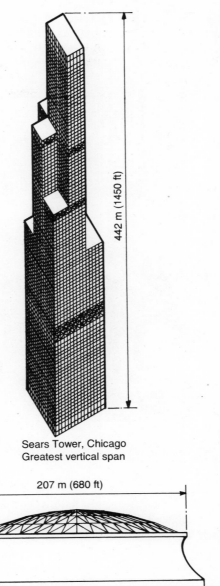

Sears Tower, Chicago
Greatest vertical span

Louisiana Superdome, New Orleans
Greatest horizontal span

2.3.6. The tallest building has a much greater span in relation to wind than any horizontal span in a building.

We have so far considered the wind pressure and the wind suction as if they are stationary, or static, forces. The wind usually does not blow with a constant velocity; high wind intensities may be of very short duration, from a few seconds to a fraction of a second. Under these conditions the building may move backward and forward, and the *dynamic* effect of the wind must then be considered. The vibrations produced in buildings by high winds can be disturbing to the occupants. The effect may be partly physiological, akin to seasickness, and partly psychological; we are at least subconsciously disturbed if the hundredth floor of a building sways noticeably in a high wind. In the World Trade Center in New York, the world's second tallest building (412 m or 1350 ft), completed in 1973, dampers that resemble the shock absorbers used in automobiles were fitted to reduce the movement due to wind (Section 8.8).

An *earthquake* is a sudden, jerky movement of the ground, which takes the foundation of the building with it but leaves the upper part of the building behind because of the high speed of the ground's motion and the high inertia of the building (Fig. 2.3.7). The effect is the same as if the building moved relative to the ground. The direction of the motion depends on the fault in the earth's crust, but it can always be resolved into a horizontal and a vertical component (Section 2.6). The vertical component of the earthquake's motion is relatively harmless, because all buildings are designed to resist large vertical loads, but the horizontal component of the movement may produce serious cracking and even collapse of the building (Fig. 2.3.8).

In designing small buildings it is normally assumed that the earthquake produces a static horizontal force, which depends on the expected intensity of the worst earthquake likely to occur and is specified in the local building code. The building must be made strong enough to resist this force, but in addition it must be able to absorb the movement of the earthquake without collapsing.

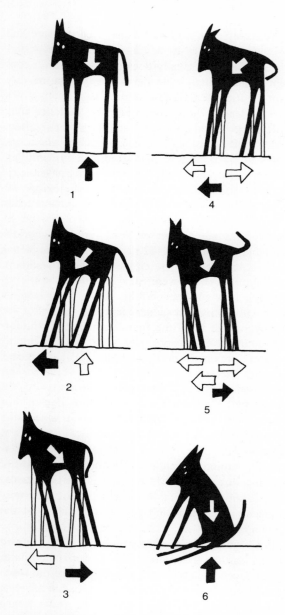

2.3.7. An earthquake is a sudden, jerky motion of the ground, which takes the foundation of the building with it but leaves the upper part of the building behind because of the high speed of the motion and the high inertia of the building.

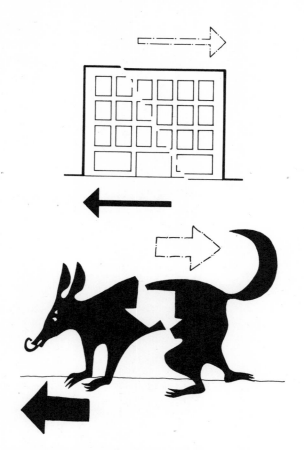

2.3.8. Unless the building is designed to resist the horizontal force of the earthquake, it will crack and may collapse.

Thus loadbearing walls built from brick, concrete block, natural stone, mud blocks or mud perform badly in earthquakes even if they are well constructed. However, brick or block walls reinforced with steel mesh are satisfactory. Timber-framed construction is suitable for small buildings, but reinforced concrete or structural steel is required for multistory buildings (Chapters 3 and 8).

Even if the earthquake produces only a single shock, its high intensity sets a tall building swinging like a pendulum. The frame of the building must be sufficiently *ductile* (see Glossary) to absorb the energy and dissipate it through its pendulum motion. For tall buildings the dynamic effect of the earthquake must therefore be considered.

The main earthquake zones are the Circum-Pacific Belt, which contains Chile, Central America, California, Alaska, Japan, Indonesia, and New Zealand; and the Alpide Belt, which runs through Iran, Turkey, Greece, Yugoslavia, Italy, and Portugal. The latter passes through the region of some of the ancient civilizations, and major earthquakes have become part of their legends.

Only minor earthquakes have been recorded in the British Isles, in Australia, and in eastern North and South America.

2.4 Fire Loads

Unlike the loads considered in the previous sections, fire loads do not impose any forces on the structure unless there is a fire. They are, however, very important for structural design.

The concept of *fire load* originated in the United States in the early twentieth century. It denotes the amount of combustible material per unit area; this can be stated as the mass of the material in kilograms (or pounds) or as the thermal energy of the material in joules (or btu).

The fire load is specified to accord with the type of occupancy, because, for example, a store selling textiles is likely to contain more combustible material than a school. For particularly hazardous occupancies, such as a warehouse for organic solvents, it may be necessary to make a special determination of the fire load by measuring the volume of combustible material and multiplying this by the thermal energy of combustible material released by a fire; the latter can be determined by burning the material in a calorimeter.

The *fire endurance* of the structure is the time taken to cause its failure in a fire. For simple structural members the fire endurance can be calculated from theoretical considerations, but for more complex members and for structural assemblies the fire endurance can only be determined by means of a fire test. This is carried out by placing a prototype of the structure in a furnace whose temperature is raised in accordance with a standard fire curve (Fig. 2.4.1). This corresponds to the temperature rise observed in actual fires. When a fire starts, the temperature rises rapidly. It reaches 500°C (930°F) in about five minutes and 900°C (1650°F) in about an hour. Thereafter the increase in temperature is much slower (Ref. 2.2).

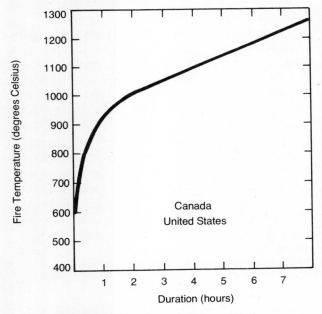

2.4.1. Standard temperature-time fire curve.

The fire endurance is reached *either* when the structure collapses under the service loads (see Section 3.10) *or* when the temperature on the far, or opposite, side of a structural member (for example, a floor slab) rises to a level at which materials on that side (for example, carpets or plastic tiles) would be set on fire.

The fire endurance of the structure specified by building codes depends on the fire load and on the height and purpose of the structure. The rules are based on two concepts: It should be possible to evacuate everyone in the building, and it should be possible for the fire to burn itself out without causing a catastrophe, even if the fire brigade is unable to extinguish it.

Thus isolated single-story buildings are not generally required to possess any fire endurance, except when the building has a high fire load, because the occupants can walk out of the building at ground level. Buildings of two or three stories need only a low fire endurance, if they contain single-family apartments; if the same building were modeled as a hotel or a small hospital, it would need a much higher fire endurance.

For the purpose of fire codes, buildings are tall if their height is greater than the tallest ladder of the fire brigade, generally 50 m or 150 ft. It is then impossible to evacuate the occupants through upper-floor windows. Fire-secure stairs and/or elevators are provided for the evacuation of the building and for giving firefighters access to the fire. However, fire stairs are not really practical for one-hundred-story structures. Recent buildings have been designed with fire-secure floors at certain levels, to which people can be evacuated; their fire endurance must be such that people would be safe in them even during an exceptionally severe fire on nearby floors. In addition, the structural frame must have a fire endurance greater than that of the most severe fire that could conceivably arise under the most adverse conditions, so that the building as a whole cannot collapse, even though individual floors may be destroyed by fire.

Although all this may not sound entirely satisfactory, very few lives are lost by fire in tall buildings. When we consider all buildings, however, the loss of life due to fire is still distressingly high, although far less than the loss of life on the roads. Most deaths and most property damage occur in relatively small and old buildings, where fire precautions have been allowed to fall below the standards required for safety.

2.5 The Effects of Temperature, Moisture, and Foundation Settlement

All materials, with a few exceptions, expand when heated and contract when cooled. The annual change in temperature from the coldest night to the hottest day is small on a tropical island, but it is quite large in places remote from the ocean, particularly in a cold climate. A change in humidity also causes expansion or contraction in timber and masonry materials, but not in metals.

Long walls and floors in buildings are provided with *expansion joints* to allow for the expansion and contraction caused by temperature and moisture movement. If these joints are not provided at suitable intervals, stresses that can cause serious damage are set up in the materials. In the case of brick and concrete these stresses are often relieved by the formation of unsightly cracks.

Expansion joints cannot be provided in structural members. The beam in Fig. 2.5.1 would simply collapse if we placed an expansion joint at mid-span. However, only a few beams in buildings are so long that the temperature or moisture movement presents a problem; these are placed on rollers (Fig. 2.5.1) or rockers. Bridge beams are frequently long enough to require rollers, rockers, or sliding joints; they can be seen on the underside of the bridge if it is accessible.

There is a particular problem in curved structures. Arches have horizontal restraints (Section 7.7), and these are responsible for their superior structural performance. If we placed the arch in Fig. 2.5.2 on rollers, like the beam in Fig. 2.5.1, it would not be as efficient a structure. During the night the arch shortens, and during the day it gets longer, particularly if the sun is shining. Similar problems, although on a smaller scale, are created by moisture movement in concrete, as the concrete absorbs water and then dries out again. The stresses caused by temperature and moisture movement in arches are often much greater than the stresses caused by the live load, and thus they cannot be ignored.

There is a similar problem in concrete shell structures (Section 10.3), particularly in domes. It is more serious than in arches, because the shells are usually thin and therefore crack easily, so that temperature stresses must always be considered.

2.5.1. Long beams must be placed on roller or rocker bearings so that the expansion and contraction due to changes in temperature and moisture can take place without causing additional stresses in the beam.

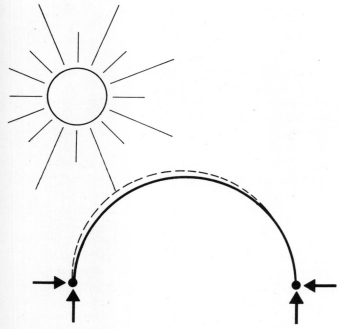

2.5.2. Since arches must be restrained at both supports, a change in temperature causes a change in the shape of the arch. This in turn produces stresses within the arch that may be greater than the stresses due to dead and live loads.

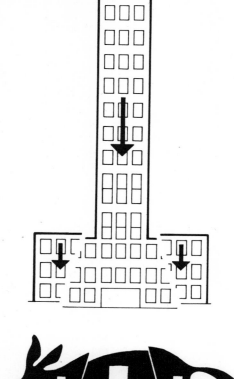

2.5.3. Cracking of building with brick walls due to uneven settlement of the foundations.

Uneven settlement of foundations can also create stresses in the superstructure. It was a common cause of trouble in Gothic cathedrals, which were often built with inadequate foundations. It should not occur today if the soil on which the building is to be erected is properly investigated (Section 9.1) and a suitable foundation designed for the building. Uniform (or even) settlement can be irritating, but it is not dangerous (Section 9.6). Several buildings in Chicago and in Shanghai (both cities have clay subsoil) have sunk so that it is now necessary to walk several steps down to the floor that was originally at ground level, but the buildings are quite safe. Nor does uneven settlement necessarily cause collapse; the Leaning Tower of Pisa, which was built 800 years ago, is still standing, although it is 5 m (16 ft) out of plumb in a height of 55 m (179 ft).

Problems arise particularly in buildings of varying height, such as the one shown in Fig. 2.5.3, which has a tower much heavier than its wings. The soil under the tower is compressed more than the soil under the wings. The uneven settlement causes serious cracks on both sides of the tower, and the building must be demolished.

Temperature and moisture movement and uneven settlement can therefore have an effect similar to that of the loads imposed on a structure. There is, however, one important difference between temperature and moisture movement on the one hand and uneven settling on the other. The former are unavoidable; we cannot stop the sun and the rain. But uneven settlement that results in cracks is the result of poor foundation design, and it can be avoided.

2.6 The Forces Produced by the Loads, and the Conditions of Equilibrium

The dead loads and the live loads always act vertically, because the weight of the building and its contents are attracted by the earth (that is, they are gravitational forces). Wind pressures and wind suction on a vertical wall act horizontally. The other loads act at some angle to the vertical, but it is convenient to resolve them into horizontal and vertical components.

We therefore have two types of load, those which act vertically (Fig. 2.6.1), and those which act horizontally (Fig. 2.6.2).

The technical term used in mechanics is *force*, and this is defined as anything that changes or tends to change the state of rest of a body. We may confine ourselves to a definition of force as something that tends to change the state of rest of a body, since buildings do not (or at least should not) move. This is because the foundations on which they sit provide a reaction (Fig. 2.6.3.a) that is exactly equal and opposite to the force and maintains the state of rest.

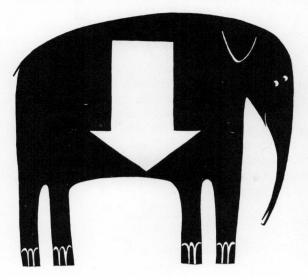

2.6.1. Vertical force.

2.6.2. Horizontal force.

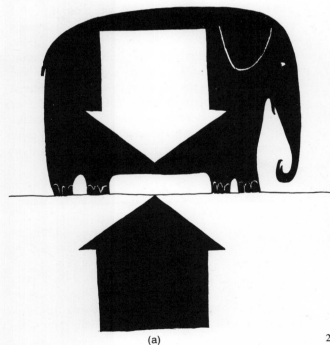

(a)

(b)

2.6.3. To every force there must be an equal and opposite reaction.

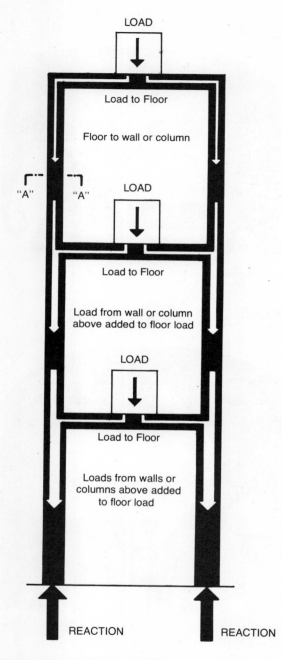

2.6.4. Transmission of forces through a building.

Let us now consider how this applies to the vertical loads acting on a building (Fig. 2.6.4). The loads are transmitted through the members of the structure to the foundation, which provides the reaction to them. In the process each structural member is acted on by a force or forces (Fig. 2.6.5). The member must be strong enough to resist the force or forces. If it is not, the member fails. If enough members fail, the structure collapses.

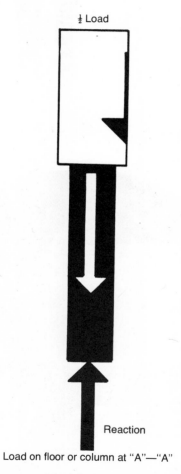

2.6.5. Forces on one of the columns in the building shown in Fig. 2.6.4.

2.7 How Traditional Structures Fail, and How Steel and Reinforced Concrete Structures Fail

Except for Roman concrete structures, traditional structures were of two kinds: those made from timber or reeds and those made from natural stone or brick.

Traditional timber joints, like the mortise-and-tenon joint or the dovetail joint, require a reduction in the cross-sectional area of pieces of timber to be connected, and the joints were therefore the weakest part of the structure. In reed or bamboo structures also the joints, whether made by binding or by cutting, tended to fail before the structural members.

Natural stone and brick were usually laid in lime mortar, which was weaker than the blocks they joined.

Thus the failure of traditional structures was generally in the joints, and the strength of the structural materials was not generally utilized to the fullest extent. This is one of the reasons why these structures were so much heavier than similar ones built today (Chapter 1).

In steel structures and reinforced concrete structures the joints are always made slightly stronger than the structural members (Section 5.7). We assume that the failure will not occur in the joints. The strength of the structure therefore depends on the strength of the material in the structural members. Next we will examine the properties of the principal structural materials.

References

2.1 SCIENCE AND BUILDING by Henry J. Cowan. Wiley, New York, 1978. 347 pp.

2.2 FIRE AND BUILDINGS by T. T. Lie. Applied Science, London, 1972. 276 pp.

Chapter 3 Structural Materials and Safety Factors

We shall never be able to get people whose time is money to take much interest in atoms.
(Samuel Butler [1835−1902])

In this chapter we examine the physical properties of the principal structural materials and the criteria for selecting the appropriate material for various types of building. Finally, we consider the criteria for safety and for serviceability.

3.1 Elastic Deformation

Solids are held together by bonds between the atoms of which they are composed. These bonds can be extended or compressed by forces acting on them, and this is called *elastic deformation* (Fig. 3.1.1).

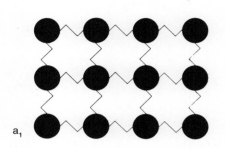

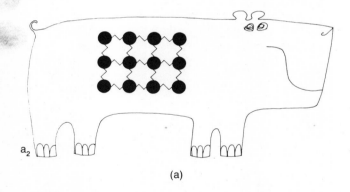

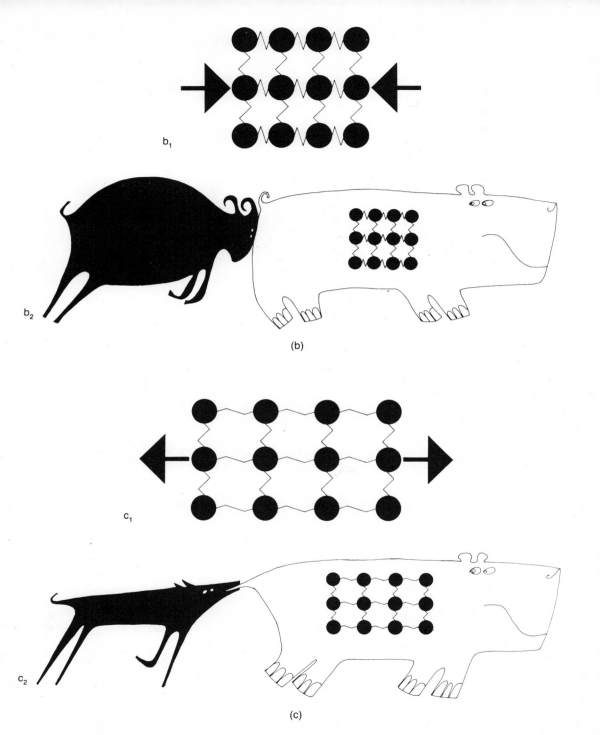

3.1.1. Bonds between atoms can be extended or compressed by a force. Provided the bonds are not broken, the deformation is elastic, that is, it is fully recovered when the force is removed.
(a) Atomic structure both before and after elastic deformation.
(b) Compression.
(c) Tension.

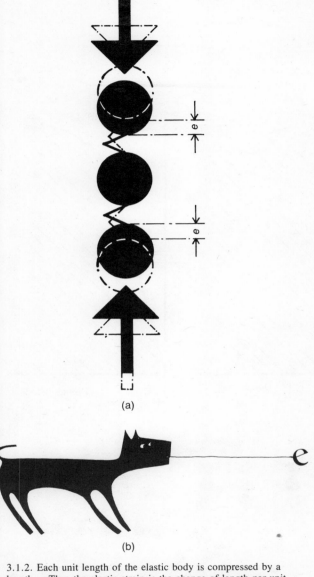

3.1.2. Each unit length of the elastic body is compressed by a length e. Thus the elastic strain is the change of length per unit length. An elastic body resumes its original shape when the load is removed, and the elastic strain is therefore fully recovered.

The elastic deformation is a characteristic property of the material and can be predicted from its atomic structure. If f is the force per unit area (which is called the *stress*) and e is the elastic deformation per unit length (Fig. 3.1.2) resulting from it (which is called the *strain*), then

$$f = e E \qquad (3.1)$$

where E is a constant for the material, called the *modulus of elasticity*. This proportional relationship follows from the atomic structure of crystalline materials, and it is confirmed by experiments on structural materials.

For steel, for example, the modulus of elasticity E is about 200 000 MPa*. Thus a steel bar 1 m long, under a stress of, say, 150 MPa (which is the maximum permissible stress in tension), extends only $1 \times 150 / 200\,000 = 0.75 \times 10^{-3}$ m $= 0.75$ mm**, and this cannot be seen with the unaided eye, although it is easily measured with an extensometer.

The fact that the elastic deformation of building materials is so small gives the impression that buildings are rigid and do not deform under load. This is not so; all materials deform under load, and without deformation there is no stress. However, the smallness of the movement is psychologically helpful: most people are disturbed by noticeable, rubberlike deformation in a structure.

As long as the atomic bonds remain unbroken, the material remains elastic. It deforms elastically under the action of forces (Figs. 3.1.1.b and c) and recovers its original shape (Fig. 3.1.1.a) when the force is removed (Fig. 3.1.2).

---

* We will use MPa (Megapascals or millions of pascals) throughout this book. 1 MPa = 1 MN/m² = 1 N/mm²
** For a steel bar 36 in. long with a modulus of elasticity of 30 000 ksi (kilopounds per square inch), under a maximum permissible stress of 22 ksi, the extension is $36 \times 22 / 30\,000 = 0.026$ in.

3.2 Ductility

If we exceed the capacity of the material for elastic deformation, we break the atomic bonds. There are two distinct ways in which this can happen, and we accordingly distinguish between those which are *plastic* or *ductile* and those which are *brittle*.

When a material fails in a ductile manner, it deforms plastically (that is, permanently), because the atomic bonds re-form many times without a significant loss of strength (Fig. 3.2.1); indeed, there may be a small increase in strength. Eventually fracture occurs, but only after the material has absorbed a great deal of energy, far in excess of the elastic energy. This is particularly important if the loads are of short duration, as during earthquakes and cyclones (hurricanes, typhoons), since the building has a better chance of still being upright when the earthquake or the wind has subsided if the structure is capable of absorbing energy. Even with permanently acting loads, ductility delays failure, and the lapse of time may be sufficient to permit repair of the structure or at least evacuation of the building (Fig. 3.2.2).

Ductility is not the only characteristic necessary for a good structural material. We also need high strength. The traditional wrought iron, which is almost pure iron, is highly ductile but not very strong, and for this reason it is no longer used for structures.

Steel, which replaced wrought iron as a structural material a century ago, is an alloy of iron and carbon; quite a small amount of carbon (about ¼%, or one part in 400) is sufficient to produce a substantial increase in strength. This is because the carbon atoms, being different from the iron atoms, tend to block the sliding action of the iron atoms over each other. This low-carbon steel is called structural steel.

3.3 Brittleness

Structural steel, in spite of its higher strength, is still a fairly ductile material. If we increase the carbon content to 1%, we further block the sliding action of the atoms and consequently obtain a great increase in

3.2.1. When a plastic material is extended beyond the elastic limit, the atomic bonds are broken and reformed with the adjacent atoms. The material is just as strong as it was before, but it is now permanently deformed. The process can be continued many times before fracture occurs. The ultimate deformation of a plastic material is much greater than the elastic deformation, and it is easily visible to the naked eye. The material is therefore also called ductile.

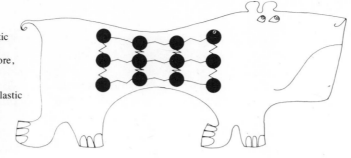

3.2.2. The energy absorbed by the material is represented by the shaded area under the load-deformation curve. Evidently a ductile material, which is capable of a great deal of plastic deformation (a), absorbs more energy before failure than a brittle material, which fractures soon after reaching the elastic limit (b). The energy due to the elastic deformation is represented by the cross-hatched region.

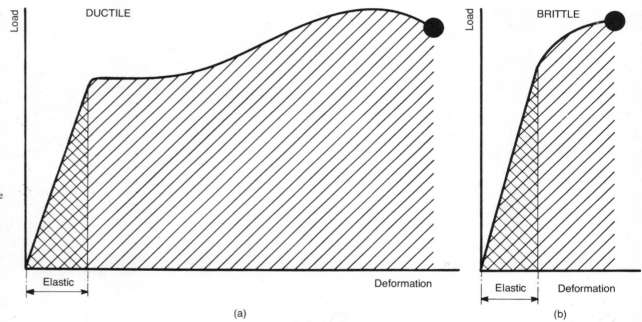

3.3.1. When a brittle material is extended beyond the elastic limit, failure occurs by a sudden breaking of the atomic bonds, at 90° to the line of tension.

strength, about 4 to 5 times that of structural steel. But since the ductile deformation is blocked, a brittle fracture results from a failure of the atomic bonds (Fig. 3.3.1). This failure occurs with little energy absorption (Fig. 3.2.2.b) and therefore with little warning. These high-carbon steels have their place in structural design (Section 5.6), as long as we remember their lack of ductility (Fig. 3.3.2).

We can remove the brittleness by reducing the carbon content of high-carbon steel, if we regard this as desirable. However, there are many important building materials that are inherently brittle. Natural stone, brick, and concrete are all in that category. They have high compressive strength but poor tensile strength and negligible ductility.

Stone, brick, and concrete have always been and still are among the most important structural materials. They are durable and very strong in compression, and there is an ample supply of the raw materials. Concrete has two important additional advantages: it can be cast into any simple shape, and its cost is low (although in many parts of the world brickwork is as cheap or cheaper).

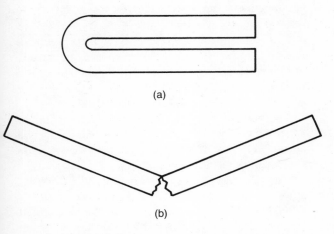

3.3.2. (a) Low-carbon steel is fairly strong and highly ductile. (b) High-carbon steel is very much stronger but lacks ductility.

3.3.3. The strength of steel is about the same in compression and in tension. The strength of concrete is much higher in compression than in tension.

In a ductile material the strength in tension is about the same as the strength in compression. In a brittle material the tensile strength is much lower than the compressive strength, because the atomic bonds are relatively easily fractured in tension, while a compression failure can occur only by sliding action, or *shear,* as the atoms are pressed together (Fig. 3.3.3).

Most traditional structures in stone, brick, and concrete were therefore designed to eliminate tensile stresses almost completely, which imposed drastic restrictions on span and form. This is why so many important buildings prior to the nineteenth century were roofed with vaults and domes (Section 10.2). The only alternative for all but the shortest spans was to use timber, which was not durable.

In the nineteenth century iron and, later, steel became available as structural materials. We now had a ductile material that was also durable, and structural form gradually changed to that used today.

3.4 The Cure for Brittleness

Reinforced concrete was invented in the second half of the nineteenth century. Steel bars are placed into the parts of the concrete that are in tension (Fig. 3.4.1). The concrete cracks but does not break apart, because the ductile steel resists the tension which the brittle concrete cannot withstand.

The strain (that is, deformation per unit length) at which concrete cracks is about 3×10^{-4}, or 3 parts in 10 000; this corresponds to a 3 mm increase in length for a beam 10 m long, or 0.1 in. in 30 ft. If the steel is cast into concrete, its strain is exactly the same as that of the surrounding concrete, because the two materials deform together. For a modulus of elasticity of 200 000 MPa, the corresponding stress is

$$f = eE = 3 \times 10^{-4} \times 200\,000 = 60 \text{ MPa*}$$

---

*For a modulus of elasticity of 30 000 ksi,
$f = eE = 3 \times 10^{-4} \times 30\,000 = 9$ ksi

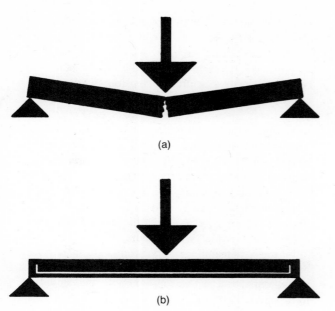

3.4.1. Plain concrete (a) and reinforced concrete (b).

This is only 15%, or about 1/7, of the strength of the most commonly used reinforcing steel, and the steel would be used very inefficiently if we did not allow the concrete to crack (Fig. 3.4.2).

It is important that water does not reach the steel, so that it does not rust. Rust has a greater volume than steel, and its formation would further crack the concrete and quickly cause its disintegration. A few large cracks would admit water, but a large number of fine cracks will not (Section 5.6). The size of the cracks depends on the adhesion between the steel and the concrete. This is improved if the bars have a deformed surface and are slightly rusty and if a large number of small-diameter bars, rather than a few large-diameter bars, are used, because this gives more surface area per unit volume of steel. Furthermore, the lime (CaO, calcium oxide) that is a component part of portland cement prevents rusting.

The steel reinforcement resists the tensile stresses wherever they occur. It turns the brittle concrete into a material with sufficient ductility. This makes it suitable for most purposes for which structural steel was previously considered necessary (Section 5.5), even though the volume of the steel is usually only about one one-hundredth of the volume of the concrete (Fig. 3.4.3).

3.4.2. Reinforced concrete beams and slabs are always cracked on the tension face. This does not matter provided that the cracks are so fine that moisture cannot penetrate to the reinforcement.

3.4.3. Although most reinforced concrete structures contain many reinforcing bars, carefully arranged to ensure that all tensile stresses in the concrete structure are resisted by steel bars, the total volume of the steel is only about 1% of the volume of the concrete. However, the cost of this reinforcement can be quite high. In most reinforced concrete structures the cost of the reinforcement is roughly the same as the cost of the concrete.

3.5 Which is the Right Structural Material?

Timber is easily cut and nailed, and this makes it a particularly useful material for domestic buildings. In some countries, such as Australia, loadbearing brick walls are to an increasing extent replaced in one- and two-story construction by timber frames clad with a nonloadbearing brick veneer on the outside and an insulating board on the inside. Timber is not fireproof, and it is not used for the frames of multistory buildings. Its strength per volume is slightly lower than that of reinforced concrete and much lower than that of steel. However, its strength per weight is not much less than that of steel, and long-span roofs have been built from laminated timber. This is made by gluing comparatively small pieces of timber with a strong waterproof glue, such as melamine formaldehyde (Fig. 3.5.2). Thus larger pieces can be produced than exist in nature.

Natural stone is now rarely used as a structural material, because of the high cost of skilled labor. It is still used as a nonstructural material. Brick and concrete blocks are used extensively for nonloadbearing partitions, and they are competitive for loadbearing walls of multistory buildings of medium height.

Today the floor structure of all but the smallest buildings is built from steel, reinforced concrete, or prestressed concrete, and so are the vertical supports of the larger buildings. Timber and aluminum have been used for a few long-span roofs, but the great majority are now also built from steel or reinforced concrete. In most countries for the past fifty years there has been a steady move to use reinforced concrete in place of steel for reasons of cost.

Although steel does not burn, it softens and loses its strength in a fire. It must therefore be protected for fire-resistant construction.

Fire resistance is not normally required in single-story buildings, because the people inside are already at ground level and can walk out of the building if a fire breaks out. Steel is therefore commonly used for single-story factories and assembly halls.

Steel rusts unless protected by zinc (galvanizing) or anticorrosive paint or covered by a fire-protective coating. The cost of regularly repainting steel can add appreciably to its lifetime cost.

This difficulty can be overcome by using Cor-Ten, a low-alloy steel that forms a protective coating through weathering. This is, in fact, a durable form of rust. The color changes from light brown (after about a month), to dark brown and purple (after one or two years); however, the coating is partly soluble in

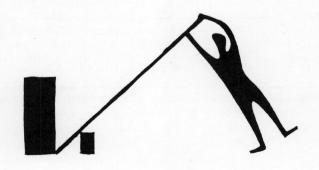

Which is the right structural material?

3.5.1.

water, and rain produces brown streaks on any material below the Cor-Ten steel. The structure must therefore be suitably detailed; in particular, it must not be supported directly on concrete, which is stained by rust. The rust washed off the steel is best absorbed in a bed of gravel at the base of the steel structure, as the gravel can be replaced from time to time until the oxidation process is complete.

3.5.2. The weakness of timber along the grain is eliminated in laminated timber.

Table 3.1. Structural materials

Material	Advantages	Disadvantages
Small timber sections	Easily fabricated; low cost	Attacked by fungi, insects, and fire
Large timber sections	Longer spans and greater durability than for smaller sections	Can only be cut from large trees, which take many years (for some species, centuries) to grow
Laminated timber	Large sections are produced from small pieces of timber	Additional cost due to fabrication and gluing
Structural steel	Easily fabricated; very large spans possible with fabricated structures; moderately low cost	Rusts unless protected; not fire-resistant without protection; fire protection adds to the cost
Cor-Ten steel	Does not need galvanizing or painting	Costs more than structural steel; building must be detailed to prevent rust staining
Aluminum	Does not need painting; very light weight	Costs more than steel; higher deflection and greater tendency to buckle than steel
Loadbearing walls of bricks or blocks	Low cost; good thermal insulation and high thermal inertia	Poor resistance to earthquakes and hurricanes unless reinforced
Plain concrete	Low cost; high thermal inertia	Limited to walls and foundations; poor resistance to earthquakes and hurricanes without reinforcement; surface finish on vertical surfaces presents problems
Reinforced concrete	Low cost; high thermal inertia; may be cast in one piece in any shape that can be produced by formwork; very large spans possible	Speed of construction limited by the need for the concrete at lower levels to gain strength; heavier than structural steel
Prestressed concrete	Lower structural depth than reinforced concrete; dead load can be compensated by load balancing; prestressing prevents the formation of fine cracks, which are unavoidable in plain and reinforced concrete	More expensive than reinforced concrete

Fire Endurance	Maintenance	Availability of Raw Materials	Demolition
Low	Requires regular painting if used externally	Regenerated indefinitely by regular reforestation	Easy
Burns initially, but forms protective layer of charcoal	Some hardwoods require no maintenance	Reforestation to produce large trees takes a long time	Easy
As for large timber sections	Some glues and protective coatings are weatherproof	Regenerated indefinitely by regular reforestation	Easy
Moderate if unprotected; excellent if protected by layer of concrete or vermiculite	Requires galvanizing or regular painting both for internal and external use	Deposits of iron ore and supply of scrap steel are limited, but sufficient for at least two centuries; turning iron ore into steel requires energy	Easy
Moderate; used without cover	None	As for steel	Easy
Low	None	Raw material plentiful, though not unlimited; energy expenditure to produce metal higher than for steel	Easy
Excellent	None	Unlimited	Easy
Excellent	None	Unlimited	More expensive than for concrete blocks
Excellent if reinforcement has adequate cover	None	As for concrete and steel (but uses only a small amount of steel)	Expensive for small structures, difficult for large structures
Excellent if prestressing steel has adequate cover	None	As for concrete and steel (but uses only a very small amount of steel)	Can be difficult and dangerous because of the energy locked in the pressing steel

3.6 The Choice of the Structural Material for Domestic Buildings

Single-story, single-family houses must carry their own vertical load and that of the people inside, but because of the small size of the building the forces and moments due to these loads are low. The sizes of the structural members are rarely calculated; the rules given in building codes are often based on experience rather than structural mechanics (Fig. 3.6.1).

Ordinary wind forces have little effect on the walls, but it is important to tie down the roof structure as a whole and the individual parts of the roof covering, such as roof sheeting or shingles, to prevent their being lifted off in a high wind (Figs. 2.3.2 and 2.3.3).

Fire endurance is not usually required for single-family houses of one or two stories, because the residents endanger only themselves, and people do not have far to go to reach outdoors.

Timber is thus an excellent material for the structural frame. It is plentiful and can be regenerated by proper reforestation. It is relatively cheap, particularly in countries that have large forests. It is easily cut and assembled, and its comparative lightness reduces the cost of handling and transportation. On the other hand, it is combustible and can be damaged by fungi and insects.

Roof structures of small buildings are almost invariably made from timber. Timber frames for the walls are also normally used in the United States, Canada, Australia, and New Zealand. In Great Britain, where most timber is imported, loadbearing brick or block walls are more common.

Bricks and concrete blocks are incombustible and very durable. But the loadbearing walls built from them generally cost more than timber-framed walls with timber siding or with brick veneer. They also lack resistance to hurricanes and earthquakes.

Light frames of steel or aluminum have been used as substitutes for timber frames.

Diagonal framing members to resist horizontal wind forces (Fig. 2.3.2) are not usually required for

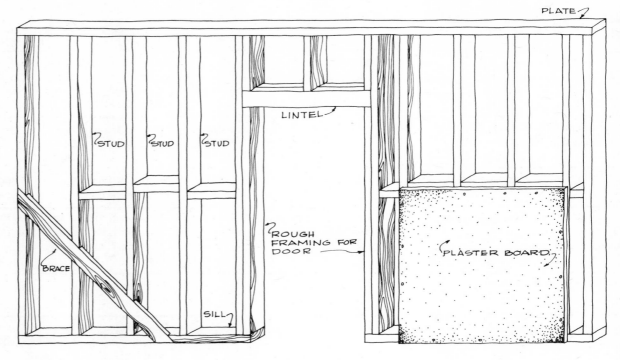

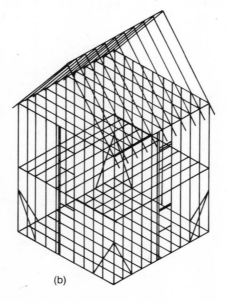

3.6.1. Timber structures for domestic buildings.

single-family timber-framed houses, provided that the cladding is securely fixed to the frame to give the necessary resistance to racking.

In some parts of the world, such as the southeastern United States and tropical Australia, storms with a very high uplift develop as a result of the effect of tropical heat on the ocean; these are called tropical cyclones, hurricanes, or typhoons. In such locations, not only must the roof structure be tied to the walls below, but holding-down bars must be taken into the foundation; the foundation itself must be a heavy slab of concrete whose weight is much greater than the force of the uplift. The various parts of the building must be tied together; depending on the location of the eye of the hurricane, the wind forces may act horizontally and upwards from any direction. If the connection between one part of the building and the rest is broken, the building is often quickly destroyed because members that were previously supported at both ends now become cantilevers (Fig. 3.6.2).

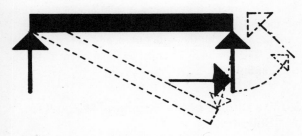

3.6.2. Structural members previously supported at both ends can become cantilevers through hurricane damage; this quickly leads to the destruction of the building.

In regions with a record of severe hurricanes it is better to use a reinforced concrete frame (Fig. 3.6.3) or reinforced block walls (Fig. 3.6.4). Steel reinforcing bars firmly anchored both in the foundation slab and the roof slab provide greater resistance to uplift than can be achieved with timber connectors.

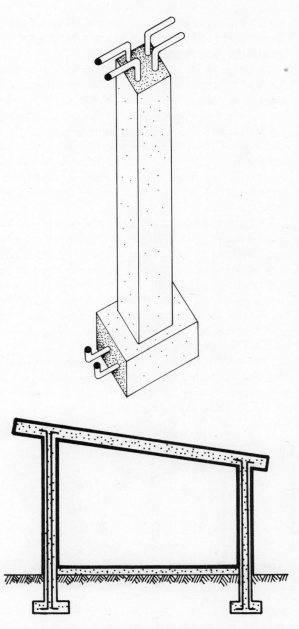

3.6.3. Reinforcement for columns in a region liable to hurricane damage.

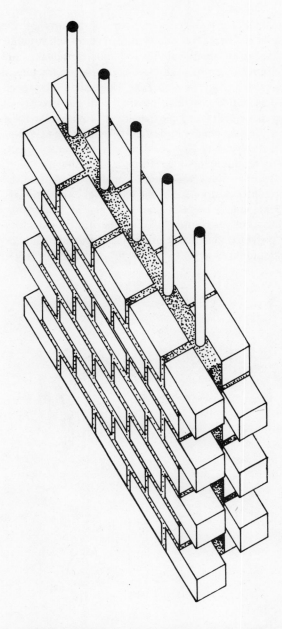

3.6.4. Reinforcement for concrete block buildings in a region liable to hurricane damage.

Minor earthquakes occur in most parts of the world, but these have no significance for the structural design of small buildings. Major earthquakes are confined to certain regions where there is a fracture in the earth's crust (Section 2.3). These should be specified in building codes, together with the severity of the earthquakes to be expected in a particular location. The design of single-family buildings in California, Alaska, and New Zealand is significantly affected by requirements for resistance to earthquakes.

Brick or block walls are incapable of resisting the horizontal component of the earthquake motion (Fig. 2.3.8) unless reinforcement is placed within the joints. Timber-framed buildings perform satisfactorily provided that the parts of the structure and the cladding are securely fixed to one another (Fig. 3.6.5).

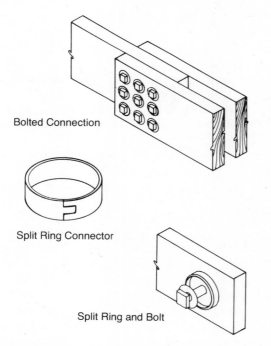

3.6.5. Timber-framed connections suitable for a region liable to earthquake damage.

3.7 The Choice of the Structural Material for Small and Medium-Sized Commercial and Industrial Buildings

Factories are often built as single-story structures for ease of internal transportation, and if the materials handled in them are not unduly hazardous, the structure generally does not require fire protection (Section 2.4). Structural steel is then a suitable material. It is cheap when there is no need for fireproofing, it is suitable for longer spans (which eliminate obstructing columns), and it is relatively easy to modify the structure if the layout of the factory requires alteration.

Multistory commercial and industrial buildings generally must have a fireproof structure. Steel does not burn, but it loses some of its strength at high temperatures. Usually, 500°C (930°F) is taken as the critical temperature (Ref. 2.2) at which the factor of safety has been dissipated, that is, at which the strength of the steel structure has been reduced to the point where it collapses under the action of service (or actual) loads (Section 3.10).

Steel must therefore be insulated against the high temperatures of the fire. The most common methods are spraying the steel structure with an insulating material, such as vermiculite, or casting it into concrete (Fig. 3.7.1).

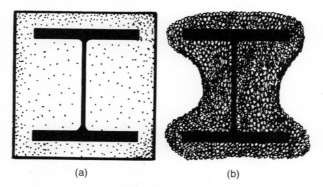

3.7.1. Fire protection of steel (a) by concrete and (b) by sprayed vermiculite, which is a mineral expanded by heating.

Concrete is widely used for protecting steel structures, but the resulting beams and columns are sometimes not much smaller than those of a reinforced concrete structure of the same strength. Reinforced concrete members are therefore generally more economical than structural steel members for medium-sized buildings that require a fire-resistant structure.

Loadbearing walls of brick or concrete block are coming back into favor, used in conjunction with reinforced concrete floors. Because the walls need to be of a substantial thickness to carry the loads, they reduce variations in the interior temperature and thus conserve energy. After a quarter-century of emphasis on structural lightness and thin curtain walls supported by a skeleton frame, there is now a new interest in thick walls for energy conservation.

In the nineteenth century loadbearing walls were designed as individual structural members, resulting in very thick walls that would be considered uneconomical today. Modern design methods consider the loadbearing wall part of a three-dimensional assembly. The structural members at right angles to it (that is, the cross walls and the reinforced concrete floors) provide the depth the structure needs to resist lateral loads. The problem is considered in more detail in Section 8.9.

3.8 The Choice of the Structural Material for Tall Buildings

At the time of writing the tallest building with loadbearing walls, in Pittsburgh, is 21 stories high (Section 8.9); the tallest building with a reinforced concrete frame, in Chicago, is 76 stories high, and the tallest steel-framed building is 110 stories high, also in Chicago. However, there are fewer than a dozen steel-framed buildings taller than 70 stories, and all are in the United States or Canada.

The designer has a choice between steel and reinforced or prestressed concrete for all but the tallest buildings. In North America and Japan steel is at present cheaper for most tall buildings, but in other countries reinforced concrete has proved more economical. The tallest buildings in South America (Bogota), Africa (Johannesburg), and Australasia (Sydney) are of reinforced concrete; the tallest building in Europe (Paris) is of mixed concrete and steel construction.

Many structural designers consider steel a more suitable material in earthquake zones, because a steel structure can absorb more energy than a concrete structure without suffering serious structural damage (Sections 3.1, 3.2, and 8.8).

Tallness is a relative term, and in most cities, particularly in Europe, Asia, and Africa, the tallest building has about twenty stories. Loadbearing walls are then a definite possibility. The subject is discussed in more detail in Section 8.9.

3.9 The Choice of the Structural Material for Long-Span Buildings

Long spans are occasionally used in tall buildings to create open plazas under buildings or monumental spaces at or immediately above ground level. Since these must conform to the fire-endurance requirements of the tall building, the structure must be built from fire-protected structural steel, reinforced concrete, or prestressed concrete.

When a building contains only one large room at ground level and some supporting rooms for foyers, refreshments, and toilets, as in a covered sporting arena or an exhibition hall, there is a much wider choice of material. Steel need not be encased in concrete or vermiculite, provided that satisfactory arrangements are made for the evacuation of the building and for fire extinguishment.

Even materials that are in themselves combustible, such as timber and fabric, can be used as structural materials if the requirements for evacuation and fire extinguishment are met. Very light and economical structures can be built from fabric, steel, or aluminum sheet supported by suspension cables and from membranes supported by pneumatic pressure. These are particularly suitable for temporary exhibition buildings, as they are easily dismantled and reerected elsewhere. Suspension cables are structurally very efficient, and in theory there is no limit to the span of a pneumatic membrane, because the weight of the structure is balanced by a small internal air pressure. However, these structures are not used for the longest spans because of practical difficulties (Sections 10.10 and 10.11).

Steel is used in the American sporting arena that holds the present record for span (207 m or 680 ft), and reinforced concrete is used in a French exhibition hall (Table 1.1) that spans almost as far (206 m or 676 ft). The subject is discussed in more detail in Chapter 10.

3.10 Safety Factors

Structures must have a margin of safety to ensure that they do not collapse (Fig. 3.10.1). There are two different philosophies of structural design. In the ultimate strength method the *service load* (which the structure is actually required to support during its useful life) is calculated and multiplied by a *factor of safety*. This factored load is called the *ultimate load*, and the structure is designed so that it would just fail if the load were increased to the ultimate load.

A factor of safety of 1.0 would mean that the structure collapses as soon as the service load is imposed on it. Evidently, factors of safety must always be greater than 1.0. At present, factors of safety for buildings vary from about 1.5 to about 2.5 (that is, a margin of safety of from 50% to 150%), depending on the structure and the material used (Fig. 3.10.2).

The factor of safety allows for imperfections in the materials. It would be too expensive to test every single piece of material used in the building, so we only test representative samples. It is therefore possible that a piece of material of lower strength than the minimum specified is used in a highly stressed part of the structure; however, the testing should be sufficiently rigorous to ensure that there is only a low probability of such an occurrence.

Since we cannot make buildings with the precision of a watchmaker, it is possible that some structural members are slightly undersized, and the factor of safety allows for this as well. Building codes specify dimensional tolerances to which buildings must conform.

The factor of safety also allows for the possibility that the service load we have assumed may be slightly exceeded and for the simplification normally made in the analysis of a complex structure. It does not allow for gross errors of judgment or mistakes in the arithmetic.

The second approach to structural design is to ensure that the maximum stresses in the structural materials under the action of the service loads (that is, the actual maximum loads imposed on the structure) are less than the *maximum permissible stresses*. These are determined by dividing the ultimate stresses (that is, the stresses at which the structural material fractures or deforms plastically) by the factor of safety. The maximum permissible stresses are invariably within the elastic range of the material, and this method of design is called the *elastic method*.

The two methods sometimes give the same answer. In the following chapter we will use the elastic method, which is the more common of the two, except for reinforced concrete sections (Section 5.5). This means that we design the structure so that the stress in the structure nowhere exceeds the maximum permissible stress for the material. The most economical, or *optimum*, structural design is one in which the stresses are as high as possible within this limitation. Thus the structure is optimized for the quantity of structural material. The optimum structural design is not always the one that produces the cheapest building, because other factors have to be considered (Chapter 11).

3.10.1. Factor of safety.

(a) DON'T BE AFRAID OF The UNKNOWN →

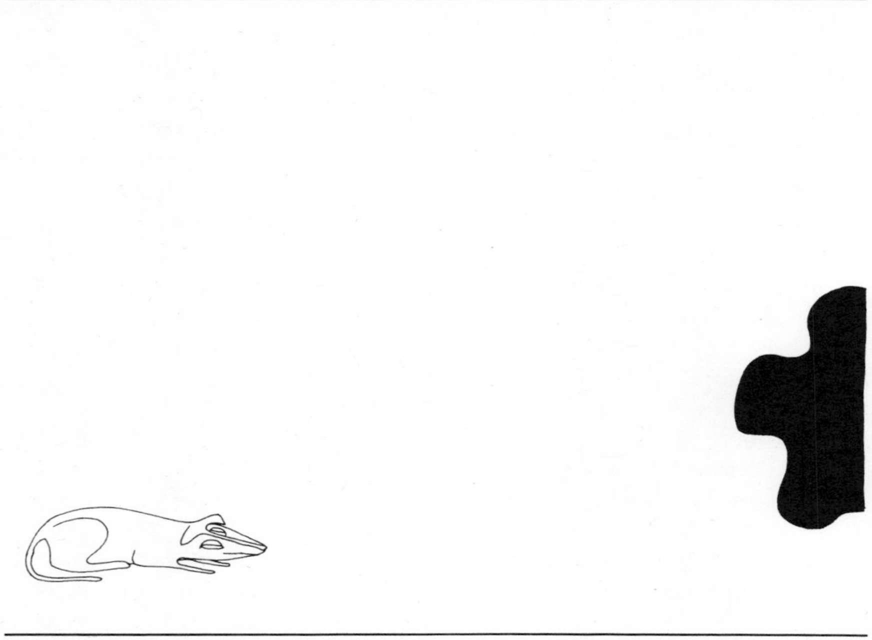

(b)

(c)

(d)

(e)

(f)

(g)

(h)

3.10.2. The factor of safety lies between 1.5 and 2.5.

3.11 Serviceability

The factor of safety ensures that the structure does not collapse. Additional calculations are needed to avoid cracking of the structure under the action of the service loads (that is, the actual maximum loads imposed on the structure) and cracking of the non-structural parts of the building.

Cracking of the structure is a problem only for reinforced concrete. Building codes include rules for the arrangement of the reinforcement within the concrete to ensure that the cracks are so fine that water cannot penetrate to the reinforcement and cause rusting (Section 5.6). Absolute avoidance of fine cracks is impossible if the reinforcement is to be effective (Section 3.4).

However, cracking of other parts of the building can be caused by the deformation of any structural material. Excessive deflection of a concrete floor slab, for example, causes cracks in brittle finishes, such as plaster, attached to the soffit of the slab; excessive deflection of steel beams produces cracks in suspended ceilings made of brittle materials (Fig. 3.11.1).

The deflection of steel beams is purely elastic under the action of the service loads. The beam deflects when it is loaded, and the deflection is fully recovered when the load is removed (Section 3.1).

Reinforced concrete, however, has a time-dependent deflection that is additional to the elastic deflection. This is called *creep,* and it is caused by the squeezing of the water in the hydrated cement from the pores of the concrete structure by the load. It occurs slowly over a period of time, and it is still significant a year after the structure has been loaded (Fig. 3.11.2). Creep deflection can be 2 to 3 times as much as the elastic deflection (Fig. 3.11.3). As a result, a building that is quite satisfactory when it is completed may show signs of distress several months later. Brittle finishes, such as plaster ceilings, and brittle partitions made from bricks or blocks may crack as a result of creep deflection. Doors and windows may jam because the creep deflection has caused a slight distortion.

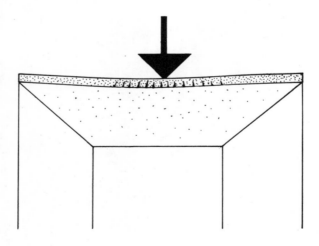

3.11.1. Cracking of brittle ceiling finish, for example, plaster.

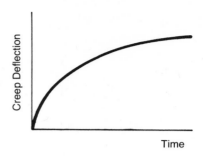

3.11.2. Creep deflection occurs over a period of time.

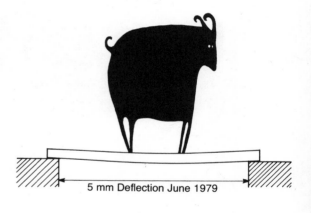

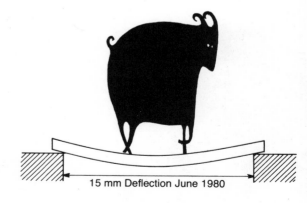

3.11.3. Creep deflection may be 2 to 3 times as much as the initial elastic deflection.

The creep deflection is directly proportional to the elastic deflection. If the elastic deflection is small, the creep deflection is also small. It is therefore necessary to keep the elastic deflection within the limits specified in the building codes, to ensure that the gradually occurring creep deflection is also kept within admissible limits. This can be done by making the floor structure sufficiently deep or the column spacing sufficiently close, or else by prestressing (Fig. 3.11.4) the floor structure (Section 5.6).

3.11.4. Creep deflection can be kept within reasonable limits by prestressing the concrete floor structure (Section 5.6).

Chapter 4 The Problem of Span

"Excellent!" I cried. "Elementary," said he. Sir Arthur Conan Doyle [Dr. Watson to Sherlock Holmes in "The Crooked Man"]

This chapter explains how to add and subtract forces and how to resolve inclined forces into their horizontal and vertical components. It then explains how to take moments and how to add and subtract them. This is best done with the aid of numerical examples.

It would be confusing to use both American units of force and length (kips and ft) and SI units (kN and m) in the examples and illustrations; we have used kN and m, partly because they are the units we will all be using soon and partly because they make a clear distinction between force (or weight) and mass. If you prefer American units, replace kN and m wherever they occur by kips and ft. The examples are equally valid, whether American or SI units are used with the same numbers.

Span is the major problem in most large buildings. It dominates the design of tall buildings and of auditoriums and assembly halls with long interior spans. In the first case horizontal loads act at right angles to the height of the building (Fig. 4.0.1), and in the second vertical loads act at right angles to the interior span (Fig. 4.0.2). Both these load systems produce bending moments (see Glossary), and it is the designer's task to select the structural system that resists these bending moments most efficiently.

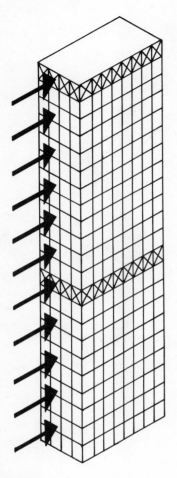

4.0.1. Vertical span.

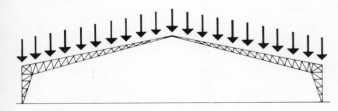

4.0.2. Horizontal span.

4.1 A Brief History of Span

The ancient civilizations were impressed by size, but there is no evidence that they had any interest in span (Fig. 4.1.1). Thus the Great Pyramid at Gizeh in Egypt held the record for the highest building for 44 centuries (Sections 1.2 and 1.4), but this great height was a byproduct of its size. It was a marvelous feat of construction, of organization, and of materials handling, but there was no structural problem to be solved. The Great Pyramid could not fail as a structure, and its height did not pose any problem of vertical span.

The seven wonders of the ancient world, listed by Antipatrus in the first century B.C., were all remarkable for their sheer size:

The pyramids of Egypt, which were solid.

The hanging gardens of Babylon, which were probably roof gardens on top of a solid ziggurat.

The statue of Zeus at Olympia.

The temple of Artemis of Ephesus, which employed 120 columns but used only short-spanning lintels between the columns.

The tomb of King Mausolus at Halicarnassus.

The lighthouse on the Island of Pharus in Alexandria.

The Colossus that stood with its legs across the harbor entrance of the Island of Rhodes.

Only the last two posed structural problems. The Colossus collapsed in an earthquake 56 years after it was completed (Ref. 1.4, p. 51). The Pharus lighthouse probably had substantial setbacks, so that the bending stresses due to wind would have been small.

This interest in size rather than span was not limited to the ancient civilizations of the Mediterranean and the Middle East. Great mounds of earth or stone have survived from other early civilizations that probably had no contact with Ancient Egypt, Babylon, and Greece; examples are the stupas of south and southeast Asia and the pyramids of Mexico.

Interior spans exceeding a few meters were first used in ancient Rome. The Pantheon, which is still impressive today, held the record for seventeen centuries (Section 1.2).

The height of buildings was restricted less by structural considerations than by lack of vertical transportation. Ancient Rome must have had apartment buildings at least seven stories high, because an edict of the Emperor Nero limited their height to 70 Roman feet (68 U.S. feet, or 21 m) after a disastrous fire. The Flavian Amphitheater, also known as the Colosseum, the tallest building surviving from ancient Rome, is 48 m (158 ft) high. However, these buildings had thick walls, and there were probably no great structural problems to be solved in their construction. Seven-story apartment buildings survive in some cities from medieval times, but the upper floors were very inconvenient for their inhabitants.

In the Middle Ages and the Renaissance a number of towers were built, some as fortified lookouts and some as spires or minarets for churches or mosques. The tallest was 142 m (465 ft) high (Section 1.2), that is, slightly less than the Great Pyramid. Many of them expressed verticality in a way that the Great Pyramid and the Colosseum do not, and some needed great structural skill to construct. However, it seems likely that traditional masons found more difficulty in building the greatest interior spans of their day than the tallest buildings.

Long interior spans are an invention of the nineteenth century. The need for them was created by railroads, exhibition buildings, and assembly halls for various forms of entertainment, and their span increased fivefold in the space of a century (Sections 1.2 and 1.4).

Tall buildings became possible only after the invention of the safety passenger elevator in 1854.

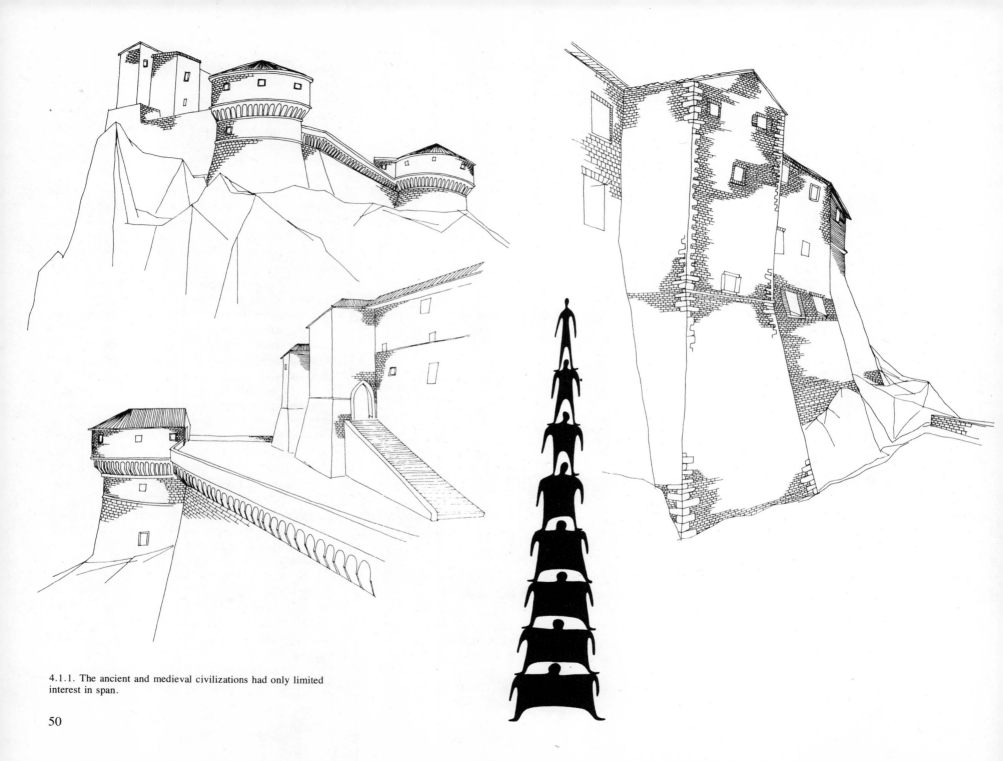

4.1.1. The ancient and medieval civilizations had only limited interest in span.

After a fire destroyed most of Chicago in 1871, many of its new buildings were designed with elevators, and the heights of buildings increased rapidly. In 1881 the maximum height reached ten stories, in 1892 twenty-one stories. The leadership then passed to New York, and by 1931 it was 85 stories. The present record is 110 stories. This is a threefold increase in absolute height in one hundred years (Section 1.4). However, if we exclude structures that were not for habitation (such as the tower of Rouen Cathedral), the increase in building height in the last century (1877-1977) is fifteenfold.

4.2 The Concept of Moment

We defined *force* in Section 2.6 as something that changes or tends to change the state of rest of a body. This is rather an abstract definition, but many words that are in common use and are generally understood are difficult to define in simple terms. (For example, what is a building? Look the term up in an unabridged dictionary!)

Thus weights and wind loads are forces, but in addition there are the reactions of the structure to these imposed forces, and the internal forces within the structure (Figs. 2.6.1 to 2.6.5). We will presently examine these in more detail.

A force tends to move a body in the direction in which it acts, and it also tends to rotate the body if it does not pass through its center of gravity (see Glossary).

A *moment* is the rotating effect of a force (Fig. 4.2.1). It is measured by multiplying force by distance (Fig. 4.2.2). The further the perpendicular distance of the force from the point under consideration, the greater the moment of that force.

4.2.1. Force × distance = moment.

(a)

(b)

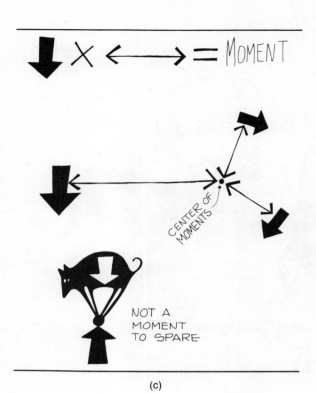

(c)

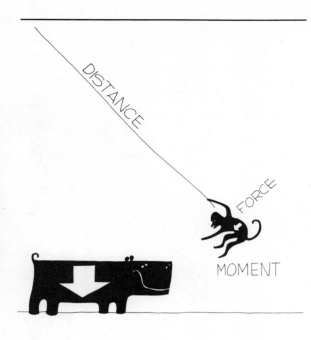

(d)

(e)

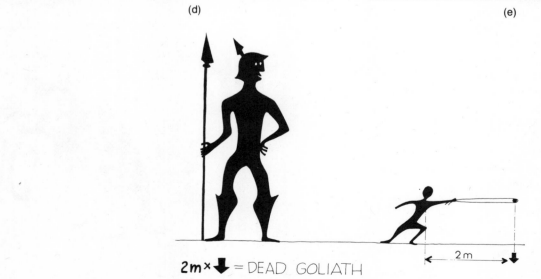

(f)

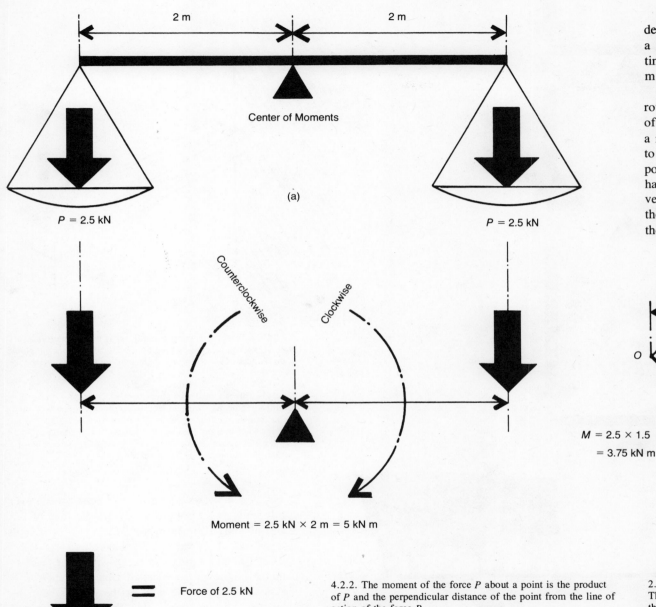

This is the *lever principle* discovered by Archimedes in Syracuse (Sicily) in the third century B.C. Thus a force 3 m from the fulcrum of the lever is three times as effective as the same force at a distance of 1 m from the fulcrum (Fig. 4.2.3).

Moments can be classified into those tending to rotate in a clockwise direction (that is, like the motion of the hands of a clock, or like the motion of driving a right-handed screw into a hole) and those tending to rotate in a counterclockwise direction. The two oppose each other. Thus the hippopotamus in Fig. 4.2.3 has a clockwise moment about the support of the lever. The dog has a counterclockwise moment about the support. If the two moments cancel one another, the lever is in balance.

4.2.2. The moment of the force P about a point is the product of P and the perpendicular distance of the point from the line of action of the force P.
(a) Two forces, each 2.5 kN, balance horizontally at a distance of 2 m each from the fulcrum of the balance.
(b) The moment of each force about the center of moments is 2.5 kN multiplied by 2 m, that is 5 kN m (kilonewton meters). The moment arm of the force is the perpendicular distance from the center of moments; it is 2 m.
(c) The moment of the arm of the inclined force is 1.5 m. Thus the moment of the inclined force 2.5 kN about O is $2.5 \times 1.5 = 3.75$ kN m.

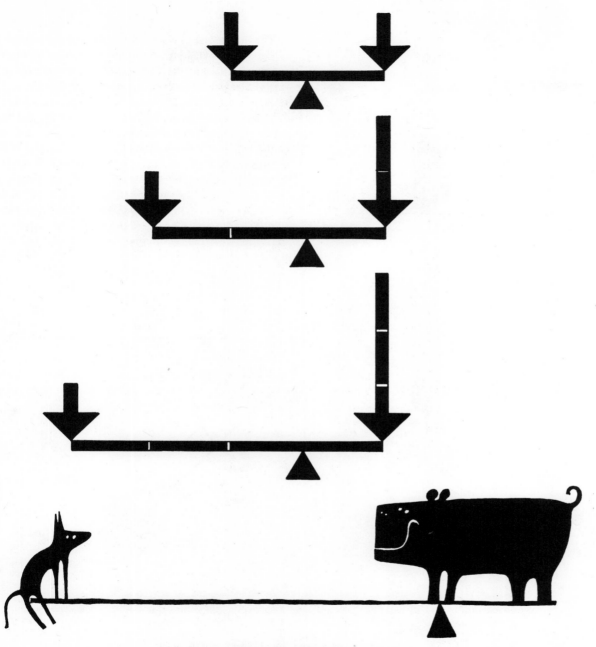

4.2.3. Forces on a lever balance if they have equal moments about the point where the lever is supported.

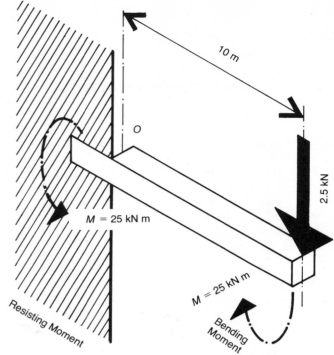

4.2.4. The reinforced concrete cantilever is built into a brick wall. At its end it carries a load of 2.5 kN, 10 m from the point O on its upper face where it is built into the wall. The force has a moment of

$$2.5 \times 10 = 25 \text{ kN m}$$

about O. This is called a bending moment, because it tends to bend the cantilever. The cantilever is restrained by the wall, and the restraining moment or resisting moment is also 25 kN m. This resisting moment balances the bending moment. (A cantilever is a beam firmly restrained at one end, and not at all restrained or supported at the other end.)

Fig. 4.2.4 shows a reinforced concrete cantilever firmly built into a brick wall, carrying a load of 2.5 kN at its end, 10 m from the point where it is built into the wall. The load of 2.5 kN has a moment of 2.5 × 10 kN m about the point 0. This moment tends to bend the cantilever and is called a *bending moment*. It also tends to bend the wall at the point where the cantilever is built into it. Unless the wall is capable of exercising a *resisting moment* of 25 kN m on the cantilever, it will break the wall.

4.3 Composition and Resolution of Forces

Forces have direction as well as magnitude. Thus a horizontal force of 1.5 kN can move a block standing on a concrete floor sideways if the friction between it and the concrete floor is less than 1.5 kN (Fig. 4.3.1.a); the larger, vertical force of 2.0 kN (Fig. 4.3.1.b) does not move the block. Thus we cannot add forces as if they were numbers.

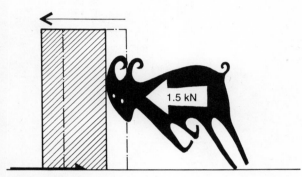

Friction less than 1.5 kN

(a)

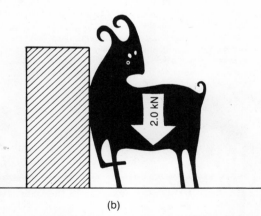

(b)

4.3.1. The ram weighs 2.0 kN, and it is also capable of pushing horizontally with a force of 1.5 kN. The ram's weight of 2.0 kN has no noticeable effect on a solid concrete floor (b), but it can push a block standing on the floor horizontally if the friction between the block and the floor is less than 1.5 kN (a).

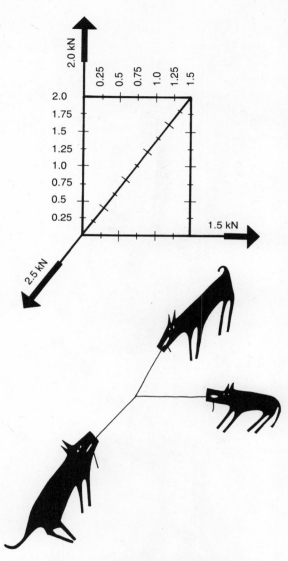

4.3.2. To combine the vertical force of 2.0 kN and the horizontal force of 1.5 kN shown in Fig. 4.3.1, we measure 2.0 vertically and 1.5 horizontally and draw the closing line of the triangle. We can obtain the answer either by measurement (2.5 kN) or by calculation, using the Pythagorean Theorem:

$$\sqrt{1.5^2 + 2.0^2} = \sqrt{6.25} = 2.5 \text{ kN}$$

Thus 2.5 kN is the *resultant* of a horizontal force of 1.5 kN and a vertical force of 2.0 kN.

We can perform the addition by using the *triangle of forces*, which was discovered by the Flemish physicist Simon Stevin in the seventeenth century. It is best illustrated by an experiment (Fig. 4.3.2).

It follows from Fig. 4.3.2 that an inclined force T_{AB}, represented by the line AB in Fig. 4.3.3, can be resolved into a vertical component force T_{AC} and a horizontal component force T_{BC}. We can proceed from A to B directly or via C; the result is the same. If θ is the angle of inclination of the line AB to the horizontal (Fig. 4.3.3), the line $AC = AB \sin \theta$, and the line $BC = AB \cos \theta$. Therefore the vertical and horizontal component forces are

$$T_{AC} = T_{AB} \sin \theta \tag{4.1}$$

and

$$T_{BC} = T_{AB} \cos \theta \tag{4.2}$$

We can thus replace all inclined forces by their vertical and horizontal components and consider the equilibrium of the structure under the action of horizontal and vertical forces only.

Conversely, a force of 2.5 kN in the direction shown can be *resolved* into a horizontal component of 1.5 kN and a vertical component of 2.0 kN.

We can prove the correctness of the triangle of forces by means of an experiment. Join a vertical spring balance, a horizontal spring balance, and a third spring balance at the angle shown by means of strings. When the third balance indicates 2.5 kN, the horizontal spring balance indicates 1.5 kN and the vertical spring balance indicates 2.0 kN. Each force balances the other two.

4.4 The Conditions of Equilibrium

In Fig. 4.3.3 the component of the vertical force T_{AC} in the direction of the horizontal force is $T_{AC} \sin 0° = 0$, and thus the vertical force has no effect on the horizontal force. Similarly, $T_{BC} \cos 90° = 0$, and the horizontal force has no effect on the vertical force.

It follows that we need at least *two* conditions to determine whether a set of forces is in equilibrium. First, all the horizontal forces must balance one another; that is, the sum of all horizontal forces must be zero (Fig. 4.4.1), or

$$\Sigma R_H = 0 \qquad (4.3)$$

4.3.3. T_{AC} is the vertical component of T_{AB}: $T_{AC} = T_{AB} \sin \theta$.
T_{BC} is the horizontal component of T_{AB}: $T_{BC} = T_{AB} \cos \theta$.

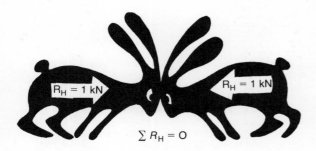

4.4.1. Horizontal equilibrium.

Second, all the vertical forces must balance one another; that is, the sum of all the vertical forces must be zero (Fig. 4.4.2), or

$$\Sigma R_V = 0 \qquad (4.4)$$

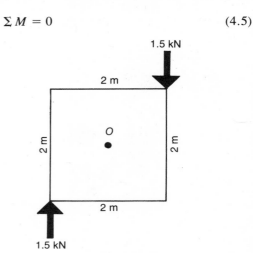

4.4.2. Vertical equilibrium.

This ensures that the structure cannot be moved either vertically or horizontally, but it still leaves the possibility that it can be rotated (Fig. 4.4.3). Evidently, we need a third condition of equilibrium, namely, that the sum total of all the moments of all the forces about any convenient point be zero (Fig. 4.4.4), or

$$\Sigma M = 0 \qquad (4.5)$$

4.4.3. Let us consider a 2 m square slab pivoted at its centroid, O. It is pushed at opposite corners by two equal forces, one upward and one downward. The two forces balance, thus satisfying the condition of vertical equilibrium. There are no horizontal forces, so that the condition of horizontal equilibrium is satisfied. Nevertheless, the slab will evidently rotate in a clockwise direction, so that it is not in equilibrium.

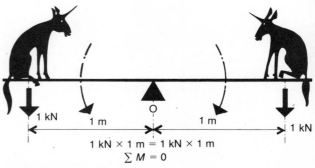

4.4.4. Moment equilibrium.

4.5 Bending Moments and Shear Forces: Why Do Structural Engineers Talk So Much About Them?

Our task is to devise a structural system that safely resists the loads discussed in Chapter 2 at the smallest possible cost. It must also produce an attractive-looking building, and it must not interfere with the building's environmental functions (Chapter 11).

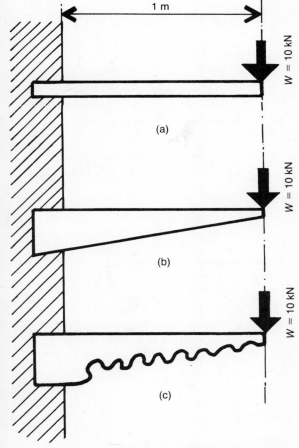

4.5.1. A cantilever carrying a single concentrated load of 10 kN at a distance of 1 m from a wall.
(a) Cantilever of uniform depth.
(b) Cantilever of uniformly varying depth.
(c) Cantilever of free shape.

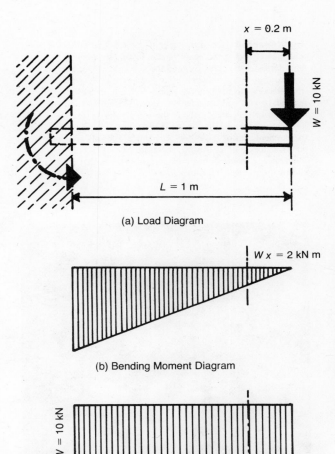

4.5.2. Bending moment and shear force diagrams for a cantilever.
(a) The cantilever of Fig. 4.5.1.a is "cut" 0.2 m from the end.
(b) First consider the moment equilibrium ($\sum M = 0$). The load of 10 kN at a distance of 0.2 m has a moment of 2 kN m about the "cut," or "section." This is the *bending moment* of the force 10 kN at that section. The cantilever must be strong enough to resist this bending moment. We can determine the bending moment at other sections; at 0.4 m from the load 10 kN the bending moment is 4 kN m, and at the support of the cantilever it is 10 kN m. We can plot the variation of the bending moment along the length of the cantilever. This is called a bending moment diagram.

In order to solve this problem we work out the shear forces (see Glossary) and bending moments produced by the loads. Let us consider a single vertical force of 10 kN at the end of a 1-meter cantilever projecting from a wall. It does not matter whether this cantilever is of uniform depth (Fig. 4.5.1.a) or of variable depth (Fig. 4.5.1.b and c); the bending moment at the wall, which is

$$M = 10 \text{ kN} \times 1 \text{ m} = 10 \text{ kN m}$$

and the connection of the cantilever to the wall must be strong enough to resist it.

Let us examine the strength required of the cantilever to carry the load of 10 kN (Fig. 4.5.2). The cantilever must be strong enough to resist the bending moment, whose variation is shown in Fig. 4.5.2.b, and strong enough to resist the shear force, which in this case is the same at every section of the cantilever (Fig. 4.5.2.c). In practice the bending moment is more likely to be critical for the sizing of structural members in a building (see Chapter 6). Since this increases uniformly, the cantilever of variable depth shown in Fig. 4.5.1.b has some merit. However, the material saved may not compensate for the extra labor required to vary the depth.

(c) Consider the vertical moment equilibrium at the section ($\sum R_V = 0$). There is an unbalanced vertical force of 10 kN, which tends to cut or shear through the cantilever. This is called the *shear force*, and it is resisted by an equal and opposite force within the cantilever. In this particular example the shear force is the same at every section (namely, 10 kN). The shear force diagram shows the shear force at each section of the cantilever.

In practice the bending moment is more likely to be critical for the sizing of the structural member in a building, and we will in future examples determine only the variation of the bending moment.

4.6 A Bevy of Beams

Cantilevers usually occur in buildings as projections (Fig. 4.6.1), and their span is mostly small. This is fortunate, because they are not an efficient type of beam. We noted in Fig. 4.5.2 that a concentrated load of 10 kN on a 1 m cantilever in the worst position for the load (at the end of the cantilever) produced a maximum bending moment of 10 kN m. If the same load is placed on a beam of 1 m span that is simply supported at its ends, the maximum bending moment is only 2.5 kN m when the load is in its worst position (in the middle of the beam or slab) (Fig. 4.6.2).

4.6.2. (a) Consider a beam or slab simply supported at its ends over a span of 1 m, carrying a concentrated load of 10 kN at mid-span. Because the beam or slab is simply supported, it can bend freely at the supports, and the support reactions are vertical only. By symmetry, each is 5 kN. (b) Now make an imaginary cut in the beam or slab 0.001 mm to the left of the load of 10 kN, and consider the resulting bending moment that the beam or slab must resist at the section $A-A$. (c) The moment is formed by a load of 5 kN (the left-hand reaction) and the moment arm of 0.499 999 m, which is for all practical purposes the same as 0.5 m. Therefore $M = 5 \times 0.5 = 2.5$ kN m. The bending moment increases uniformly up to that point and then decreases uniformly to zero (d).

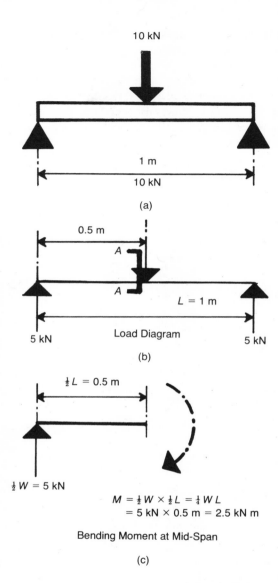

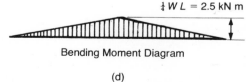

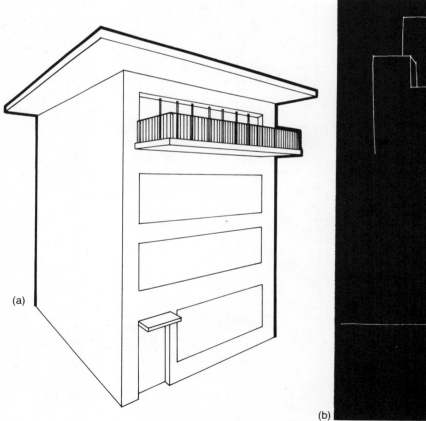

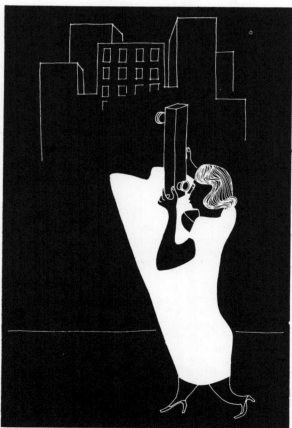

4.6.1. Cantilevers usually occur in buildings as projections.

In actual fact, most loads are distributed (Fig. 4.6.3), although not necessarily uniformly. Thus a 150 mm thick concrete slab is 150 mm thick over the entire span, and it is therefore a uniformly distributed load. The people standing on the concrete slab are not normally uniformly distributed, but for the sake of simplicity we assume that they are. Thus we treat all live loads (except for some special heavy loads; see Section 2.2) as if they had been put through a mincer and the mince had been spread uniformly over the entire span.

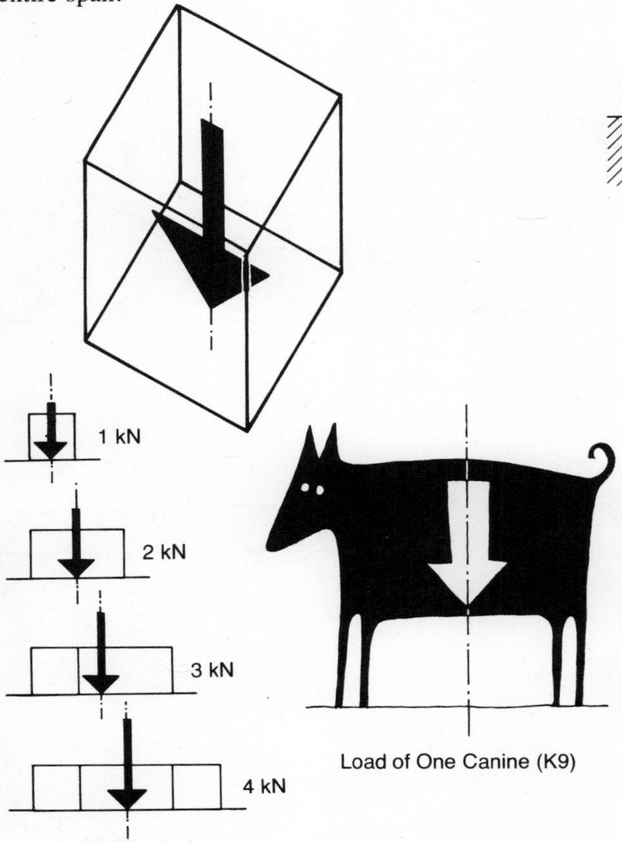

4.6.4. A simply supported beam or slab carrying a concentrated load (a) and a uniformly distributed load (b). The (maximum) bending moments at mid-span are calculated in (c) and (d). The bending moment diagram for a concentrated load is linear (e). The bending moment diagram for a uniformly distributed load is parabolic (f). The maximum bending moment due to a uniformly distributed load is only half as much as that for the same total load concentrated at mid-span.

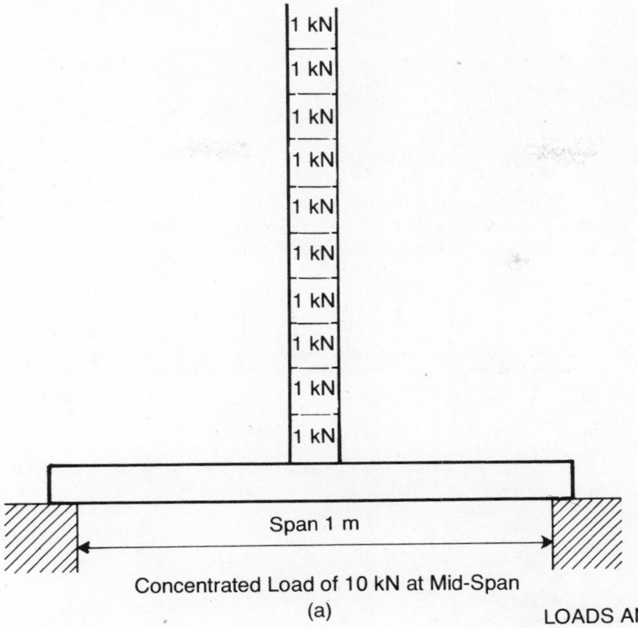

Concentrated Load of 10 kN at Mid-Span
(a)

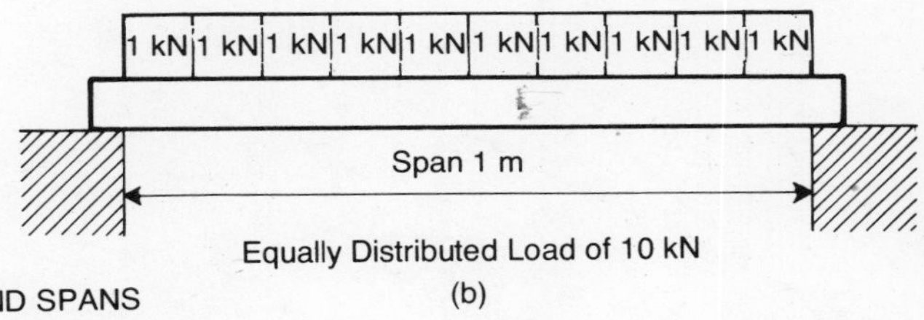

Equally Distributed Load of 10 kN
(b)

LOADS AND SPANS

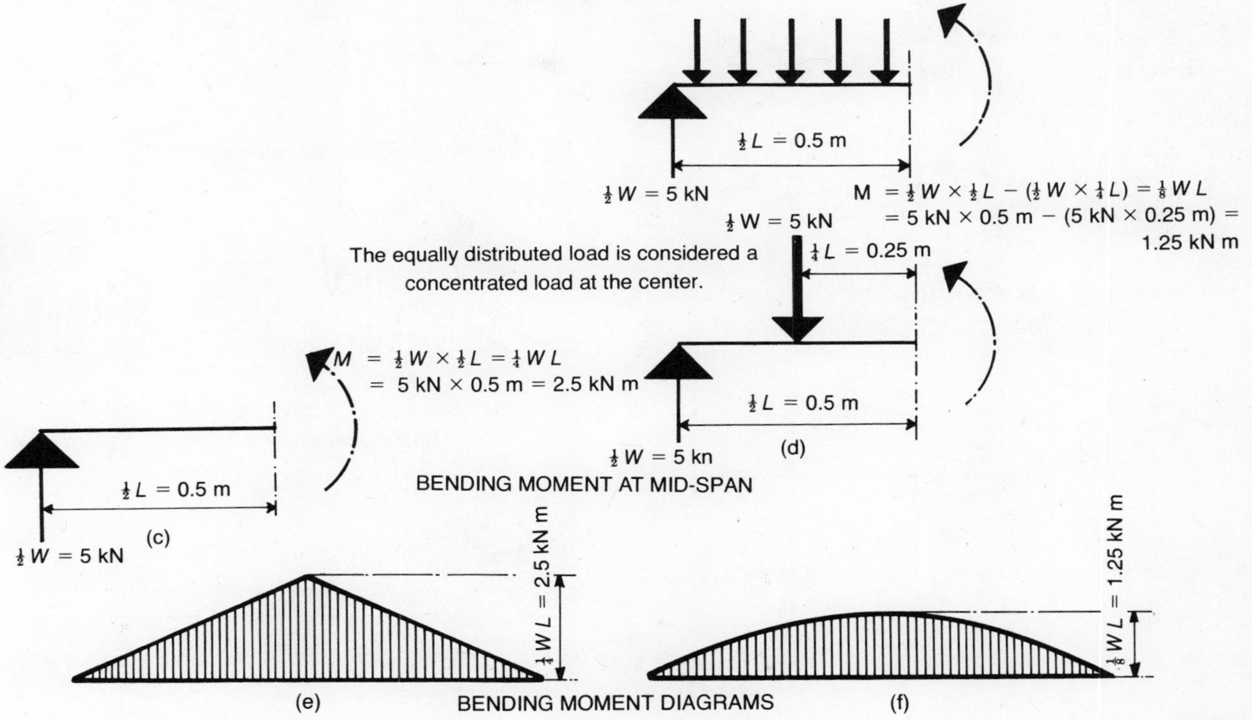

$\frac{1}{2}W = 5$ kN, $\frac{1}{2}L = 0.5$ m

$M = \frac{1}{2}W \times \frac{1}{2}L = \frac{1}{4}WL$
$= 5$ kN $\times 0.5$ m $= 2.5$ kN m

(c)

$\frac{1}{2}W = 5$ kN, $\frac{1}{4}L = 0.25$ m, $\frac{1}{2}L = 0.5$ m

$M = \frac{1}{2}W \times \frac{1}{2}L - (\frac{1}{2}W \times \frac{1}{4}L) = \frac{1}{8}WL$
$= 5$ kN $\times 0.5$ m $- (5$ kN $\times 0.25$ m$) = 1.25$ kN m

The equally distributed load is considered a concentrated load at the center.

(d)

BENDING MOMENT AT MID-SPAN

$\frac{1}{4}WL = 2.5$ kN m

$\frac{1}{8}WL = 1.25$ kN m

(e) BENDING MOMENT DIAGRAMS (f)

Load of One Canine (K9)

1 kN
2 kN
3 kN
4 kN

4.6.3. Loads (W) are considered to act at their center in moment diagrams.

59

The maximum bending moment due to a uniformly distributed load is only half that due to a concentrated load of the same magnitude in its worst position (Figs. 4.6.4 and 4.6.5). Even greater reductions in bending moment can be achieved by building the ends of the beam or floor slab firmly into the supports (Fig. 4.6.6) or by making the beam or floor slab continuous over the supports (Figs. 4.6.7 and 4.6.8).

4.6.6. (a) A beam or slab carrying a uniformly distributed load with simple (that is, purely vertical) supports. The deflected shape is shown in (c) and the bending moment diagram in (e). (b) Beam or slab carrying a uniformly distributed load with built-in supports, which provide both moment and vertical restraints as in a cantilever. The deflected shape is shown in (d), and the bending moment diagram in (f). The maximum bending moment is only two-thirds that of a simply supported beam or slab. The built-in beam or slab curves concavely near the supports and convexly at mid-span (d). The concave curvature is called *negative*, and it produces a negative bending moment. The convex curvature is called *positive*, and it produces a positive bending moment (f). In a reinforced concrete beam or slab the reinforcement is required where the beam is in tension (Fig. 3.4.1.), that is, at the bottom of the beam or slab for a positive bending moment and at the top of the beam or slab for a negative bending moment (g and h).

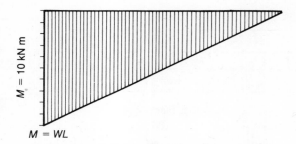

(a) Cantilevered Beam with Concentrated Load at End
$L = 1$ m $\quad\quad W = 10$ kN

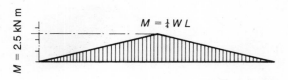

(b) Concentrated Load at Center of Simply Supported Span
$L = 1$ m $\quad\quad W = 10$ kN

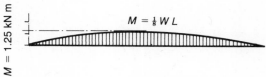

(c) Uniformly Distributed Load at Center of Supported Span
$L = 1$ m $\quad\quad W = 10$ kN

4.6.5. Comparison of bending moment diagrams.

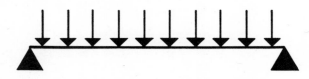

(a) Load Diagram: Uniformly Loaded Beam with Vertical Supports

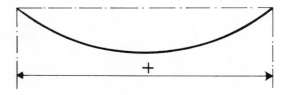

(c) Deflected Shape (exaggerated) and Sign of Curvature

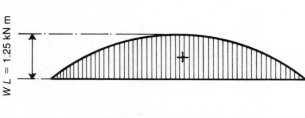

(e) Bending Moment Diagram

(g) Layout of Reinforcement in Concrete Beam (not to scale)

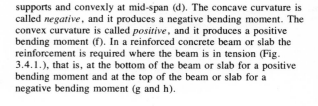

(b) Load Diagram: Uniformly Loaded Beam with Fixed Ends

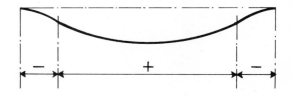

(d) Deflected Shape (exaggerated) and Sign of Curvature

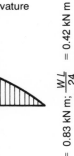

(f) Bending Moment Diagram

(h) Layout of Reinforcement in Concrete Beam (not to scale)

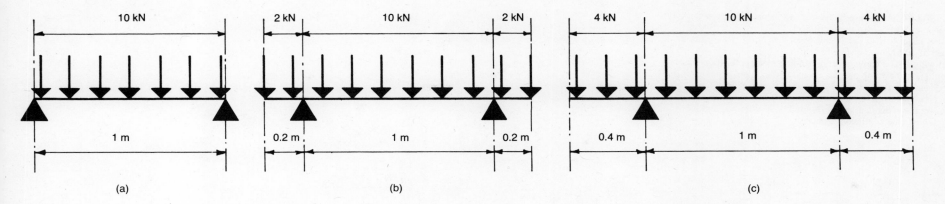

Load Diagrams

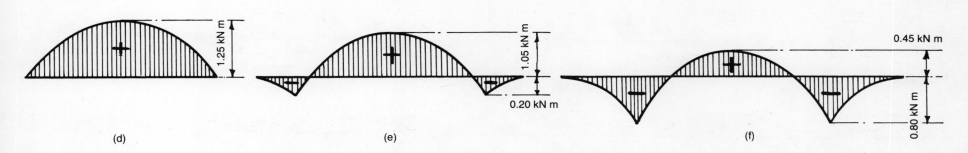

Bending Moment Diagrams

4.6.7. A simply supported beam or slab (a) without overhangs, (b) wth cantilever overhangs each equal to 20% of the span, and (c) with cantilever overhangs each equal to 40% of the span. The corresponding bending moment diagrams (in kN m) are shown in (d), (e), and (f). The cantilever moments are negative; that is, reinforcement in concrete beams or slabs is required on top. They consequently reduce the positive moment in the simply supported span. The 40% cantilever overhang actually reduces the positive bending moment at mid-span so much that the negative bending moment, at the supports, is now bigger. The biggest bending moment determines the size of the beam.

4.6.8. (a) Three simply supported beams or slabs, each carrying a load of 10 kN over a span of 1 m.

(b) A beam or slab continuous over three spans, each carrying a load of 10 kN over a span of 1 m.

(c) and (d) The corresponding bending moment diagrams. The continuous beam or slab has a smaller maximum bending moment, but it varies from positive to negative. In a concrete beam or slab, reinforcement is required at the bottom only for the simply supported beams (e), and both at the top and the bottom for the continuous spans (f).

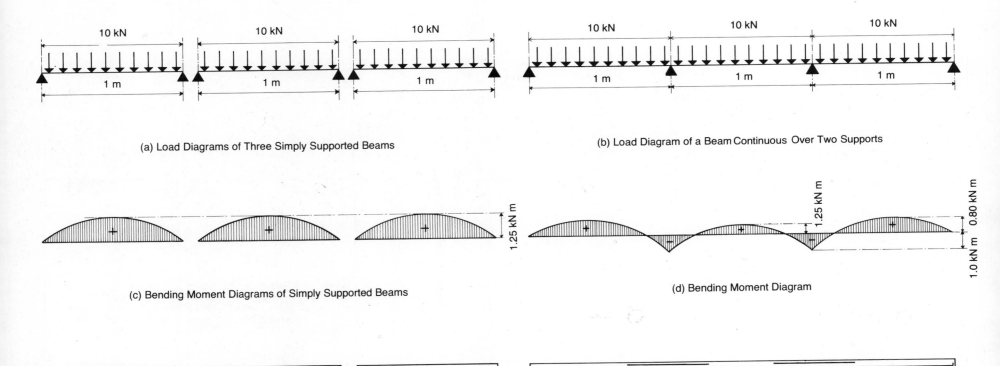

(a) Load Diagrams of Three Simply Supported Beams

(b) Load Diagram of a Beam Continuous Over Two Supports

(c) Bending Moment Diagrams of Simply Supported Beams

(d) Bending Moment Diagram

(e) Layout of Reinforcement in Simply Supported Beams or Concrete Slabs Without Restraints (not to scale)

(f) Layout of Reinforcement in Continuous Beam or Concrete Slab (not to scale)

Table 4.1. Maximum Bending Moments for Single-span Beams

	Loading Diagram Cantilevers	Maximum Bending Moment Is*	Occurs at
1	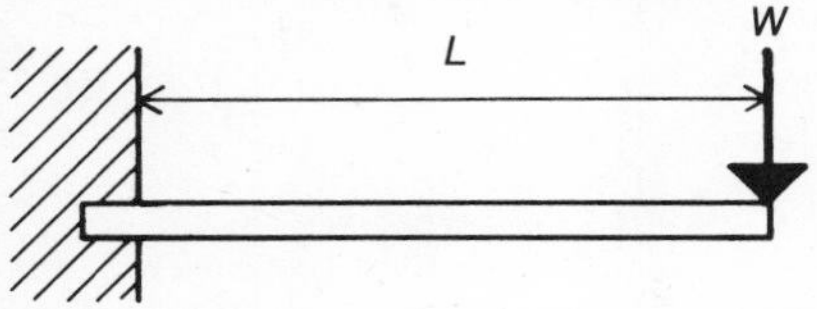 Point load W at free end, span L	$-WL$	Support
2	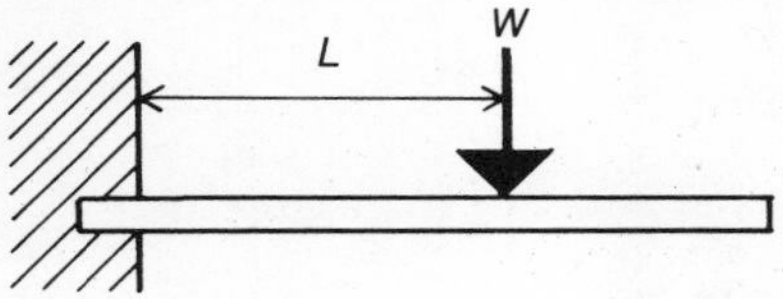 Point load W at distance L from support	$-WL$	Support
3	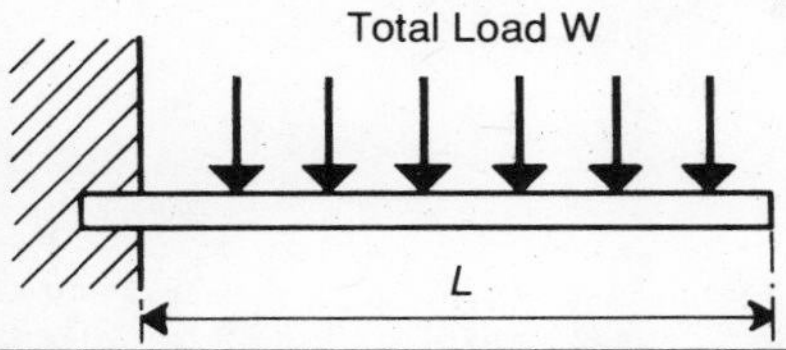 Total Load W uniformly distributed, span L	$-\tfrac{1}{2}WL$	Support
4	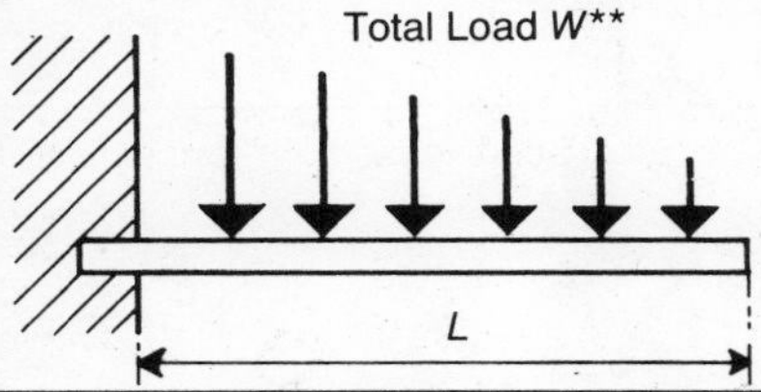 Total Load W** varying uniformly (max at support), span L	$-\tfrac{1}{3}WL$	Support
5	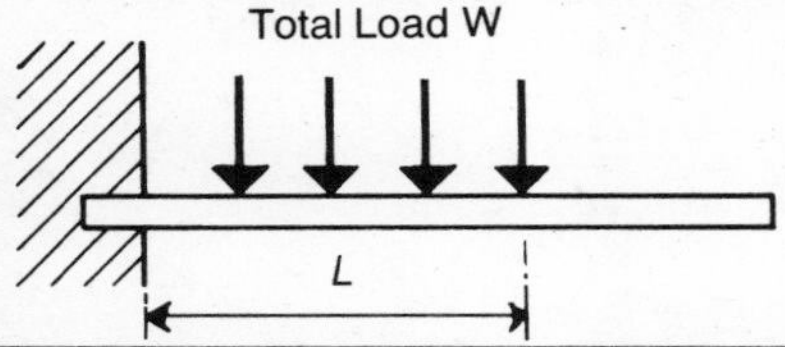 Total Load W uniformly distributed over part of span	$-\tfrac{1}{2}WL$	Support

Table 4.1. Continued

	Loading Diagram Simply Supported Beams	Maximum Bending Moment Is*	Occurs at
6	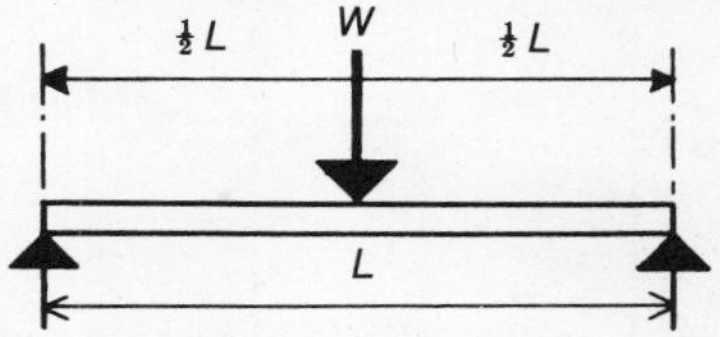 Point load W at mid-span ($\tfrac{1}{2}L$ each side), span L	$+\tfrac{1}{4}WL$	Mid-span
7	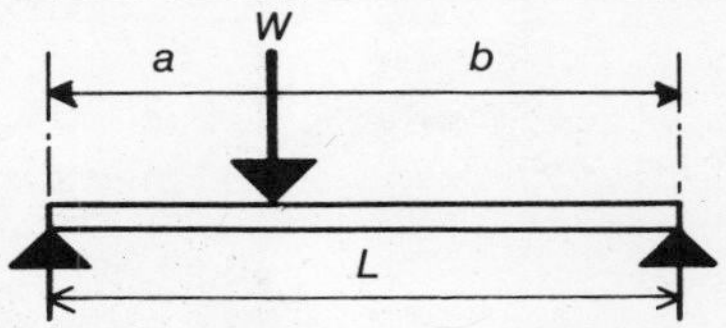 Point load W at distances a and b from supports, span L	$+\dfrac{ab}{L}W$	Under the Load
8	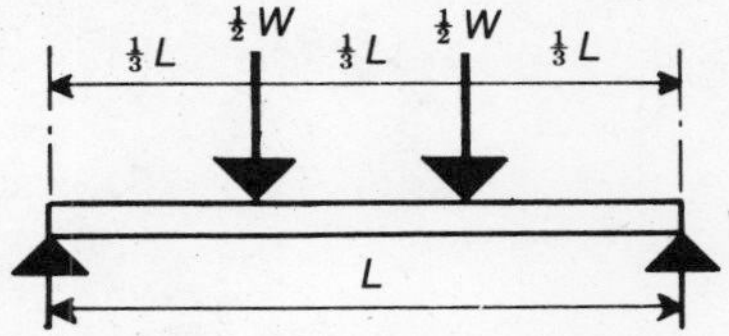 Two loads of $\tfrac{1}{2}W$ at third-points ($\tfrac{1}{3}L$ spacing), span L	$+\tfrac{1}{6}WL$	Mid-span and Under the Load
9	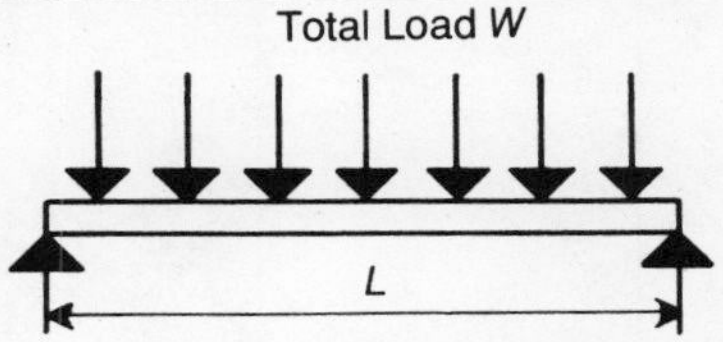 Total Load W uniformly distributed, span L	$+\tfrac{1}{8}WL$	Mid-span
10	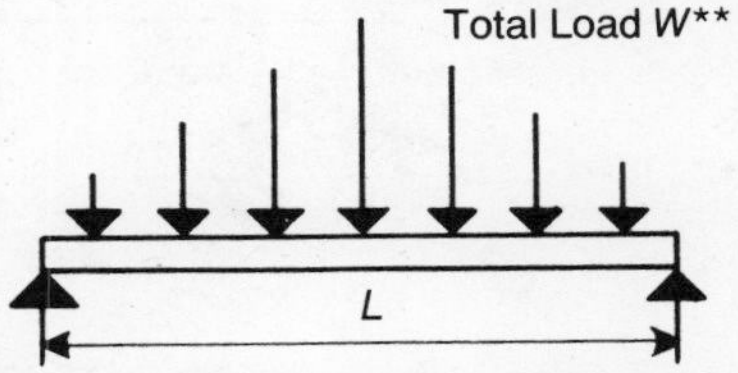 Total Load W** varying uniformly in magnitude, span L	$+\tfrac{1}{6}WL$	Mid-span

*The terms "positive bending moment" and "negative bending moment" are defined in the Glossary.
**Varying uniformly in magnitude as shown above.

Table 4.1. Continued

	Loading Diagram Built-in Beams	Maximum Bending Moment Is*	Occurs at
11	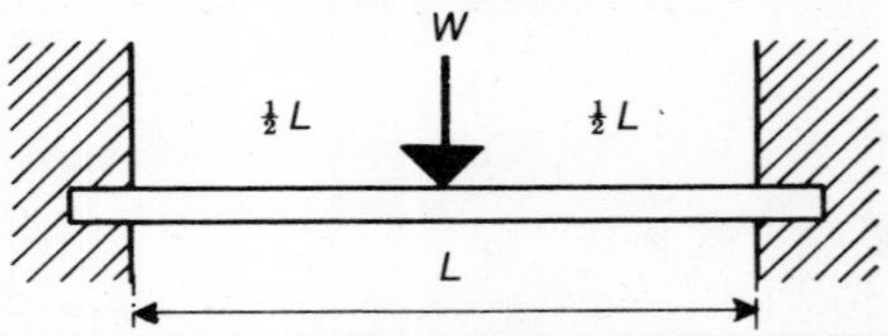 W at mid-span ($\frac{1}{2}L$, $\frac{1}{2}L$), span L	$-\frac{1}{8}WL$ $+\frac{1}{8}WL$	Supports Mid-span
12	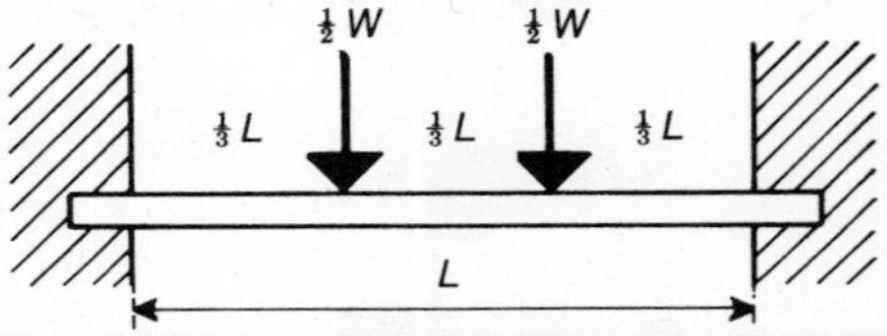 Two loads of $\frac{1}{2}W$ at third points ($\frac{1}{3}L$, $\frac{1}{3}L$, $\frac{1}{3}L$)	$-\frac{1}{9}WL$	Supports
13	Total Load W, 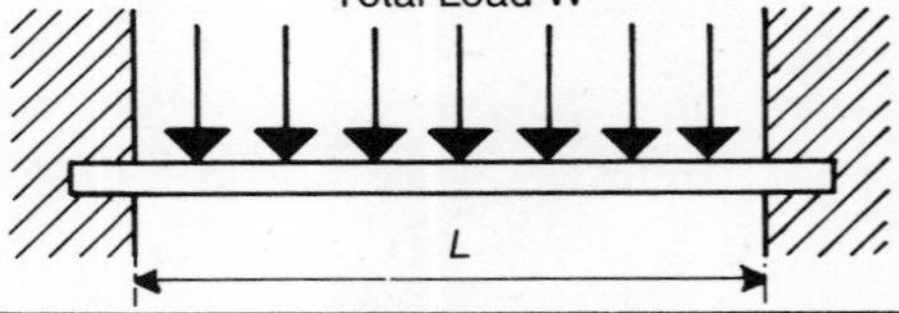 uniformly distributed, span L	$-\frac{1}{12}WL$	Supports
14	Total Load W**, 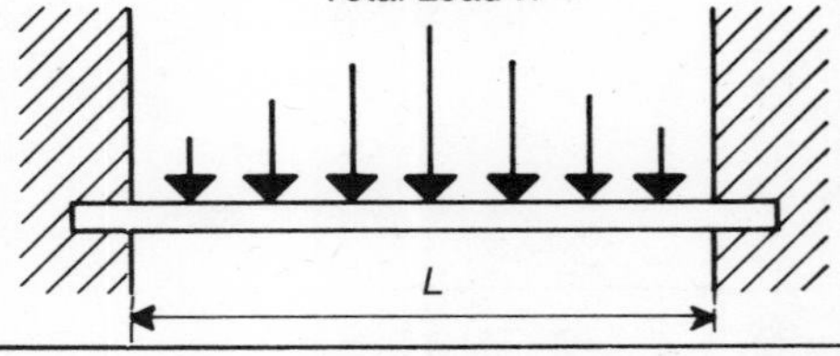 varying uniformly, span L	$-\frac{5}{48}WL$	Supports

*The terms "positive bending moment" and "negative bending moment" are defined in the Glossary.
**Varying uniformly in magnitude as shown above.

Table 4.2. Maximum Bending Moments for Continuous Beams

	Loading Diagram	Maximum Bending Moment Is*	Occurs at
	Load W at Center Point of each Span of Length L		
1	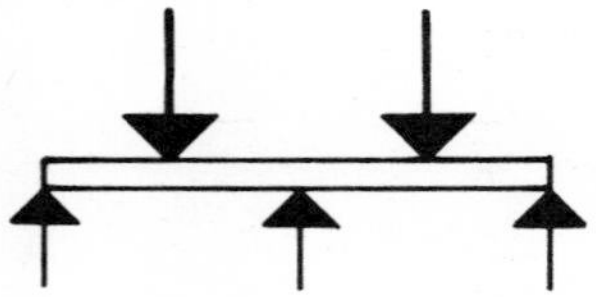 Two-span continuous beam	$-0.188\,WL$	Interior Support
2	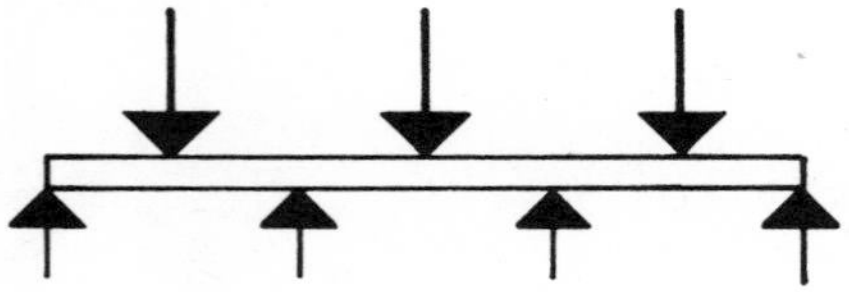 Three-span continuous beam	$-0.150\,WL$	Interior Supports
3	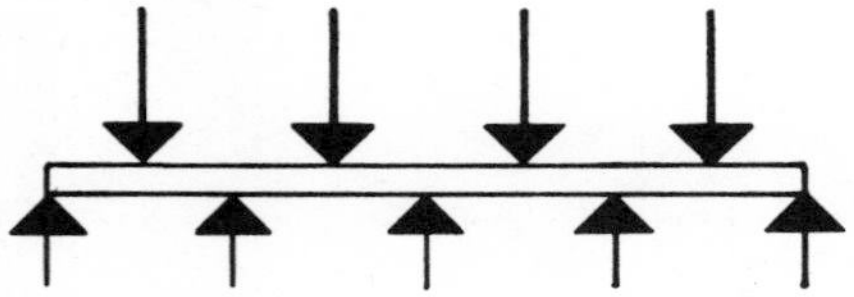 Four-span continuous beam	$-0.161\,WL$	First Interior Supports

*The terms "positive bending moment" and "negative bending moment" are defined in the Glossary.

Table 4.2. Continued

	Loading Diagram	Maximum Bending Moment Is*	Occurs at
	Loads W at Third Points of Each Span Total Load W per Span of Length L		
4		−0.167 W L	Interior Support
5		−0.134 W L	Interior Supports
6		−0.143 W L	First Interior Supports
	Total Load W Uniformly Distributed Over Each Span of Length L		
7		−0.125 W L	Interior Support
8		−0.100 W L	Interior Supports
9		−0.107 W L	First Interior Supports

*The terms "positive bending moment" and "negative bending moment" are defined in the Glossary.

4.7 Curved Structures and Trusses Use Less Material for Horizontal Spans

The beams we have considered so far are suitable for small and medium spans. For really long spans we require a structure that is either deep or curved, and we will discuss these long-span roof structures in more detail in Chapter 10. In theory the most efficient structural form is the suspension cable, although in practice it presents waterproofing and vibration problems (Chapter 10).

A cable offers no resistance to bending whatever (Fig. 4.7.1), because it is flexible. It carries its load by virtue of its shape (Fig. 4.7.2), and it is purely in tension. The cable tension T at mid-span (Fig. 4.7.2.c) is equal to the simply supported bending moment for the uniformly distributed load, $\frac{1}{8} WL$, divided by the sag of the cable, s:

$$T = \frac{WL}{8s} \tag{4.6}$$

4.7.1. A flexible cable offers no resistance to bending whatever.

(a)

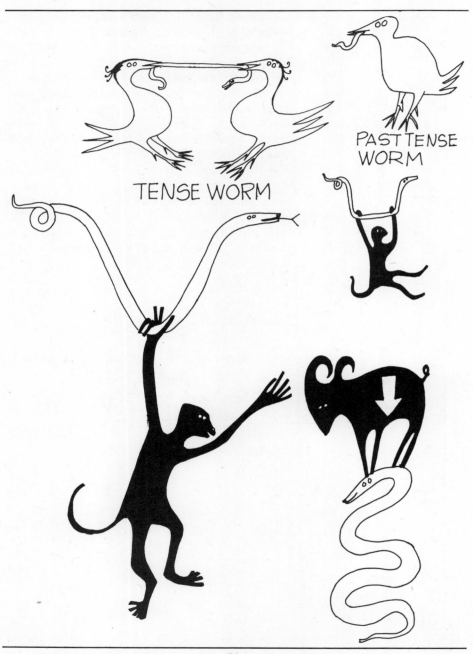

(b)

4.7.1(b). A flexible cable offers no resistance to bending whatever.

4.7.2. A suspension cable carrying a uniformly distributed load.
(a) A suspension cable carrying a load uniformly distributed along its span hangs in a parabolic curve. It requires both the vertical reactions needed for a simply supported beam and horizontal reactions to stop it from collapsing inwards.
(b) We can divide this structural system into two parts: the first comprises the uniformly distributed load, W, and the vertical reactions, R_V, and the second the horizontal reactions R_H acting on the cable with a sag s.
(c) We cut each of these systems at mid-span. The uniformly distributed load W over a span L produces a bending moment $M_1 = \frac{1}{8}WL$, as in a simply supported beam. Let us assume that the cable tension at mid-span is T. Then for horizontal equilibrium the horizontal reaction is

$$R_H = T$$

and its moment about the lowest point in the cable at mid-span is

$$M_2 = T s$$

where s is the sag of the cable.

We know that the cable cannot resist a bending moment because it is flexible. Therefore

$$M = 0 \text{ and } M = M_1 + M_2 = 0$$

Consequently

$$T s = \tfrac{1}{8} W L$$

and the cable tension is

$$T = \frac{WL}{8\,s} = \frac{100 \times 100}{8 \times 20} = 62.5 \text{ kN}$$

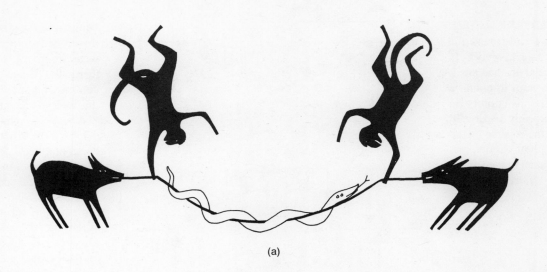

(a)

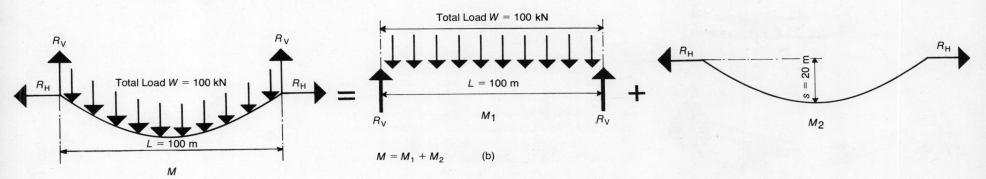

$M = M_1 + M_2$ (b)

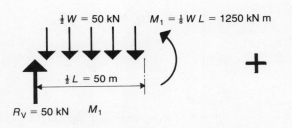

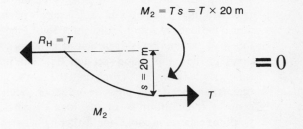

$M_1 + M_2 = 0$ (c)

We can turn the flexible cable upside down (Fig. 4.7.3). If we are making a model, we must freeze the shape first with a lacquer (such as hairspray). In building practice, we use a stiff material (such as steel, laminated timber, or reinforced concrete) to obtain a parabolic arch or arched vault that is entirely in compression under the action of a uniformly distributed load (Fig. 4.7.4). This is precisely the reverse of the situation with a suspension cable, and the arch compression C at mid-span (Fig. 4.7.4.b) is equal to the simply supported bending moment, $\frac{1}{8} WL$, divided by the rise of the arch, r:

$$C = \frac{WL}{8r} \quad (4.7)$$

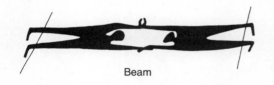
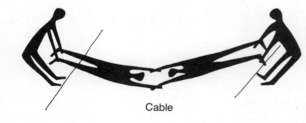
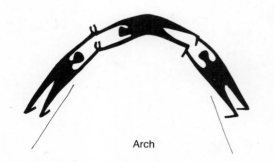

4.7.3. Beam, cable and arch.

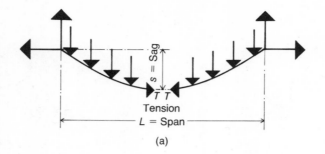
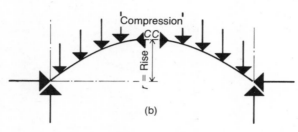

4.7.4. The parabolic arch (b) is the reverse of the parabolic suspension cable (a). Under the action of a uniformly distributed load, the arch is entirely in compression.

Because traditional masonry construction was incapable of resisting substantial tensile stresses, the longer-spanning roofs in Roman, medieval, and Renaissance architecture were designed to be in pure compression, as nearly as was possible with the empirical methods of the time (Section 1.2, Ref. 1.4). The resulting domes and cross-vaults are among the most beautiful structures ever built.

Even when modern structural theory is used, compression structures cannot ordinarily be entirely free from bending. The parabolic arch is purely in compression if the load is uniformly distributed along the span. If this load distribution is disturbed, bending is superimposed on the compression (Section 7.7). This happens, for example, if workers climb onto the roof to do some maintenance or if wind pressure or suction (Section 2.3) disturbs the uniformity of the load distribution. A flexible suspension cable would alter its shape to adjust itself to the new load distribution (Fig. 4.7.5) and would thus remain purely in tension; an arch or arched vault, being rigid, must maintain its shape. On the other hand, the arched vault, because of its rigidity, is easier to waterproof and easier to insulate acoustically and thermally (Chapter 10). It remains predominantly a compression structure, and we will see in Chapter 7 why this uses so much less material than an ordinary beam.

A bending moment is produced by a load system (Fig. 4.7.6). If the structure that supports the load system is a horizontal beam, then the bending moment produces bending stresses in the structural material (Section 5.4). The longer the span of the beam, the more structural material is required. The weight of this material constitutes a load on the beam, and this increases rapidly as the span increases. The resulting structure is heavy and expensive. As we have seen, the bending moment can be resisted purely by tension *or* by compression, and this requires far less material.

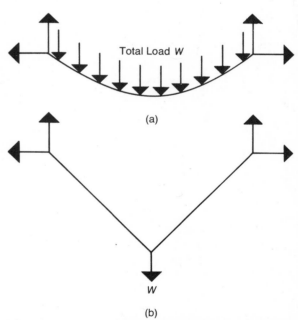

4.7.5. A flexible suspension cable alters its shape to adjust itself to a new load distribution.
(a) Uniformly distributed load.
(b) Central concentrated load.

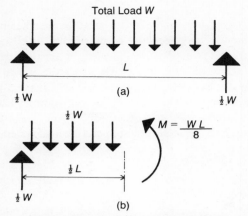

4.7.6. A bending moment is produced by the load system, not by the structural system. The bending moment due to a uniformly distributed horizontal load and two vertical reactions produces a bending moment $M = \frac{1}{8}WL$, irrespective of the shape of the structure.

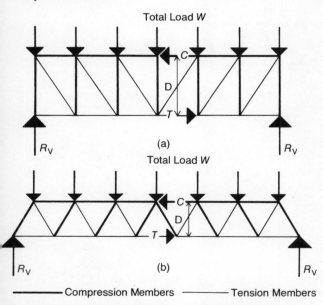

—— Compression Members   ——— Tension Members

4.7.7. A truss of straight tension and compression members is an efficient structure for carrying a uniformly distributed load over a long span. The top chord is in compression and the bottom chord in tension.
(a) Pratt truss with vertical compression and inclined tension members.
(b) Warren truss with inclined tension and compression members.

An economical structural system can also be produced from a mixture of tension and compression members in the short connecting members (Fig. 4.7.7). If the truss carries a total load W uniformly distributed over a span L, then the maximum bending moment at mid-span is $\frac{1}{8}WL$. This is resisted by a compressive force in the middle of the top chord, C, and a tensile force in the middle of the bottom chord, T. If the overall depth of the truss is D, then the resistance moment M formed by the compressive and the tensile force is

$$M = \tfrac{1}{8} WL = CD = TD \qquad (4.8)$$

Thus the *structural system* resists a bending moment, but the individual members of the system are either in tension or in compression; none of the *members* of the system are in bending.

4.8 A Note on Vertical Spans

In the earliest iron-framed and steel-framed buildings erected in Chicago and other American and European cities (Section 4.1), the main problems in the design of the superstructure were the beams that carried the vertical load of the floors and the columns that transmitted this load to the foundations. In the 1880s the effect of wind was considered for some of the taller buildings, and diagonals were inserted to resist this horizontal load (Fig. 4.8.1.a). For most buildings, however, the external walls were sufficient to resist wind forces (Fig. 4.8.1.b).

In the 1890s building heights increased to an extent where it became necessary to consider the horizontal wind loads on the building (Fig. 2.3.5) as if they were acting on a vertical cantilever. The problem (Fig. 4.8.2) is in principle the same as for the cantilever in Figs. 4.5.1 and 4.5.2, but there are two important differences. First, the height of really tall buildings is much greater than the greatest interior spans (Fig. 2.3.6), so that we are dealing with larger bending moments. This is more than compensated by the second difference, the fact that the entire width of the building is available to resist these bending moments, whereas for an interior span it is usually necessary to keep the depth of the roof structure small. In practice therefore the structure is rarely a limiting factor in the height of a building.

The various methods used for the design of tall buildings are discussed in detail in Chapter 8.

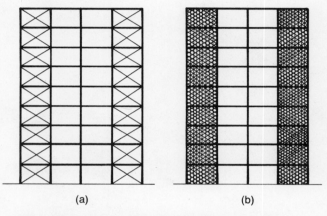

4.8.1. Wind bracing for multistory buildings in the late nineteenth century, (a) by diagonal bracing and (b) by solid wall panels or shear walls.

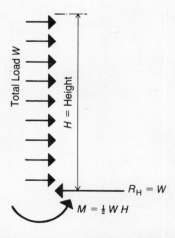

4.8.2. The wind loads produce a cantilever bending moment $M = \frac{1}{2}WH$ that is independent of the structure of the building.

PART 2 PRELIMINARY STRUCTURAL DESIGN

Chapter 5 Structural Members

I have the simplest tastes. I am always satisfied with the best. (Oscar Wilde)

Complex structures are assembled from members with comparatively simple properties. We consider the design problems of ties, columns, and beams, the choice of various structural materials for each, and the methods of jointing them. Shear and torsion are discussed in Chapter 6.

5.1 Structural Members and Structural Assemblies

Structural members (Fig. 5.1.1) may be subjected to:

Pulling or tension (a force)

Pushing or compression (a force)

Cutting or shear (a force)

Bending or flexure (a moment)

Twisting or torsion (a moment)

or to any combination of these. For example, most columns in structural frames are subject to some bending (Sections 8.3 and 8.4), and they are therefore not under pure compression but under combined compression and bending. The spandrel beams or edge beams of a floor are subject to some torsion, and they are therefore under combined bending and torsion. Tension, compression, and bending are considered in this chapter, shear and torsion in Chapter 6.

Because stress is proportional to strain (Section 3.1), stresses are usually proportional to the forces and moments that cause them. It is then possible to work out each stress separately and add all of them up, that is, superpose them on one another. This is called the *principle of superposition*, and it greatly simplifies structural calculations. We will therefore discuss each force and moment in turn, and we can then superpose them as required.

The forces or moments to which structural assemblies are subjected may be quite different from those acting on their members. Let us consider as a simple example the truss shown in Fig. 4.7.7.b. The truss carries a vertical load, which produces a bending moment (Fig. 4.7.6). The truss is therefore subjected to bending. However, the individual members of the truss are either in tension or in compression; none are in bending. The mechanics of the structure can therefore be examined at two levels: The truss can be considered as a whole (in bending), or the members of the truss can be considered individually (in tension or compression).

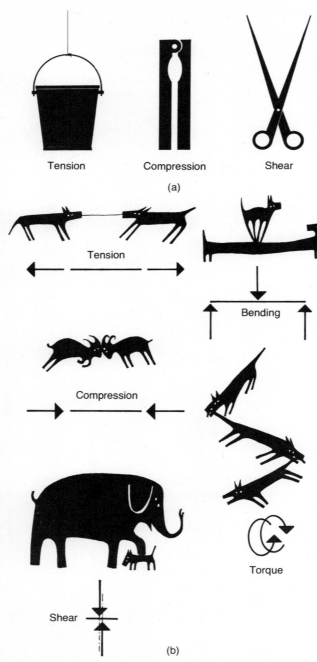

5.1.1. Basic forces and moments acting on structural members.

This book is concerned with the choice of structure, not with the detailed design of structures. We will therefore consider structural members only inasmuch as they affect structural decisions. A more detailed consideration of structural members can be found in any of a number of more advanced textbooks; this particular problem, for example, is solved by four different methods in Ref. 1.1, pp. 83–99.

For readers who wish to take the problem a little further, we will determine two of the forces. The following problem can be omitted without loss of continuity. This topic is considered further in Section 7.9.

Problem 5.1 We will determine the forces in two of the members in the truss shown in Fig. 4.7.7.b. These are shown enlarged in Fig. 5.1.2.

Let us assume that the truss carries a load of 16 kN/m over a span of 15 m. Then the vertical reaction $R_V = 16 \times 7.5 = 120$ kN (half the total load). From Eq. (4.1) in Section 4.3,

$$R_V = T_{AB} \sin 60°$$

which makes

$$T_{AB} = \frac{120}{0.866} = 138.6 \text{ kN}$$

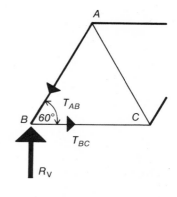

5.1.2. Enlargement of left-hand end, Fig. 4.7.7.b *(Problem 5.1)*.

This is the "internal resistance" of the member. The external forces acting on the member are equal and opposite (Fig. 5.1.3). The member AB is therefore subject to a compressive force of 138.6 kN.

From Eq. (4.2)

$$T_{BC} = T_{AB} \cos 60° = 138.6 \times 0.5 = 69.3 \text{ kN}$$

This force acts in the opposite sense from T_{AB}, so that the member BC is subject to a tensile force of 69.3 kN.

For a load of 1 kip/ft and a span of 50 ft the solution is as follows:

$$R_V = 1 \times 25 = 25 \text{ kips}$$

$$T_{AB} = \frac{25}{.866} = 28.87 \text{ kips}$$

$$T_{BC} = 28.87 \times 0.5 = 14.43 \text{ kips}$$

The member AB is subject to a compressive force of 28.87 kips, and the member BC is subject to a tensile force of 14.43 kips.

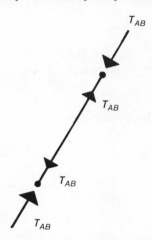

5.1.3. Internal resistance forces in the compression member AB, and the equivalent external forces.

5.2 Tension Members

We noted in Section 5.1 that a truss, as a structural assembly, is subjected to bending but that its members are in tension or compression. Half the members are in tension, and the other half are in compression. The same applies to bracing members in rectangular frames (Fig. 4.8.1.a). Under any system of horizontal loads (wind or earthquake), half the members are in tension and half are in compression. However, as we will see presently, slender compression members tend to buckle. A small amount of sideways buckling does no damage to a slender compression member, but it ceases to carry any further compressive force. Consequently the bracing acts effectively only in tension; half the members are needed when the wind blows from the left, and the other half are needed when it blows from the right.

Bracing members are normally hidden within the outer walls or the roof covering (Figs. 2.3.2 and 4.8.1.a), but occasionally they are made a conspicuous feature on the facade of the building (Fig. 5.2.1).

The structure of the floor of a multistory building is sometimes hung from ties, instead of being supported on columns (Figs. 5.2.2 and 5.2.3). At first sight this does not seem to make structural sense: We first carry the loads through the ties to an upper floor, which then transmits them by bending to columns or to a central service core, and the loads are then transmitted all the way down again and past the originating floor to the foundation (Fig. 5.2.4). There are, in fact, relatively few structures with hung floors. The advantages are not structural but constructional; the floor structure is precast and assembled and then lifted straight into position. For the system illustrated in Fig. 5.2.2, the construction of the floor structure of the building is started from the top, lifted straight up from the ground below. For the system illustrated in Fig. 5.2.3 it is possible to work on three different levels at the same time, once the main structural frame has been completed.

The design calculations for pure tension members are very simple. If the maximum permissible tensile stress for the structural materials if f, then the cross-sectional area A needed for tensile force P is (Fig. 5.2.5)

$$A = \frac{P}{f} \tag{5.1}$$

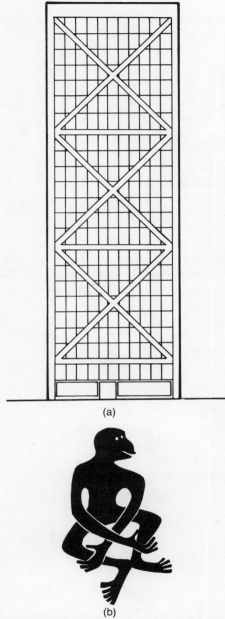

5.2.1. Diagonal braces are sometimes made a conspicuous feature on the facade of the building.

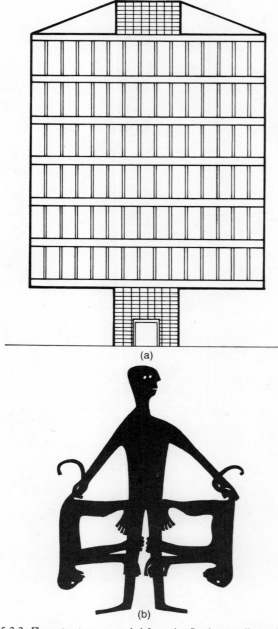

5.2.2. Floor structure suspended from ties fixed to cantilevered brackets.

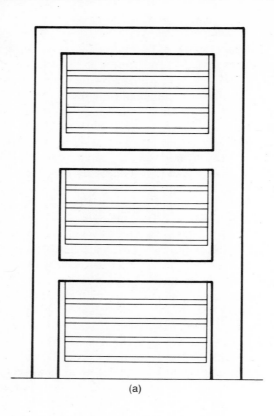

5.2.3. Floor structure suspended from giant portal frames; the supporting structure is utilized for the equipment rooms, which need a greater height and require no windows.

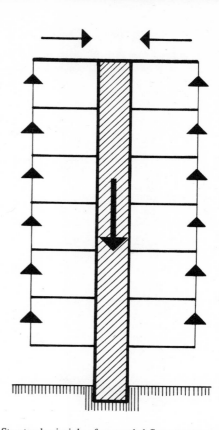

5.2.4. Structural principle of suspended floor structure.

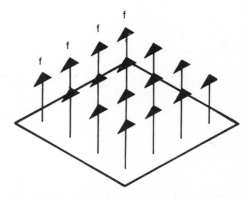

5.2.5. $P = fA$: Tensile force = Maximum permissible stress × Cross-sectional area.

5.3 Compression Members and Buckling

The *forces* acting on a compression member are the exact opposite to those acting on a tension member, but the *behavior* of the member is not the exact opposite, because a compression member may buckle if incorrectly designed. This is easily demonstrated with a thin rod or a piece of cardboard. If loaded in compression, it bends sideways; when the load is removed, it recovers its shape (Fig. 5.3.1). This is called *elastic buckling*. The member is not damaged by elastic buckling, but evidently it ceases to be a useful compression member. Buckling therefore constitutes a failure.

In addition, a compression member can fail (Sections 3.2 and 3.3) by plastic deformation, by brittle fracture, or by splitting of the material (Fig. 5.3.2). The load P'_S that causes failure is

$$P'_S = f' A \quad (5.2)$$

where f' is the compressive strength of the material, that is, the stress at which it fails, and A is the cross-sectional area.

The theory of buckling, which is quite complicated, is derived in more advanced textbooks (for example, Ref. 1.1, pp. 188–190). The column fails by buckling when the load

$$P'_l = k \frac{\pi^2 E A r^2}{L^2} \quad (5.3)$$

where π is the circular constant (approximately 3.142), E is the modulus of elasticity of the material, L is the length of the column, and r is its radius of gyration. The radius of gyration is a geometric property of the section, which is given in structural handbooks; a solid square section has a low value of r, and a thin tube of the same cross-sectional area has a much higher value of r. For rectangular sections, $r = d/\sqrt{12} = 0.29 d$, where d is the depth in the direction in which the column buckles.

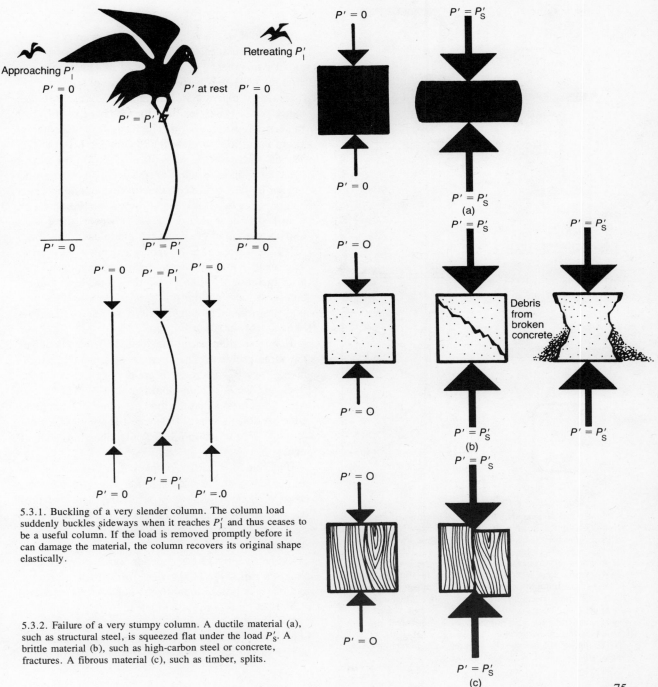

5.3.1. Buckling of a very slender column. The column load suddenly buckles sideways when it reaches P'_l and thus ceases to be a useful column. If the load is removed promptly before it can damage the material, the column recovers its original shape elastically.

5.3.2. Failure of a very stumpy column. A ductile material (a), such as structural steel, is squeezed flat under the load P'_S. A brittle material (b), such as high-carbon steel or concrete, fractures. A fibrous material (c), such as timber, splits.

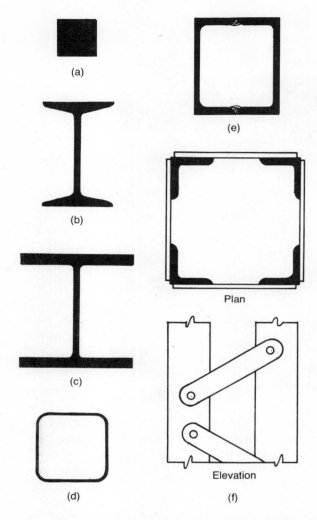

5.3.3. A metal column should have a large radius of gyration, r, particularly if high-strength steel or aluminum is used. All the columns illustrated have the same cross-sectional area, A, but different radii of gyration.
(a) A square section has a low r.
(b) A standard steel section has a high r in the vertical direction (maximum r), but a low r in the horizontal direction (minimum r).
(c) A wide-flange section has a higher r in the horizontal direction.
(d) A square tube has a still higher r in the horizontal direction.
(e) A tube formed by welding from two channel sections.
(f) The highest r is obtained by forming an open tube from four angle sections joined by diagonal lattice bars.

The load P_s' is characteristic of a very stumpy member, and the load P_l' is characteristic of a very slender member. The behavior of practical compression members lies between these two extremes.

The two equations (5.2) and (5.3) deserve some study because they explain the behavior of different materials in compression. The formula for P_s' contains f' but not E; the formula for P_l' contains E but not f'. We can increase the strength of steel, f', by adding small quantities of alloying material, by cold-working, or by heat treatment; thus we can obtain a higher compressive (and tensile and flexural) strength (more than 100%) at some extra cost. However, there is no known method of changing the modulus of elasticity, E, of steel; therefore P_s' increases, but P_l' does not. The result is that as we use higher-strength steel, we also increase the likelihood of a buckling failure, unless we take special precautions such as substituting compression members like (d) or (e) for members like (a), (b), or (c) in Fig. 5.3.3.

If we use aluminum instead of steel, we encounter the same problem. The strength of structural aluminum alloys, f', is almost as great as that of structural steel, but the E of aluminum is only one third that of steel. This means that the load at failure P_s' of comparable sections is almost the same for both metals, but the load P_l' for aluminum is only one third that of steel.

Buckling is a minor problem for timber and for reinforced concrete (see Section 8.3). It must be considered in the design of structural steel and prestressed concrete, is a major problem in high-strength steel, and is a very important problem in aluminum structures.

Evidently, buckling depends on the ratio L/r, which is called the *slenderness ratio;* for rectangular sections r is proportional to the depth of the section, d, and the ratio L/d is used instead.

The slenderness ratio L/r (or L/d) critical for buckling is that with the highest value; that is, buckling occurs in the direction for which the radius of gyration, r (or the depth, d) is lowest. Thus the compression member in Fig. 5.3.3.b may be expected to buckle by deflecting left or right, rather than up or down.

However, if the compression member is restrained in one direction, buckling in that direction is prevented, and it can buckle only in the direction in which it is not restrained. We then use the slenderness ratio in that direction, even if it has a lower value. For example, if a steel section forms part of the facade of the building, it is normally restrained in line with the facade but able to buckle at right angles to the facade.

Structural codes give tables that specify the maximum permissible stress, f, for various values of the ratio L/r or L/d. The cross-sectional area required for the column service load, P, is then

$$A = \frac{P}{f} \qquad (5.4)$$

Where there is a buckling problem, the first step is to obtain, for the same amount of material, the highest possible radius of gyration. Solid square or rectangular sections, which are normally used for timber and reinforced concrete columns, are not suitable for steel; wide flange shapes are more suitable than standard shapes because, for the same amount of material, they have a higher minimum radius of gyration. Square tubes, tubes formed from two channels or four plates, and open tubes formed from four braced angles also have high radii of gyration and therefore higher permissible stresses (Fig. 5.3.3).

5.4 Lintels and Beams: Simple in Appearance but Complex in Theory

Lintels and beams are the most common structural members, and they have been used since time immemorial. However, their theory is quite complicated. Its solution was first attempted by Leonardo da Vinci in the fifteenth century, but the problem was not finally solved until the early nineteenth century, by L.M.H. Navier, a professor at the *École des Ponts et Chaussées* in Paris.

If we bend a beam, as in Fig. 5.4.1, it becomes shorter on top and longer at the bottom. Somewhere near the middle of the beam (exactly at the middle if the section is symmetrical) there is a surface that gets neither longer nor shorter, and this surface is called the *neutral axis*. The beam in Fig. 5.4.1 is in compression above the neutral axis and in tension below it. The greatest tension occurs at the very bottom of the beam, the greatest compression at the very top. These are the greatest tensile and compressive stresses, and they must not exceed the maximum permissible stresses specified in the structural code.

The basis of the theory is an experimental observation that all elastic straight beams, when bent by a uniform bending moment M, assume the shape of a circular arc (Fig. 5.4.2). The vertical sections, which were originally plane and parallel, remain plane but converge on a center of curvature.

From the geometry of the bent beam we can derive expressions for the strain (see Glossary) at the very top and the very bottom of the beam (Section 3.1). Using Eq. (3.1), we then convert these strains into stresses (see Glossary). By multiplying each stress by its appropriate unit area we obtain the forces at the various depths of the beam, and we take the moment of each force about the neutral axis. Finally, we integrate (or add up) these individual moments over the entire cross section of the beam, thus obtaining the resistance moment of the beam. This must be equal to the bending moment applied to the beam, a condition that yields

$$M = fS \tag{5.5}$$

where M is the bending moment, f is the maximum permissible flexural stress, and S is a geometric property of the beam, called the *section modulus*. For a rectangular beam we have

$$S = b\,d^2 / 6 \tag{5.6}$$

where b is the width and d the depth of the beam.

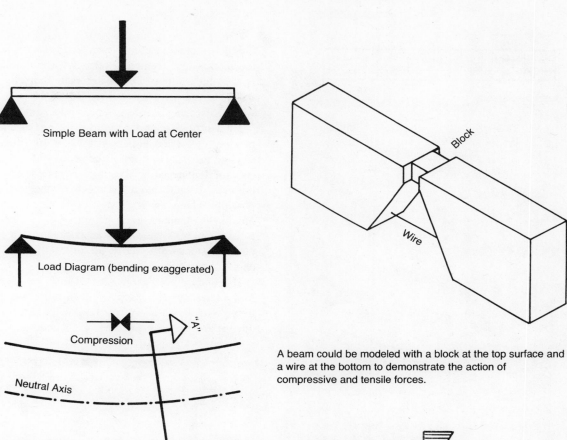

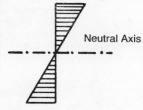

A beam could be modeled with a block at the top surface and a wire at the bottom to demonstrate the action of compressive and tensile forces.

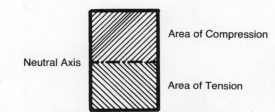

Bending stresses in the beam section move from the maximum at the surfaces furthest from the neutral axis to zero at the neutral axis.

5.4.1. Bending of a beam.

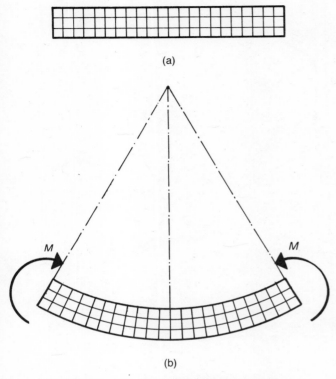

5.4.2. Experimental basis of the theory of bending. The originally parallel straight vertical lines (a) remain straight but converge on a center of curvature (b). The originally straight horizontal lines are bent into circular arcs by a uniform bending moment M. The stresses are compressive on top and tensile at the bottom, and there is a neutral axis where there is no stress.

Timber beams are usually rectangular, because timber is easily sawed along straight lines. It should be noted, however, that the stated "nominal" size of the timber is sometimes the size of the sawn timber before dressing and sometimes the finished size. If the former, b and d must be reduced by the amount of material lost in dressing; this can reduce the section modulus by as much as 40%. Section moduli of standard timber sections are listed in structural handbooks.

Steel is a more expensive material than timber, and the sections are formed differently. The molten steel is cast into billets, which are then thinned out by rolling at red heat. By shaping the rollers in different ways, different sections can be produced.

Evidently rectangular sections are not efficient for bending, since the highest stresses occur at the very top and the very bottom of the beam and most of the structural material is thus understressed. It is economical to concentrate as much of the material as possible where the highest stresses occur. Thus we obtain the standard steel section (Fig. 5.3.3.b). The rolls are expensive, and there is therefore a limited range of standard sections, up to 914 mm (36 in.) in depth.

If a deeper beam is needed, it must be assembled from plates and angles (Fig. 5.4.3). It is often cheaper to use a truss, which may be regarded as a girder with open spaces in the web (Figs. 4.7.7 and 5.4.4).

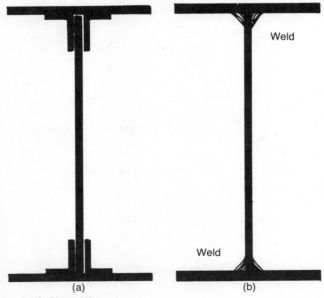

5.4.3. Plate girder.
(a) Plates joined through angles.
(b) Plates joined by fillet welds only.

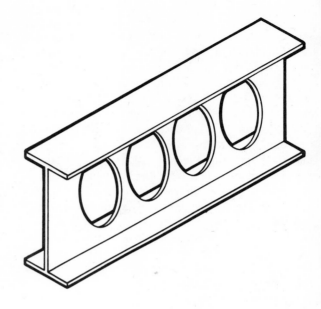

5.4.4. A parallel-chord truss requires more fabrication but uses less material than a plate girder. It could be regarded as a plate girder into which holes have been cut, like the one above.

Steel manufacturers list the section moduli of standard sections and also of some standard combinations of sections for plate girders.

Aluminum softens at a much lower temperature than steel, and therefore the machinery required for hot-rolling is not as heavy. Moreover, small beams of aluminum can be produced by extruding the metal through a die (or orifice), as toothpaste is squeezed from a tube (Fig. 5.4.5). Since dies are not expensive, it is unnecessary to restrict the range of aluminum sections to the same extent as steel sections. Section tables that list the section moduli of standard sections are available from their manufacturers.

For light construction, beams can be produced from thin, cold-rolled steel or aluminum sheet, which is cold-bent to form I-shapes or Z-shapes (Fig. 5.4.6).

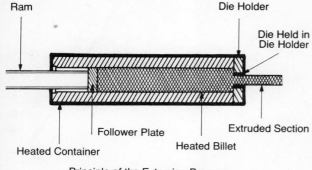

Principle of the Extrusion Process

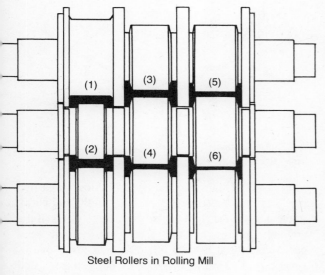

Steel Rollers in Rolling Mill

5.4.5. Producing beam sections by extrusion (aluminum only) and by rolling (steel or aluminum).

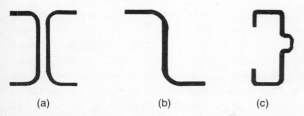

5.4.6. Beams produced from cold-formed steel or aluminum: (a) I-section, (b) Z-section, (c) Doorbuck.

5.5 Reinforced Concrete: Concrete with Flexural Strength

It would not be impossible to use plain concrete for roofs and floors of short span. The strength of concrete is comparable to that of natural stone. A soft sandstone has roughly the same strength in bending as a weak concrete, and a good granite or diorite is comparable to high-strength concrete. Stone lintels have been used extensively since ancient Egypt and Greece, and they continued to be used until the end of the nineteenth century. Many can still be seen spanning door and window openings in old buildings. Stone slabs for roofs have been used to a lesser extent and not as recently; a few survive, for example, in Gizeh, Karnak, and Athens.

However, unreinforced stone and concrete are extremely uneconomical materials when used in bending, because both are weak in tension; their compressive strength is about ten times as great. The beam in Fig. 5.4.1 would fail at the bottom in tension long before the ultimate compressive strength of the stone or concrete was reached at the top (Section 3.4). By adding a comparatively small amount of reinforcement, say 1% of the beam's volume, we can increase its flexural strength by at least 500%.

Metal reinforcement for natural stone has been used from time to time since ancient Rome, but it is difficult to fix and to protect from corrosion. Reinforced concrete was invented in the 1860s, but it did not become a commonly used structural material until the 1920s.

Reinforced concrete is particularly useful for floor structures; it is easy to form and cast concrete floors. Steel plate floors need a concrete topping both as a wearing surface and for fire protection. Timber floors are not fireproof, and they have been used in the twentieth century only for houses and other small buildings.

Concrete can be brought to the place where it is required in liquid form and cast on the site (*site-cast*), or it can be *precast* on the ground or in a factory a long distance away and then hoisted into the position where it is required.

Precasting has particular advantages in severe climates. The chemical reaction of cement is inhibited below 0°C (32°F); in hot weather, above approximately 30°C (86°F), concrete also does not gain its proper strength. Weather continuously above 30°C rarely lasts for more than a few days, but in some parts of the United States, the U.S.S.R., and Canada, temperatures may remain continuously below freezing for several weeks. Under those conditions precasting is clearly economical.

It is easier to obtain a good surface finish on precast concrete components than on site-cast concrete, and this is an important reason for precasting wall panels (Sections 5.7 and 8.9), but surface finish is not as important a consideration for concrete floor structures.

The great majority of concrete floor structures are therefore site-cast. This not only is more economical but also gives important structural advantages. Because the entire floor is cast *monolithically* (see Glossary), it has greater rigidity, and the maximum bending moments are lower (Fig. 4.6.8). The greater rigidity of the structure increases its resistance to hurricanes, earthquakes, and explosions (for example, from gas appliances or small bombs; no structural defense is possible against a megaton bomb).

The simplest and most recently developed type of reinforced concrete floor is the *flat plate* (see Glossary and Fig. 5.5.1). Although the reinforced concrete slab is fairly thick, the overall depth of the floor structure is that of the slab only, and the connection between the slab and the column is not very stiff. The slab therefore has a relatively high deflection relative to its span. This creates a special problem because of the creep deflection of the concrete (Section 3.11).

It is possible to compensate for the elastic deflection that occurs as soon as the structure is loaded (Fig. 5.5.2) by giving it an upward camber that disappears under the load. However, creep deflection poses a more serious problem, because it is much larger than the elastic deflection, usually 100% to 250% more; as the creep continues for a year or more, the structure either retains an upward camber for some months or acquires a permanent deflection.

Since deflection is proportional to the fourth power of the span*, the spans of flat plates should be kept low, 5 m to 8 m (16 to 26 ft). Alternatively, the deflection can be counteracted by load balancing (see Glossary), but this can only be accomplished by prestressing the steel (Section 5.6).

It is also advisable to provide suitable constructional details for windows and doors supported on flat reinforced concrete plates, because these could jam as a result of excessive deflection. There should be a space between the top of a partition and the underside of a flat plate (filled with a flexible material such as mastic, if necessary) to allow for the creep deflection of the plate.

Flat plates are most economical for buildings that require partitions but no false ceilings. The formwork is extremely simple. The ceiling can be left as it comes off the form, or it can be given a coat of acoustic plaster or some similar treatment. When false ceilings are required, beam-supported floors are often cheaper, because they have a greater depth and therefore permit larger spans with less concrete in the floor structure.

Beamless floors supported on enlarged column heads (Fig. 5.5.3) are called *flat slabs* (see Glossary). They are particularly suitable for buildings with open spaces, such as factories, warehouses, and parking garages; partitions do not fit easily between the enlarged column heads. Because these floors are stiffer than flat plate floors, spans can be greater.

---

*This means that a 20% increase in span produces a more than 100% increase in deflection for the same depth.

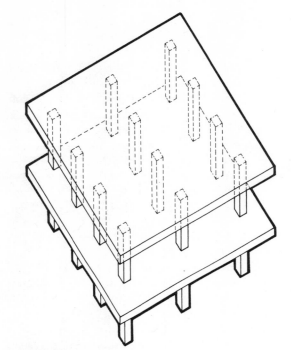

5.5.1. Flat plate floor.

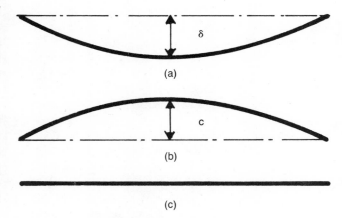

5.5.2. (a) Elastic deflection produced by the load, maximum δ at mid-span.
(b) Structure is built with an upward camber c at mid-span such that $\delta = c$.
(c) Under load, the camber cancels out the elastic deflection.

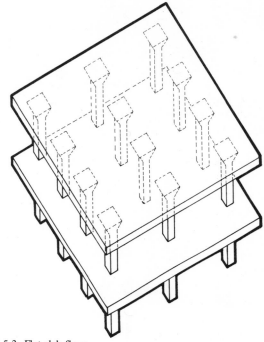

5.5.3. Flat slab floor.

In both flat slabs and flat plates the load is carried partly in one direction (say, north−south) and partly in the direction at right angles (say, east−west). If the column spacing is the same north−south and east−west (as it usually is), then as a result of this symmetry, half the load carried by the slab is resisted by the reinforcement in the north−south direction and the other half by the reinforcement in the east−west direction.

A beam-supported slab needs to span only from beam to beam (Fig. 5.5.7), and the load is then transmitted by the beams to the columns. In flat plates and flat slabs the floor slab transmits the load directly to the columns. The floor slab thus performs a double function, and the bending moments are increased accordingly; the bending moments in flat slabs and flat plates are approximately double those in beam-supported slabs. Thus in a square flat plate or flat slab the bending moment in *each direction* (say, north—south and east—west) is approximately equal to that shown in Tables 4.1 and 4.2 for slabs spanning in one direction only; that is, the sum of the bending moments in a beamless slab in both the north—south and east—west directions is twice the amount shown in tables 4.1 and 4.2. In a beam-supported square slab (Fig. 5.5.7.a) carrying the same load over the same span, the bending moment is approximately half the value for a similar beamless slab. This assumes, naturally, that the beams are made strong enough to transmit the reaction of the floor slabs to the columns.

If the concrete slabs are cast in one piece continuously over all the columns, there are negative bending movements over the columns, requiring reinforcement on top, and positive moments near mid-span, requiring reinforcement at the bottom of the slab (Fig. 4.6.8). The negative moments over the columns are larger than the positive moments at mid-span.

It is therefore possible to cut away some of the concrete at the bottom of the slab, because it is cracked in tension (Fig. 3.4.2) and consequently does not contribute to strength, and concentrate the reinforcement in the ribs (Fig.5.5.4). It is not cut away around the column, where the concrete in compression is at the bottom and where a greater thickness of concrete is needed because of the tendency of the column to punch through the slab (Fig. 5.5.5.).

The slab is constructed on flat formwork, to which plastic pans are nailed to form the ribs. After the soffit formwork has been removed, the flexible pans can be withdrawn. Because of the shape of the slab and the enlarged column heads, this type of construction is sometimes called a waffle slab on mushroom columns (Fig. 5.5.4).

Reinforced concrete *slabs supported on beams* use less concrete for the same column spacing than flat plates, and therefore weigh less (Fig. 5.5.6). This is an important consideration for multistory buildings, where column size can be critical; the ground floor columns must carry the weight of all the floors above them.

Building services can often be fitted between the beams (Fig. 5.5.8), so that the overall height of the building is not necessarily increased by the use of beams. A flat plate has a neater appearance than a slab supported on beams, but this does not matter if a false ceiling is used.

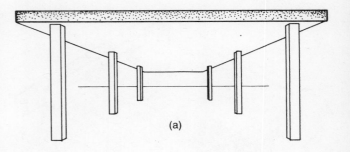

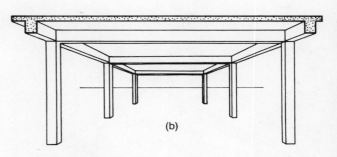

5.5.6. The flat-plate floor (a) and the beam-and-slab floor (b), suitably reinforced, have about the same strength. The floor slab (b) is thinner than the floor slab (a), but the overall depth of the floor structure (b) is greater than that of the floor structure (a). Because of its greater overall depth, (b) has less creep deflection than (a). The volume of concrete, and therefore the weight of concrete, is less for (b) than for (a). The cost of the formwork is greater for (b) than for (a).

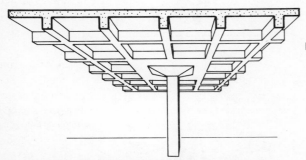

5.5.4. Underside of flat slab floor with positive tension reinforcement concentrated in ribs (waffle slab on mushroom columns).

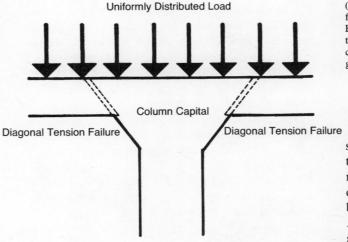

5.5.5. Columns have a tendency to punch through the slab.

There are two reinforcing systems for beam-and-slab floors. The concrete slab can span directly between primary beams, so that the reinforcement in the north—south and the reinforcement in the east—west direction of the slab each take their share of the bending resistance. This is called a *two-way system*. Alternatively, a *one-way system* can be used; the slab spans in one direction between secondary beams, and these in turn span at right angles between primary beams (Fig. 5.5.7).

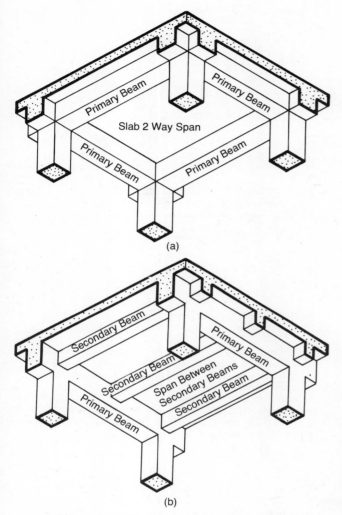

5.5.7. (a) A two-way slab spans in two perpendicular directions, like a flat slab or flat plate. The reinforcement in both directions is main reinforcement, and the beams in both directions are primary beams. If the column spacing is the same in both directions, then half the total bending moment is resisted by the reinforcement in each direction.
(b) A one-way slab spans in one direction only, and the entire load is carried in that direction. On the other hand, the span, and consequently the bending moment, is much smaller. The slab is supported by secondary beams, which in turn are supported by primary beams.

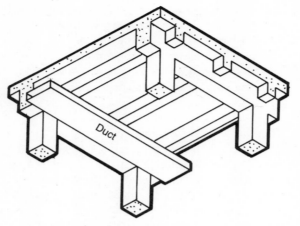

5.5.8. Building services can be fitted between the beams of a beam-and-slab floor.

When a one-way system is used, it is still necessary to provide reinforcement in both directions (Fig. 5.5.9), even though the main reinforcement is needed only in one direction to resist bending stresses. The secondary reinforcement resists the tensile stresses produced by the shrinkage and temperature movement of the concrete (Section 2.5); these could produce large and unsightly cracks unless bridged by a small amount of reinforcement.

The concrete *ribs* in two-way slabs can be produced only with removable flexible pans, or with stiffer pans which are left as permanent formwork. In one-way slabs ribs can be formed more easily (Fig. 5.5.10.a). It is also possible to use hollow tiles of terra cotta or concrete and arrange the primary reinforcement in the ribs between them (Fig. 5.5.10.b). The hollow tiles give a flat ceiling, greater stiffness, and improved insulation to the floor. If hollow tiles are used, the topping above the tiles can be reduced to 50 mm (2 in.).

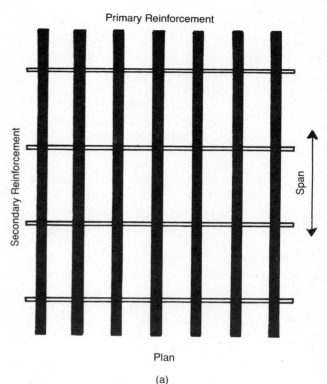

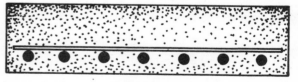

5.5.9. In one-way slabs reinforcement is needed only in one direction to resist the bending moment due to the loads; but reinforcement is required in two directions at right angles to one another to prevent tension cracks due to shrinkage contraction and temperature movement (Section 2.5) of the concrete. This secondary reinforcement (at least 0.2%) also serves to distribute load concentrations to the main reinforcement. It is placed inside the main reinforcement. Concrete cover, at least 20 mm (¾ in.), must be provided above the main reinforcement to protect it against fire and corrosion.

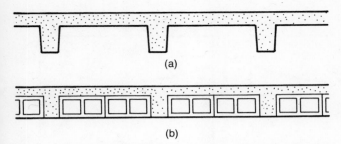

5.5.10. (a) Ribbed construction for reinforced concrete one-way slab. The main reinforcement is concentrated in the ribs. Only nominal reinforcement is required in the topping above the ribs. (b) Hollow tile construction for reinforced concrete one-way slab. The main reinforcement is concentrated in the ribs, and the space between the ribs is filled with hollow tiles of terra cotta or concrete. No reinforcement is required in the concrete topping. The tiles provide a flat ceiling and improve the stiffness and insulation of the floor.

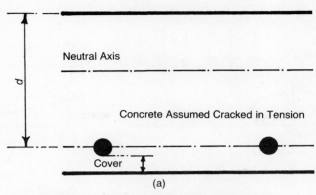

5.5.11. The force in the tension reinforcement when the slab fails is $f_y A_s$, where f_y is the stress at which the steel deforms plastically (Section 3.2) and A_s is the cross-sectional area of the steel. The resistance moment of the slab, M', is the product of that force and the length of its lever arm, z, about the center of the compressive force in the concrete. Approximately $z = 0.9\,d$, where d is the effective depth (see Glossary).
(a) Cross section of beam.
(b) Moment and forces in cross section.

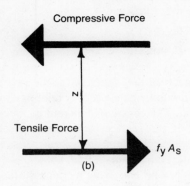

There are two ways in which a reinforced concrete slab can *fail in flexure*: The concrete can fail in compression, usually by crushing, or the reinforcing steel can fail in tension, by a gradual yielding (see Glossary) of the steel. The concrete failure is brittle and therefore sudden; the steel failure is plastic and therefore takes time (Sections 3.2 and 3.3). A sudden failure is undesirable, since it does not allow time for remedial measures. In the event of a gross overload, such as might occur during an earthquake, hurricane, or explosion, it does not allow time to evacuate the building before collapse.

A concrete (brittle) failure occurs when there is a great deal of reinforcement. Because there is plenty of steel, the stress in it is relatively low, and thus it is the concrete that fails first. The minimum percentage of reinforcing steel necessary to produce a brittle concrete failure can be worked out. The American and Australian concrete codes require that the tension reinforcement be no more than 75% of that critical percentage, to ensure that the steel deforms plastically before the concrete is crushed and therefore that the failure occurs gradually. The British concrete code gives equations that serve the same purpose.

These provisions limiting the percentage of steel reinforcement ensure that it is the steel stress which is critical for the design of a reinforced concrete slab; therefore the strength of the concrete does not enter the calculations. Reinforced concrete calculations are now always expressed in terms of the conditions at failure (Section 3.10). The equations are derived in more advanced textbooks (for example, Ref. 1.1, pp. 202–14 or Ref. 5.7, pp. 31–82). The ultimate resistance moment of the reinforced concrete slab, which must be equal to the ultimate bending moment applied to the slab, is

$$M' = f_y A_s z \qquad (5.7)$$

where M' is the ultimate bending moment (see Glossary) at failure, f_y is the stress at which the steel reinforcement yields (see Glossary), A_s is the cross-sectional area of the steel reinforcement, and z is the length of the lever arm between the compressive and the tensile force in the section; it is approximately $0.9\,d$, where d is the effective depth of the concrete section, measured to the center of the reinforcing steel (Fig. 5.5.11).

5.6 Prestressed Concrete: Concrete Without Cracks

Like all brittle materials, concrete cracks when subjected to tension (Sections 3.3 and 3.4). Experiments show that although strong concrete cracks at a higher tensile stress (see Glossary) than weak concrete, it cracks at about the same tensile strain (see Glossary), namely 3×10^{-4} (or 3 parts in 10 000). This is because strong concrete is more brittle than weak concrete. Since the steel reinforcement is fully bonded to the concrete, its strain is the same as that of the surrounding concrete. Therefore the strain in the reinforcement when the concrete cracks is also 3×10^{-4}.

From Eq. (3.1) the stress in the reinforcement is

$$f = e\,E$$

The modulus of elasticity of steel is 200 000 MPa, or 30 000 ksi. Therefore the stress at which cracks first form in reinforced concrete is

$$f = 3 \times 10^{-4} \times 200\,000 = 60\text{ MPa}$$

or

$$f = 3 \times 10^{-4} \times 30\,000 = 9\text{ ksi}$$

The actual stress in reinforced concrete under the action of the service loads (see Glossary) is 2 to 3 times as much as these figures for the two principal grades of reinforcement used in reinforced concrete construction. The concrete is therefore cracked under the action of the service loads; however, it is not cracked very much. Reinforcing steel does not rust, as long as the cracks remain fine, about 0.3 mm (0.01 in.) in width. The bars have a rough surface before they are cast into the concrete, and they are also deformed to improve the bond between the steel and the concrete; this ensures that a large number of very fine cracks, rather than a few larger cracks, form.

However, unacceptably wide cracks would occur if we increased the steel stress under the action of the service loads beyond about 200 MPa (30 ksi). This means that we cannot use high-strength steel in reinforced concrete.

In prestressed concrete the use of high-strength steel is not only possible but also necessary. The first attempts to prestress concrete were made in the 1880s, not long after the development of reinforced concrete. They were totally unsuccessful. Most of the prestress mysteriously disappeared after a few weeks, because the concrete contracted as a result of shrinkage and creep (see Glossary), and the prestressing steel contracted with it. Because of the low steel stresses used, the losses were almost equal to the original prestress. Unless the original prestress is much higher than the losses to be expected, prestressing is uneconomical. Steel with a tensile strength of 1000 to 2000 MPa (150 to 300 ksi) is therefore used for prestressing tendons (see Glossary). This extra-high-strength steel is used only for prestressing and for suspension cables. The great strength is achieved by increasing the carbon content of the steel and by distortion of its crystal structure (for example, by drawing the steel wires through a die).

Because of the very high strength of the steel, we need far less of it. Eq. (5.7) also gives the ultimate bending moment of a prestressed concrete beam. If

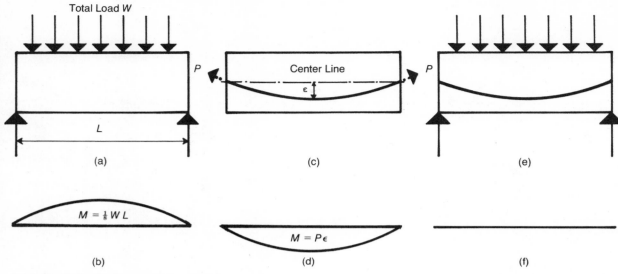

5.6.1. Load balancing in prestressed concrete beam.
(a) The beam carries a total load W over a simply supported span L.
(b) The positive bending moment varies parabolically, and at mid-span it is $\frac{1}{8}WL$.
(c) The same beam prestressed by a parabolic cable, whose eccentricity varies from zero at the supports to ϵ at mid-span. This flexes the beam upwards.
(d) The negative bending moment due to the prestress varies parabolically from zero to $P\epsilon$.
(e) To balance the load, we use the highest eccentricity that can be accommodated within the depth of the beam and adjust the prestressing force P so that
$$P\epsilon = \tfrac{1}{8}WL$$
(f) The positive and negative bending moments exactly cancel one another, and the beam is free from bending moment. However, it is compressed by the prestressing force P.

we compare a reinforced and a prestressed concrete beam of the same depth and carrying the same ultimate bending moment, then z and M' are the same and the value of $f_y A_s$ must be the same for both beams. Steel used for prestressing has a tensile strength, f_y, about five times as great as that of steel used for reinforcing concrete and therefore only one fifth the area of steel, A_s, is required.

Saving 80% of the steel can be important, particularly if steel is in short supply, as it was after World War II. Prestressed concrete had been developed as a practical structural material shortly before the war, and during and after the war it was employed extensively because of its economical use of steel. However, the prestressing operation adds to the cost, and in recent years its use in buildings has been confined mainly to long spans. Because it uses high-strength concrete as well as high-strength steel, the dead weight of the beam is lower than for ordinary reinforced concrete. Furthermore, it is possible to reduce the bending moment by *load balancing* (see Glossary).

Prestressed concrete (see Glossary) differs from all other structural systems, because very high stresses are locked in it even when it is not carrying a high load.

If the prestressing tendons are suitably shaped, they can completely counteract the bending moment of a load system, and the beam then is entirely free from bending moment. It is compressed along its axis, but this causes no deflection and only relatively low stresses (Fig. 5.6.1). In practice the beam can be balanced for one load system only, and it is therefore subject to some (but a very much lower) bending moment under different load systems.

The beam carries a dead load due to it own weight and to other permanent loads. In addition there is a live load, which may or may not be acting. It is best to balance the full dead load and half of the live load (Fig. 5.6.2).

Load balancing greatly reduces the weight of the beam or slab. It reduces the deflection even more. We noted in Section 5.5 that creep deflection is a problem in flat plates. The creep deflection is directly proportional to the elastic deflection; if the elastic deflection is zero, as it is in Figs. 5.6.1.f and 5.6.2.b, the creep deflection also remains zero. In practice the load cannot be perfectly balanced all the time, and there is some deflection, but it is smaller with load balancing than without it. Quite long spans are therefore possible with flat plates by prestressing them.

The other principal application of prestressing to buildings is its use in beams and shells of long span, for which reinforced concrete would be too heavy. Prestressed shells are discussed in Section 10.9.

The prestress is produced by means of one or more hydraulic jacks, which pull on the tendons (wires, cables, or bars). The prestress is then anchored by a patent device. There are a number of systems in use; each has advantages and disadvantages for particular applications. The suppliers of these devices will be pleased to give the necessary information.

The demolition of prestressed concrete is unlike that of other structural materials. The prestress remains locked in the structure, and this can amount to a force of hundreds of tons. If the tendons are cut inadvertently with an oxyacetylene torch during demolition, the anchorages could be released and hit other parts of the structure, and perhaps people with a devastating force, if the tendons are unbonded or inadequately bonded. It is therfore important that the structural drawings of a prestressed concrete structure be preserved until it is demolished, so that the demolisher can be informed of the locations of the tendons and the nature of the anchorages.

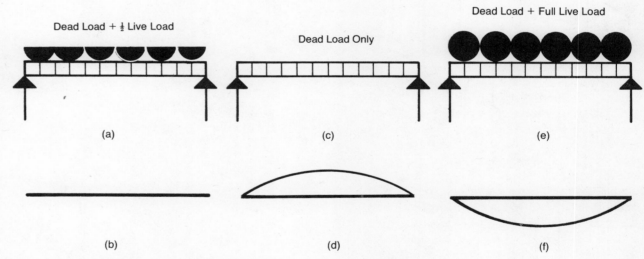

5.6.2. Load balancing for the full dead load and half the live load.
(a) Beam carrying dead load and half the live load.
(b) Bending moment is zero.
(c) Beam carrying dead load only.
(d) Maximum bending moment is negative, as a result of half the live load acting upwards.
(e) Beam carrying dead load and full live load.
(f) Maximum bending moment is positive, as a result of half the live load acting downwards.

5.7 Jointing of Structural Members

We noted in Section 2.7 that steel and reinforced concrete structures are designed so that failure does not occur in the joints. We calculate the structural sizes of the members and then design joints that are at least as strong as the members. Thus welds should be of sufficient length, bolts should be sufficiently numerous, reinforcing bars should have a sufficient overlap for bonding, and so on.

A great deal of *timber* is joined with ordinary nails and a hammer, and this is one reason for the popularity of timber for small structures (Fig. 5.7.1). Stronger timber joints can be made with nail plates, gang nails, or timber connectors.

Steel can be connected by welding, by riveting, by ordinary bolting, or by high-strength bolting.

Welds are of three types (Fig. 5.7.2); all are made by liquid metal, protected by a flux from oxidation, that is fused with the parts to be joined. The two pieces should thus be joined together by metal as strong as the pieces themselves. The strength of the joint can be reduced by a flaw in the weld metal, or a crack may be induced in the parent metal by overheating or distortion. These cracks can initiate a brittle fracture (Sections 3.3 and 6.4); this may occur without warning, particularly in cold weather, which makes metals more brittle. A well-known example is provided by the failure of welds and subsequent sinking of several Liberty Ships in the Arctic Ocean during World War II. Flaws and cracks are not normally visible to the naked eye, but they can be detected with apparatus using X-rays or ultrasonic waves, and important welds are therefore inspected with this equipment. However, X-ray inspection would be too expensive for all welds, and reliance is placed on testing the operator of the welding equipment from time to time to make sure that he has not lost his skill.

Welding is now commonly used for connections made in the factory. Welding can also be used to joint the factory-produced steel assemblies on the building site, but it is much more difficult to produce satisfactory welds under those conditions. Site connections are therefore normally made with bolts.

Ordinary low-carbon steel *bolts* have a tensile strength of 380 MPa (55 ksi). They are tightened with a spanner until they feel tight, and as a result their

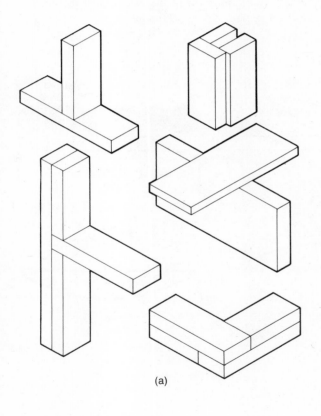
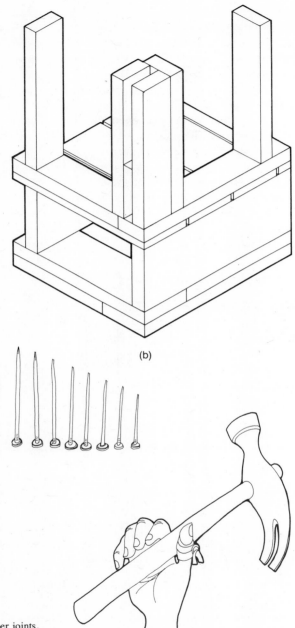

5.7.1. Timber joints.

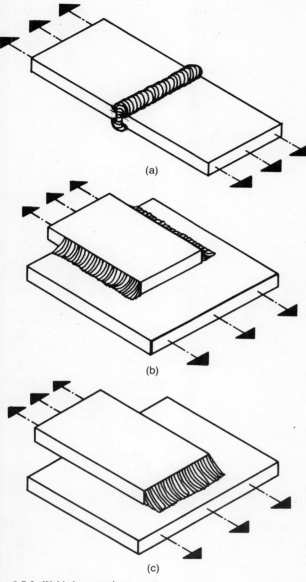

5.7.2. Welded connections.
(a) Groove weld, which acts in tension or compression.
(b) Side-fillet weld, which acts in shear.
(c) End-fillet weld, which acts partly in shear and partly in tension.

strength varies from bolt to bolt. High-strength bolts are made from a medium-carbon steel, and their tensile strength is twice as great. High-strength bolts are tightened with a torsion wrench, which is calibrated every few hours to apply a precise amount of tightening. As a result the strength of a high-strength bolted joint is more reliable than that of a joint with ordinary bolts. Bolting can easily be done on the building site.

Riveting was a common method of jointing in the United States and Great Britain (but not in Australia) prior to the 1950s, but it is now on the wane. The red-hot rivets are pressed into holes in the two pieces of steel, they expand into these holes, and on cooling they press the two pieces of steel together, so that some of the connection is due to friction, as it is in high-strength bolting. Riveting also is much easier to perform in the factory than on the building site.

Because the metal oxidizes easily, welding of *aluminum* must be performed under a protective covering of argon or helium, which makes it expensive. However, aluminum rivets can be driven cold, and this makes riveting a simple method of jointing aluminum. Bolting and gluing are other possible methods.

Although the objective of *site-cast concrete* construction is to make the structure monolithic (see Glossary), it is necessary to make a number of joints. It is important that each concrete section have some reinforcement and that this reinforcement be carried some distance beyond the point where it is required as is shown in Fig. 3.4.3. The slab reinforcement is turned down into the outer beam to make a proper connection between the slab and the beam.

Joints must be made at each floor level of a multistory structure to allow the concrete to gain strength before proceeding to the higher floors. The lower column is cast with the end of its reinforcing cage projecting. The floor is then cast. Next the reinforcing cage of the upper column is pushed over that of the lower column, and the upper column is cast. The connection between the reinforcing bars is by bond through the concrete. This is cheaper and quicker than welding the reinforcement or joining it with screw couplings. However, jointing of reinforcement by welding or screwing is sometimes necessary where there is insufficient room for bars to be overlapped for a bonded joint.

To an increasing extent the detailing of the connections of both steel and reinforced concrete structures is now done by computer, with the use of standard details (Ref. 5.6).

Precast concrete is made in the factory or on the ground and joined on the building site after it has been hoisted into position. It is particularly suitable for curtain walls and loadbearing walls, because the surface finish of a wall is a prominent feature of the building, and a good concrete surface finish is much easier to produce on a horizontal surface than on a vertical one. Thus there are great advantages in having the wall panels cast flat in a factory and then transporting them to the building site. The panels may serve as permanent formwork, or they may be fixed by bolting; the bolts are often designed to fit into slots (Fig. 5.7.3) to allow some tolerance in the manufacture of the concrete.

Quite a strong joint between two components of precast concrete can be obtained by friction. Because the coefficient between two pieces of concrete is high and concrete components are heavy, the frictional force between them is large. Until recently this has been considered sufficient for buildings that are not in an earthquake or a hurricane zone. However, joints that rely only on friction would probably not withstand even a small explosion. Since a number of such explosions have occurred during the last few years, frictional joints should always be given the added security of a bolted connection (Section 8.9).

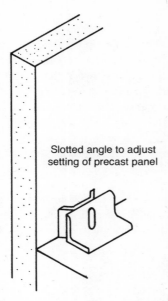

5.7.3. Fixing of precast concrete panels to steel or reinforced concrete structure.

5.8 Summary of Results for Ties, Columns, Beams, and Slabs

For *ties* the maximum permissible load is

$$P = fA \qquad (5.8)$$

where f is the maximum permissible stress in tension specified by the structural code and A is the cross-sectional area required for the tie.

For *columns* the maximum permissible load is

$$P = fA \qquad (5.9)$$

where f is the maximum permissible stress and A is the cross-sectional area. The maximum permissible stress in compression is not a fixed quantity but depends on the slenderness ratio L/r or L/d (L is the effective length of the column, r is the radius of gyration of the column section, and d is the depth of the column section).

The maximum permissible values of f for the various ratios L/r or L/d are specified in structural codes. The effective length of the column may be greater or smaller than its actual length, depending on the stiffness of the connections at the end of the column, and this also is specified in the structural code. The cross-sectional area and the radius of gyration of the various structural sections in timber, steel, and aluminum are listed in section tables.

For a rectangular section $A = b\,d$, where b is the width and d the depth of the section; $r = 0.29\,b$ in one direction, and $r = 0.29\,d$ in the other direction (at right angles).

Reinforced concrete columns are considered in Section 8.3.

For *beams* of timber, steel, or aluminum, the *maximum permissible bending moment* is

$$M = fS \qquad (5.10)$$

where f is the maximum permissible flexural stress specified in the structural code for a particular beam and S is the section modulus, which is listed in timber, steel, and aluminum section tables. For a rectangular section $S = b\,d^2/6$, where b is the width and d is the depth of the section.

For *beams and slabs of reinforced concrete* the *ultimate* bending moment is

$$M' = f_y A_s z \qquad (5.11)$$

where f_y is the specified yield strength of the reinforcing steel, A_s is its cross-sectional area, z is the lever arm of the resistance moment, which may be taken as $0.9\,d$, and d is the effective depth of the beam or slab measured from the compression face of the concrete to the center of the tension reinforcement. The depth of reinforced concrete slabs and plates is usually determined by the limitations on deflection specified in the structural code.

5.9 A Note on All Problems

You can use this book as a descriptive text. If you wish to do so, omit the remainder of this chapter and the problems in subsequent chapters. This causes no loss of continuity.

Some of the problems are very simple, and their solutions are exact. For more complicated problems brief approximate solutions are given, and these are marked with an asterisk (*).

Calculations are rounded off to four significant figures. Greater accuracy serves no purpose in preliminary calculations. It takes extra time and increases the probability of writing down an incorrect number. Figures are rounded off so as to increase the margin of safety, which usually means upwards. For example, if the calculator shows that the section modulus, S, is 642 841.74, we look for a section with S greater than 642 900. In all problems the weight of the structural member is included in the load.

All problems are solved in both systems of measurement. In the statement of the problem the metric units are given in Roman type, and the customary American units follow in italics, enclosed within parentheses. **The dimensions in italics are not conversions from meters to feet, but alternative dimensions for the examples in customary American units.** If you are using metric units, ignore the numbers in italics; if you are using customary American units, ignore the numbers in Roman type.

The actual numerical solutions are printed in a smaller type, in Roman type for metric calculations and italics for customary American units. These are alternative solutions. You need read only one of them, depending on the units you prefer.

The abbreviations are explained near the end of the book under the heading Notation. They are followed by a conversion table from metric to American units and vice-versa. When solving numerical examples it is advisable to convert all numerical data to one set of units, as the information is usually given in a variety of units.

When employing metric units, it is common practice to give stresses in MPa, loads in kN or kPa, spans in m, and section properties in mm. In the following problems we have chosen MN and m as the basic units. Thus

$$1 \text{ MPa} = 1 \text{ MN/m}^2$$
$$1 \text{ kN} = 1 \text{ MN} \times 10^{-3}$$
$$1 \text{ mm} = 1 \text{ m} \times 10^{-3}$$
$$1 \text{ mm}^2 = 1 \text{ m}^2 \times 10^{-6}$$
$$1 \text{ mm}^3 = 1 \text{ m}^3 \times 10^{-9}$$

In Australia, Canada, New Zealand, and South Africa all section properties are given in mm, and the same is proposed for the United States. In Great Britain some section properties are given in cm (centimeters). In this book only mm are used.

In customary American units, it is common practice to give stresses in ksi, loads in kips or psf, spans in ft, and section properties in in. In the following problems we have chosen kips and in. as the basic units. Thus

$$1 \text{ ft} = 12 \text{ in.}$$
$$1 \text{ lb} = 1 \text{ kip} \times 10^{-3}$$
$$1 \text{ psf} = 1 \text{ lb/ft}^2 = \frac{1}{12^2} \text{lb/in.}^2$$
$$1 \text{ psi} = 1 \text{ lb/in.}^2 = 1 \text{ kip/in.}^2 \times 10^{-3} = 1 \text{ ksi} \times 10^{-3}$$

The design is in accordance with American structural codes, but in the metric problems Canadian standard sections are used. Tables 13.1 and 13.2 give the properties of the new standard steel sections used in the problems. However, the reader is advised to acquire standard tables for steel and timber sections and for concrete reinforcing bars; they are generally available free of charge from the suppliers of the materials.

At the time of writing (1979) the United States has not determined metric maximum permissible stresses. Hence the American maximum permissible stresses have been converted from ksi to the nearest round number in MPa for use in the metric problems. As the problems deal only with preliminary design, a small variation in the maximum permissible stress is not significant.

5.10 Problems for Chapter 5

The problems for this chapter are divided into three sections. Section 5.11 contains a number of problems on steel and timber beams; Section 5.12 deals with reinforced concrete beams and slabs; and Section 5.13 gives solutions for ties and columns.

Although the theory of tension and compression members is simpler than the theory of beams, the practical design of beams is simpler than the practical design of tension hangers and of columns. The sequence of the problems is therefore the reverse of the order of discussion in the text.

Tension and compression members in trusses are covered in Section 7.9, and reinforced concrete columns in Section 8.11.

5.9.1

5.11 Problems on Steel and Timber Beams

We first solve two simple cantilever problems, one involving a construction hoist and the other involving a set of balcony supports. We then design the steel beams for the two most common types of floor structure used in small and medium-sized steel frames. The reinforced concrete slabs for both structures are considered in Problems 5.8 and 5.9 (Section 5.12).

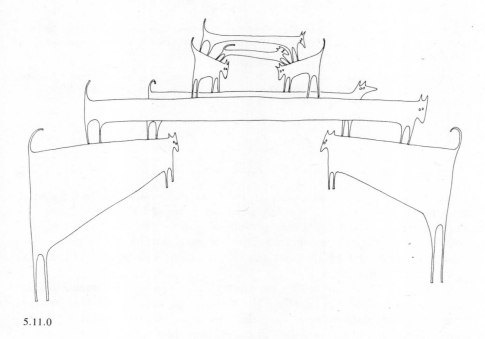

5.11.0

Problem 5.2 A steel beam 150 mm deep with a mass of 18 kg/m (*6 in. deep with a mass of 12 lb/ft*) projects 1 m (*3 ft*) beyond the facade of a building (Fig. 5.11.1). It is used to hoist loads to the top of the building while alterations are made. What is the biggest load permitted?

The beam acts as a cantilever with a single concentrated load at its end. The maximum bending moment (Fig. 4.5.2 or Table 4.1, line 1) is

$$M = -WL$$

where W is the maximum permissible load, to be determined, and L is the span.

The minus sign becomes important when we design or analyze reinforced concrete beams, because it tells us that the reinforcement must be placed on the top face. However, steel has nearly the same strength in tension and in compression, and it therefore does not matter, in determining the loadbearing capacity of the beam, whether the bending moment is positive or negative. Consequently

$$W = \frac{M}{L} \tag{5.12}$$

From Eq. (5.10) the maximum permissible bending moment is

$$M = fS \tag{5.10}$$

where f is the maximum permissible steel stress, specified by the structural steel code, and S is the section modulus listed in Table 13.1 (*Table 13.2*).

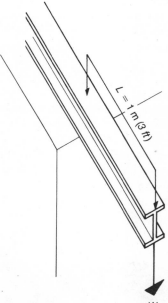

5.11.1. Span and load for hoist in Problem 5.2.

To solve the example in metric units, let us summarize the data. From Table 13.1 the section modulus is
$$S = 120\,000 \text{ mm}^3 = 120 \times 10^{-6} \text{ m}^3$$
Using a maximum permissible stress for A 36 steel (the steel most commonly used in the United States for architectural structures), we have

$$f = 165 \text{ MPa} = 165 \text{ MN/m}^2$$

From Fig. 5.11.1 the span is

$L = 1\,\text{m}$

From Eq. (5.10)

$M = fS = 165\,\text{MN/m}^2 \times 120 \times 10^{-6}\,\text{m}^3 = 19\,800 \times 10^{-6}\,\text{MN m} = 19.80\,\text{kN m}$

From Eq. (5.12)

$W = \dfrac{M}{L} = \dfrac{19.8\,\text{kN m}}{1\,\text{m}} = 19.8\,\text{kN} = 19\,800\,\text{N}$

This is the maximum load that can be hoisted. To obtain the equivalent mass we divide by g, the acceleration due to gravity:

$\dfrac{W}{g} = \dfrac{19\,800\,\text{N}}{9.8\,\text{m/sec}^2} = 2000\,\text{kg}$

To solve the example in customary American units, let us summarize the data. From Table 13.2, the section modulus is

$S = 7.31\,\text{in.}^3$

From the American (AISC) structural steel code (Ref. 5.9) the maximum permissible stress for A 36 steel is

$f = 24\,\text{ksi} = 24\,\text{kips/in.}^2$

From Fig. 5.11.1 the span is

$L = 3\,\text{ft} = 36\,\text{in.}$

From Eq. (5.10)

$M = fs = 24\,\text{kips/in.}^2 \times 7.31\,\text{in.}^3 = 175.4\,\text{kip in.}$

From Eq. (5.12)

$W = \dfrac{M}{L} = \dfrac{175.4\,\text{kip in.}}{36\,\text{in.}} = 4.87\,\text{kips}$

This is the maximum load (and the maximum mass) that can be hoisted.

Problem 5.3 A balcony cantilevers 1.5 m (*5 ft*) beyond the facade of a building. It is supported on steel beams at 2 m (*6 ft*) centers (Fig. 5.11.2). The uniformly distributed load due to the weight of the balcony and to the people and furniture on it is 7 kPa (*150 psf*). Select suitable steel beams.

From Table 4.1, line 3, the maximum bending moment for a cantilever carrying a uniformly distributed load is

$$M = -\tfrac{1}{2}WL \tag{5.13}$$

where W is the total load carried by the cantilever and L is its span. The minus sign is of no significance in the design of steel beams.

From Eq. (5.10) the section modulus required is

$$S = \dfrac{M}{f} \tag{5.14}$$

where f is the maximum permissible steel stress, which is the same as in Problem 5.2.

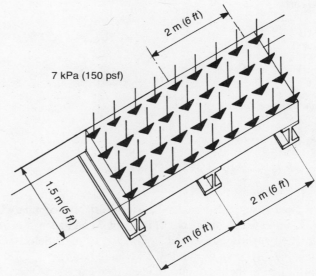

5.11.2. Span and load for steel beams in Problem 5.3.

The steel beams are spaced at 2 m centers, and the balcony cantilevers 1.5 m beyond the facade. Therefore each steel beam carries a load of

$W = 7\,\text{kPa} \times 2\,\text{m} \times 1.5\,\text{m} = 21\,\text{kN}$

The remaining data are

$L = 1.5\,\text{m}$ and $f = 165\,\text{MPa}$

The bending moment, from Eq. (5.13), is

$$M = \tfrac{1}{2} \times 21 \times 1.5 = 15.75 \text{ kN m} = 15.75 \times 10^{-3} \text{ MN m}$$

The section modulus required, from Eq. (5.14), is

$$S = \frac{15.75 \times 10^{-3} \text{ MN m}}{165 \text{ MN/m}^2} = 0.095\,5 \times 10^{-3} \text{ m}^3 = 95\,500 \text{ mm}^3$$

We select from Table 13.1 a steel beam 150 mm deep, with a mass of 18 kg/m, which has $S = 120\,000$ mm³.

In customary American units, the steel beams are spaced 6 ft apart, and the balcony cantilevers 5 ft beyond the facade. Therefore each steel beam carries a load of

$$W = 150 \text{ psf} \times 6 \text{ ft} \times 5 \text{ ft} = 4500 \text{ lb} = 4.5 \text{ kips}$$

The remaining data are

$$L = 5 \text{ ft and } f = 24 \text{ ksi}$$

The bending moment, from Eq. (5.13), is

$$M = \tfrac{1}{2} \times 4.5 \times 5 = 11.25 \text{ kip ft} = 135 \text{ kip in.}$$

The section modulus required, from Eq. (5.14), is

$$S = \frac{135 \text{ kip in.}}{24 \text{ ksi}} = 5.625 \text{ in.}^3$$

We select from Table 13.2 a steel beam 6 in. deep with a mass of 12 lb/ft, which has $S = 7.31$ in.³

Problem 5.4 A floor is supported on parallel steel joists, or secondary beams. The joists are simply supported at their ends (Fig. 5.11.3) and span 5 m (*16 ft*). The uniformly distributed superimposed load is 7 kPa (*150 psf*). Select suitable steel beams, assuming that their own weight is 0.3 kN/m (*20 lb/ft*).

From Table 4.1, line 9, the maximum bending moment is

$$M = +\tfrac{1}{8} W L$$

The procedure is otherwise the same as in Problem 5.3.

The superimposed load per joist is

$$7 \times 2.5 \times 5 = 87.5 \text{ kN}$$

and the weight of the joist itself is

$$0.3 \times 5 = 1.5 \text{ kN}$$

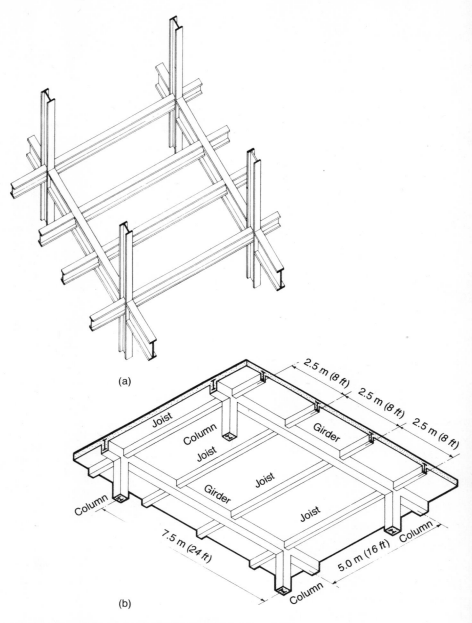

5.11.3. Layout of floor structure for Problems 5.4, 5.5, and 5.8.
(a) Top view of uncased steel frame.
(b) Upward view of steel frame encased in concrete for fire protection, with reinforced concrete floor slab in position.

The total load per joist is

$$W = 87.5 + 1.5 = 89.0 \text{ kN}$$

and the maximum bending moment is

$$M = \tfrac{1}{8} \times 89.0 \times 5 = 55.63 \text{ kN m}$$

The section modulus required is

$$S = \frac{M}{f} = \frac{55.63 \times 10^{-3}}{165} = 0.337\,2 \times 10^{-3} \text{ m}^3 = 337\,200 \text{ mm}^3$$

From Table 13.1 we select a steel beam 310 mm deep with a mass of 28 kg/m, which has $S = 351\,000$ mm^3.

The superimposed load per joist is

$$150 \times 8 \times 16 = 19\,200 \text{ lb} = 19.2 \text{ kips}$$

and the weight of the joist itself is

$$20 \times 16 = 320 \text{ lb} = 0.3 \text{ kips}$$

The total load per joist

$$W = 19.2 + 0.3 = 19.5 \text{ kips}$$

and the maximum bending moment

$$M = \tfrac{1}{8} \times 19.5 \times 16 = 39.0 \text{ kip ft} = 468 \text{ kip in.}$$

The section modulus required is

$$S = \frac{M}{f} = \frac{468}{24} = 19.50 \text{ in.}^3$$

From Table 13.2 we select a steel beam 12 in. deep with a mass of 19 lb/ft, which has $S = 21.3$ in.^{3}.

Problem 5.5 The steel joists of Problem 5.4 are supported on steel girders, or primary beams, as shown in Fig. 5.11.3. Select suitable steel beams, assuming that their own weight is 0.8 kN/m (*50 lb/ft*).

The total load carried by each steel joist was calculated in Problem 5.4. Each girder supports the end reactions of four joists. The end reaction of each joist is half the total load carried by the joist. Therefore, each girder carries the load of two regularly spaced joists. The remaining joists are carried directly by the columns.

From Table 4.1, line 8, the bending moment for a simply supported beam carrying two symmetrically spaced concentrated loads is

$$M = +\tfrac{1}{6} W L$$

The procedure is otherwise the same as in Problems 5.3 and 5.4.

The superimposed load per girder is

$$2 \times 89.0 = 178 \text{ kN}$$

and the weight of the girder itself is

$$0.8 \times 7.5 = 6 \text{ kN}$$

The total load per girder is

$$W = 178 + 6 = 184 \text{ kN}$$

The span is $L = 7.5$ m. The maximum bending moment is

$$M = \tfrac{1}{6} \times 184 \times 7.5 = 230.0 \text{ kN m}$$

The section modulus required is

$$S = \frac{M}{f} = \frac{230.0 \times 10^{-3}}{165} = 1.394 \times 10^{-3} \text{ m}^3 = 1\,394\,000 \text{ mm}^3$$

From Table 13.1 we select a steel beam 460 mm deep with a mass of 74 kg/m, which has $S = 1\,460\,000$ mm^3.

The superimposed load per girder is

$$2 \times 19.5 = 39.0 \text{ kips}$$

and the weight of the girder itself is

$$50 \times 24 = 1200 \text{ lb} = 1.2 \text{ kips}$$

The total load per girder is

$$W = 39.0 + 1.2 = 40.2 \text{ kips}$$

The span is $L = 24$ ft. The maximum bending moment is

$$M = \frac{1}{6} \times 40.2 \times 24 = 160.8 \text{ kip ft} = 1929.6 \text{ kip in.}$$

The section modulus required is

$$S = \frac{M}{f} = \frac{1929.6}{24} = 80.40 \text{ in.}^3$$

From Table 13.2 we select a steel beam 18 in. deep with a mass of 50 lb/ft, which has $S = 88.9$ in.³.

Problem 5.6 In this problem the steel joists are eliminated and the floor is supported only on steel girders, which span 6.5 m (*21 ft*) in both directions. The superimposed uniformly distributed load is 7 kPa (*150 psf*). Select suitable steel beams, assuming that their own weight is 0.8 kN/m (*50 lb/ft*).

The load carried by one girder is shown hatched in Fig. 5.11.4. It increases uniformly from zero at the column to a maximum at mid-span and then decreases uniformly to zero at the other column. The load carried by each girder is therefore equal to half the load carried by each floor panel 6.5 m (*21 ft*) square.

From Table 4.1, line 10, the maximum bending moment for this loading pattern is

$$M = +\tfrac{1}{6} W L$$

The procedure is otherwise the same as in the previous example.

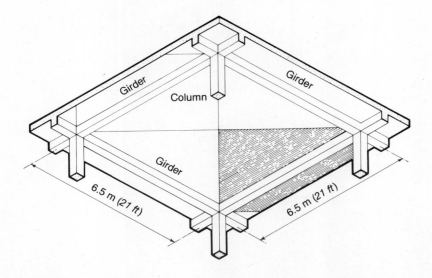

5.11.4. Layout of floor structure for Problems 5.6, 5.9, 5.10, 8.1, 8.2, 9.2, 9.3, 9.4, and 9.5.

The superimposed load per girder is

$$\tfrac{1}{2} \times 7 \times 6.5^2 = 147.9 \text{ kN}$$

The weight of the girder itself is

$$0.8 \times 6.5 = 5.2 \text{ kN}$$

The total weight per girder is

$$W = 147.9 + 5.2 = 153.1 \text{ kN}$$

The span is $L = 6.5$ m. The maximum bending moment is

$$M = \tfrac{1}{6} \times 153.1 \times 6.5 = 165.9 \text{ kN m}$$

The section modulus required is

$$S = \frac{M}{f} = \frac{165.9 \times 10^{-3}}{165} = 1.006 \times 10^{-3} \text{ m}^3 = 1\,006\,000 \text{ mm}^3$$

From Table 13.1 we select a steel beam 460 mm deep with a mass of 60 kg/m, which has $S = 1\,120\,000$ mm³.

The superimposed load per girder is

$$W = \frac{1}{2} \times 150 \times 21^2 = 33\,075 \text{ lb} = 33.08 \text{ kips}$$

and the weight of the girder itself is

$$50 \times 21 = 1050 \text{ lb} = 1.05 \text{ kips}$$

The total load per girder is

$$W = 33.08 + 1.05 = 34.13 \text{ kips}$$

The span is $L = 21$ ft. The maximum bending moment is

$$M = \frac{1}{6} \times 34.13 \times 21 = 119.5 \text{ kip ft} = 1434 \text{ kip in.}$$

The section modulus required is

$$S = \frac{M}{f} = \frac{1434}{24} = 59.8 \text{ in.}^3$$

From Table 13.2 we select a steel beam 18 in. deep with a mass of 40 lb/ft, which has $S = 68.4$ in.³.

Problem 5.7 A timber beam is simply supported over a span of 3 m (*10 ft*). It carries a uniformly distributed total load of 5 kN (*1 kip*). Select a suitable cross section.

Unlike steel, which is manufactured to a standard specification, timber is a naturally grown product, and its properties therefore vary greatly. In addition to differences between species there is a significant difference between green (or unseasoned) and dry (or seasoned) timber. Timber sizes are also highly variable, because certain traditions have grown up in different timber-producing regions. Some timber sizes are those of the actual dressed cross section, others are the nominal sizes of the rough-cut timber, which are subsequently reduced by the width of the sawcuts and by dressing. Tables of available timber sizes, which also give the section modulus, are available from the suppliers. For reasons of space they are not reproduced in this book. Timber sections are normally rectangular, and we can either use a narrow, deep beam, or one with a shallower section. A shallow section uses more material because the section modulus is proportional to $b\,d^2$, where b is the width and d is the depth (Section 5.4). However, a very deep and narrow beam needs cross-bracing to prevent sideways buckling, and this also adds to the cost.

We will use Douglas fir with a maximum permissible stress of

$f = 14$ MPa *(2 ksi)*

From Table 4.1, line 9, the maximum bending moment is

$M = +\frac{1}{8}WL$

As is the case for steel, the sign of the bending moment is not significant in the design of timber beams.

The total load is $W = 5$ kN, and the span is $L = 3$ m.
The maximum bending moment is

$M = \frac{1}{8} \times 5 \times 3 = 1.875$ kN m

The section modulus required is

$S = \dfrac{M}{f} = \dfrac{1.875 \times 10^{-3}}{14} = 0.133\,9 \times 10^{-3}$ m^3 = 133 900 mm^3

We select a rectangular section 140 mm deep by 45 mm wide (nominal and actual size). From section tables (not reproduced in this book) or by calculation (Section 5.4) it has $S = 147\,000$ mm^3.

The total load is $W = 1$ kip, and the span is $L = 10$ ft.
The maximum bending moment is

$M = \frac{1}{8} \times 1 \times 10 = 1.25$ *kip ft* = *15 kip in.*

The section modulus required is

$S = \dfrac{M}{f} = \dfrac{15}{2} = 7.5$ *in.*^{3}

We select a rectangular section 8 in. deep by 2 in. wide nominal size. After cutting and dressing this nominal size is reduced to $7\frac{1}{2}$ in. $\times\ 1\frac{5}{16}$ in. From section tables (not reproduced in this book) or by calculation (Section 5.4) it has $S = 12.30$ in.^{3}.

5.12 Problems on Reinforced Concrete Slabs and Beams

We determine the depth of the reinforced concrete slabs for the steel-framed buildings considered in Section 5.11, and we also redesign one of them as a reinforced-concrete framed structure.

All the examples in this section involve simplifications to obtain quick approximate answers, and they are therefore marked with an asterisk.

5.12.0.

Problem 5.8 Determine the depth of the reinforced concrete floor slab supported by the steel beams in Fig. 5.11.3.

We noted in Section 5.5 that the design of reinforced concrete sections is based on ultimate strength; the amount of tension reinforcement in reinforced concrete slabs is small and automatically ensures that the steel deforms plastically before the concrete is crushed. Since the stress in the concrete is not critical for design, we have only one equation, namely, Eq. (5.7).

There are two unknown quantities: the depth of the concrete section and the area of reinforcement. Since two unknown quantities cannot be calculated from one equation, we determine the depth of the concrete section by other means.

There are three considerations restricting the thickness of concrete slabs. In residential and office buildings, floor slabs should not be too thin because of the need for adequate sound insulation (Section 11.8), and in most buildings concrete roof slabs should not be too thin in order to provide an adequate thermal inertia (Section 11.5). The criterion most commonly used for determining the depth is that for limiting deflection. Brittle finishes are cracked if the deflection is too large, and eventually excessive cracks occur in the concrete slab itself. Since deflection is inversely proportional to (depth)^3, a small reduction in depth can greatly increase the deflection. For example,

$$100^3 = 1\,000\,000 \text{ and } 125^3 = 1\,953\,125$$

which is 95% more.

$$4^3 = 64 \text{ and } 5^3 = 125$$

which is 95% more.

This means that a 20% reduction in depth from 125 mm to 100 mm, *or from 5 in. to 4 in.*, produces a 95% increase in deflection.

All concrete codes have rules limiting the depth of reinforced concrete slabs to avoid excessive deflections. These range from simple empirical rules for simple problems to more elaborate, theoretically based rules that are used when deflection is critical.

The simple rule of the American (ACI) concrete code (Section 9.5 in Ref. 5.10) limits the ratio of span to overall depth of continuous reinforced concrete slabs, spanning in one direction only, to 28. The British (Ref. 5.11) and Australian concrete codes set limits that depend on the amount of reinforcement in the slab, and they give slightly higher ratios.

In metric units $L/28$ is $2.5 \text{ m}/28 = 89$ mm; we choose a reinforced concrete slab with an overall depth of 100 mm.

In customary American units $L/28$ is $8 \text{ ft}/28 = 3.43$ in. Since a 3½ in. concrete slab is rather thin, we choose a slab with an overall depth of 4 in.

It is not necessary in a preliminary design to know the amount of reinforcement required, but the architect should be aware of its location. The primary reinforcement runs at right angles to the supporting steel beams. The negative bending moment occurs over the steel beams (Fig. 4.6.8), and the reinforcement is needed on the top face of the slab. The positive bending moment occurs near mid-span, and the reinforcement is needed on the bottom face of the slab.

Secondary reinforcement is required at right angles to the primary reinforcement to resist shrinkage and temperature stresses. This varies with the strength of the steel used (Ref. 5.10, Section 7.12), but it is usually less than ¼% of the volume of the concrete. The layout of the reinforcement is shown in Fig. 5.12.1

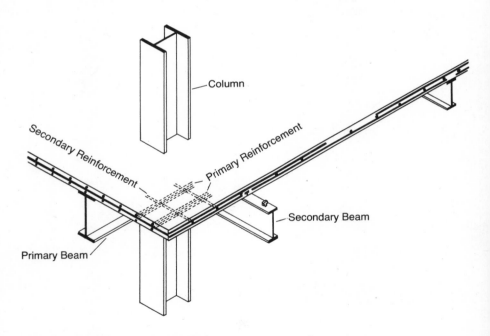

5.12.1. Layout of reinforcement for Problem 5.8.

Problem 5.9 Determine the depth of the reinforced concrete floor slab supported by the steel beams in Fig. 5.11.4.

The problem is similar to Problem 5.8, except that the bending moment is equally divided between the two spans because the slab spans in both directions. The spans are longer, but as the slab spans in both directions, it is stiffer. The American (ACI) concrete code (Ref. 5.10, Section 9.5), the British concrete code (Ref. 5.11, Section 3.3.8), and the Australian concrete code give different rules, which produce a range of ratios for span / depth depending on the type and amount of reinforcement. An average value is $L / D = 40$.

Using this approximate ratio, we find that the overall depth is $L / 40 = 6.5 \text{ m} / 40 = 163 \text{ mm}$. We choose a reinforced concrete slab with an overall depth of 175 mm.

Using this approximate ratio, we find that the overall depth is $L / 40 = 21 \text{ ft} / 40 = 6.3 \text{ in}$. We choose a reinforced concrete slab with an overall depth of 7 in.

The layout of the reinforcement is shown in Fig. 5.12.2.

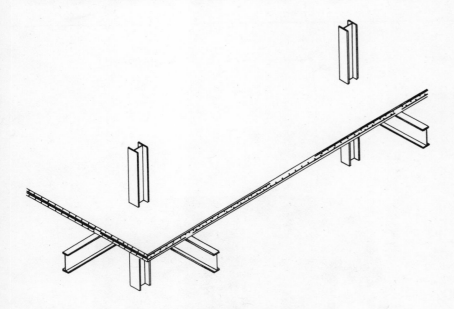

5.12.2. Layout of reinforcement for Problem 5.9.

*Problem 5.10 We wish to build the floor structure shown in Fig. 5.11.4 entirely in reinforced concrete.

We noted in Problem 5.8 that the amount of tension reinforcement in reinforced concrete slabs is small and automatically ensures that the steel deforms plastically before the concrete is crushed. In reinforced concrete beams it is possible to have sufficient reinforcement for a concrete crushing failure to occur. Because concrete is a brittle material (Section 3.3), this is a sudden failure that could occur without warning. Concrete codes based on ultimate strength (Refs. 5.10 and 5.11) therefore limit the amount of reinforcement, to ensure that this does not happen. The critical percentage of reinforcement, depending on the strength of the concrete and of the steel, varies from 2% to 7%.

In practice the beam is likely to fail the deflection criterion of the concrete code before this percentage of reinforcement can be reached, and deflection is therefore a useful basis for determining the minimum depth.

The minimum American (ACI) concrete code requirement (Ref. 5.10, Section 9.5) for the overall depth of a continuous beam is $L / 21$. It is advisable to choose a depth appreciably above the minimum; otherwise, the beam may have an excessive amount of reinforcement, and it may also deflect too much.

The width of the beam should be about half the depth.

For a span of 6.5 m we have

$$\frac{L}{21} = 310 \text{ mm}$$

We will use a depth of 600 mm and a width of 300 mm for the beams. These figures are based on judgment, not calculation.

For a span of 21 ft we have

$$\frac{L}{21} = 12 \text{ in.}$$

We will use a depth of 24 in. and a width of 12 in. for the beams. These figures are based on judgment, not calculation.

The reinforcement in continuous beams (Fig. 4.6.8) is at the bottom face near mid-span and at the top face over the supports. In addition, shear reinforcement (Section 6.1) is required. The layout of the reinforcement is shown in Fig. 5.12.3

Alternatively, we can delete the supporting beams altogether and use a flat plate (Fig. 5.5.1). Since creep deflection is a problem in flat plates, it is advisable to reduce the column spacing slightly (Section 5.5) or, alternatively, to prestress the reinforcement (Section 5.6).

For a reinforced concrete flat plate, a concrete thickness about 10% higher than that of a beam-supported slab is appropriate.

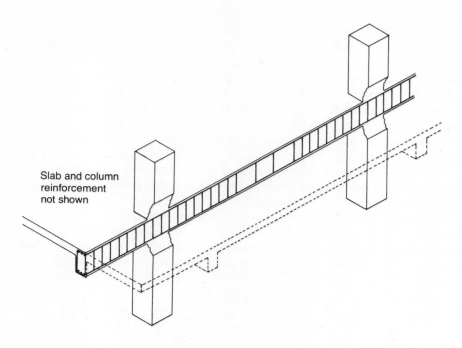

5.12.3. Layout of reinforcement for the beams only for Problem 5.10.

5.13 Problems on Ties and Columns

We calculate the size of the tension hangers required for a building whose floors are suspended from the roof. We next determine the approximate sizes of a steel column and a timber column. Reinforced concrete columns are considered in Problem 8.2.

Problem 5.11 The floor structure is suspended from ties fixed to cantilevers at roof level, as shown in Fig. 5.2.4. The weight on each tension hanger at the top story is 250 kN *(50 kips)*. Determine the size of the tension hanger.

It is good practice to encase the hanger in concrete, partly to give it fire and weather protection and partly to dampen vibrations due to wind. However, as concrete has negligible tensile strength, the tensile force is resisted entirely by the steel.

5.13.0.

The tension hanger alters its length with change in temperature. On the side of the building where it is exposed to the maximum solar radiation, the change in length is appreciable. It is essential for the safety of the structure that the hanger not become slack on a hot day, and the steel is therefore prestressed. Since prestressing is used, it is possible to use high-tensile steel with a maximum permissible stress of 700 MPa *(100 ksi)*. This reduces the size of the hangers and also makes it easier to ensure that there is a residual stress even at the highest temperature.

The thermal expansion of steel produces a strain (or an increase in length per unit length) of

$$e = 6 \times 10^{-4}$$

for a temperature increase of 50°C *(90°F)*. Since the modulus of elasticity of steel, E, is 200 000 MPa *(30 000 ksi)*, a change in length of 6×10^{-4} is produced by a change in stress (Section 3.1):

$$f = eE = 6 \times 10^{-4} \times 200\ 000 = 120 \text{ MPa}$$

$$f = eE = 6 \times 10^{-4} \times 30\ 000 = 18 \text{ ksi}$$

We therefore design the hangers with a slightly larger residual stress to ensure that they do not become slack as a result of an increase in temperature of 50°C *(90°F)* above that at which the hangers were installed. We will make that residual stress 140 MPa *(20 ksi)*, so that the available prestress is

$$700 - 140 = 560 \text{ MPa}$$

$$100 - 20 = 80 \text{ ksi}$$

Consequently each hanger requires a cross-sectional area of

$$\frac{250 \text{ kN}}{560 \text{ MPa}} = \frac{0.25 \text{ MN}}{560 \text{ MN/m}^2} = 446.4 \times 10^{-6} \text{m}^2 = 447 \text{ mm}^2$$

$$\frac{50 \text{ kips}}{80 \text{ ksi}} = 0.625 \text{ in.}^2$$

This cross-sectional area is supplied by one stranded cable 30 mm (1 ⅛ in.) in diameter or by two stranded cables 20 mm (¾ in.) in diameter. Sectional properties of cables, which are available from the suppliers, are not reproduced in this book.

*Problem 5.12 A steel column is required to carry an axial load of 2 MN *(450 kips)*. Select a suitable steel section.

It is a simple matter to determine the loadbearing capacity of a given column section. Since we know the effective length of the column and its section properties (including the radius of gyration, Section 5.3) we can calculate its slenderness ratio. We then ascertain from the structural steel code the maximum permissible stress, f, corresponding to that ratio. The cross-sectional area, A, of the column section is given in section tables (Table 13.1 and 13.2), and the product fA is the permissible column load.

When we design a column, we do not know the slenderness ratio, and hence the permissible stress, until we have selected a column section, and we cannot select a column section until we know the maximum permissible stress. The design process is thus a matter of making an intelligent guess and then correcting it if the first attempt was too much in error.

For a preliminary design it is sufficient to use an approximate value of f. The maximum permissible stress for an appropriately designed column section with a reasonable slenderness ratio is unlikely to be less than 80% of the maximum value. Hence we take f at 80% of the value applicable to a zero slenderness ratio.

We will take this stress for A 36 steel as 140 MPa *(20 ksi)*. Therefore we design the section for a maximum permissible stress of

$$0.8 \times 140 \text{ MPa} = 112 \text{ MPa}$$

$$0.8 \times 20 \text{ ksi} = 16 \text{ ksi}$$

5.13.1. Cross section of steel column for Problem 5.12. The actual dimensions differ slightly from the nominal dimensions; for example, this column has a nominal depth of 310 mm *(12 in.)* and an actual depth of 323 mm *(12.71 in.)*; its actual width is 309 mm *(12.16 in.)*.

Therefore the cross-sectional area required is

$$A = \frac{2 \text{ MN}}{112 \text{ MPa}} = 0.017\ 86 \text{ m}^2 = 17\ 860 \text{ mm}^2$$

$$A = \frac{450 \text{ kips}}{16 \text{ ksi}} = 28.13 \text{ in.}^2$$

From Table 13.1 we select a 310 mm wide flange shape, with a mass of 143 kg/m (Fig. 5.13.1), which has $A = 18\,200$ mm^2.

From Table 13.2 we select a 12 in. wide flange shape, with a mass of 96 lb/ft (Fig. 5.13.1), which has $A = 28.2$ in.^{2}.

*Problem 5.13 A timber column is required to carry an axial load of 220 kN *(50 kips)*. Select a suitable solid timber section.

The problem is similar to that of Problem 5.12. However, timber columns are not as efficient in resisting buckling as steel columns, which are shaped into I-sections to give a high radius of gyration in both directions for a given cross-sectional area. Solid timber columns are usually made square, and their radius of gyration for a given cross-sectional area is thus lower than for steel. Consequently the slenderness ratio is higher, and the maximum permissible compressive stress is reduced to a greater extent. We will use a reduction factor of 0.6, but this is only approximate, and for many satisfactory timber columns the factor could be higher or lower.

The maximum permissible compressive stress (Ref. 5.1) for a Douglas fir column with zero slenderness ratio is 9.0 MPa *(1.3 ksi)*.

Using a reduction factor of 0.6, which is only a rough approximation, we find that the cross-sectional area required is

$$A = \frac{0.22 \text{ MN}}{0.6 \times 9.0 \text{ MN/m}^2} = 40\,740 \times 10^{-6} \text{ m}^2 = 40\,740 \text{ mm}^2$$

A section with nominal and actual dimensions of 191 mm by 241 mm has $A = 46\,030$ mm^2.

$$A = \frac{50 \text{ kips}}{0.6 \times 1.3 \text{ ksi}} = 64.10 \text{ in.}^2$$

A section with nominal dimensions of 8 in. by 10 in. has actual dimensions (after dressing) of $7\frac{1}{2}$ in. by $9\frac{1}{2}$ in., and $A = 71.25$ in.^{2}.

Structural Handbooks for the Design of Structural Members

5.1 TIMBER CONSTRUCTION MANUAL published for the American Institute of Timber Construction. Wiley, New York, 1966. ca. 600 pp.

5.2 STEEL BUILDINGS: ANALYSIS AND DESIGN by Stanley W. Crawley and Robert M. Dillon. Wiley, New York, 1970. 397 pp.

5.3 LIGHT GAUGE COLD-FORMED STEEL DESIGN MANUAL. American Iron and Steel Institute, New York, 1956. 91 pp.

5.4 ALUMINUM STANDARDS AND DATA. The Aluminum Association, New York, 1968. 173 pp.

5.5 DESIGN HANDBOOK. American Concrete Institute, Detroit, 1973. 403 pp.

5.6 MANUAL OF STANDARD PRACTICE FOR DETAILING REINFORCED CONCRETE STRUCTURES. American Concrete Institute, Detroit, 1974. 167 pp.

5.7 REINFORCED CONCRETE FUNDAMENTALS, Fourth Edition, by P.M. Ferguson. Wiley, New York, 1979. 724 pp. (S.I. Edition, 1981)

5.8 SPECIFICATIONS FOR THE USE OF STRUCTURAL STEEL IN BUILDING, Part 2, Metric Units, BS 449. British Standards Institution, London, 1969. 120 pp.

5.9 MANUAL OF STEEL CONSTRUCTION, Seventh Edition. American Institute of Steel Construction, New York, 1970, which contains the AISC Code. 985 pp.

5.10 BUILDING CODE REQUIREMENTS FOR REINFORCED CONCRETE (ACI 318-77). American Concrete Institute, Detroit, 1979. 103 pp.

5.11 CODE OF PRACTICE FOR THE STRUCTURAL USE OF CONCRETE, Part 1, Design, Materials and Workmanship, CP 110. British Standards Institution, London, 1972. 155 pp.

Chapter 6 Some Secondary Problems in the Design of Beams

Depend upon it, Sir, when a man knows he is to be hanged in a fortnight, it concentrates his mind wonderfully.
(Samuel Johnson)

This chapter can be omitted by readers who do not wish to concern themselves with secondary aspects of structural design. It considers the effect of shear, torsion, and stress concentrations on the design of beams.

6.1 Shear

It is possible to have a beam subject to a bending moment and not to a shear force, but in buildings this is very unusual (Fig. 6.1.1). Most beams carry a vertical load, and this not only bends the beam but also tends to cut through it (Fig. 6.1.2). This is called the shear force, and the resistance to the cutting action is the *shear resistance* of the beam. Thus for a beam simply supported at its ends and carrying a uniformly distributed load, the shear force is greatest at its ends and equal to the support reaction; it is then gradually reduced to zero at mid-span and increases again toward the other end of the beam (Fig. 6.1.3).

All beams, whether supported at one end only (that is, cantilevers, Fig. 4.5.2) or at both ends (Fig. 4.6.2), and whether they are continuous over their supports (Figs. 4.6.7 and 4.6.8) or built into their supports (Fig. 4.6.6), have a basically similar distribution of shear force: the shear force is greatest near the supports.

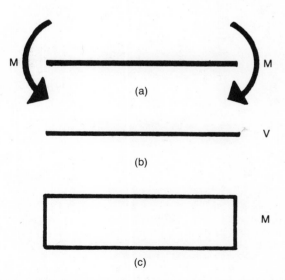

6.1.1. A uniform bending moment acting on a beam (a) produces a zero shear force V (b) and a constant bending moment M (c) along the beam.

6.1.2. Bending moment (a) and shear force (b).

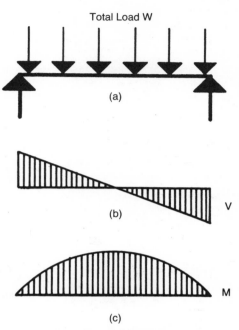

6.1.3. A uniformly distributed load W (a) produces in a simply supported beam a shear force V (b) uniformly varying from $+\frac{1}{2}W$ to $-\frac{1}{2}W$ and a bending moment M (c) varying parabolically. The maximum shear force occurs at the supports, the maximum bending moment at mid-span.

We did not consider shear forces in Chapters 4 and 5 because they are of secondary importance in the design of buildings. The magnitude of the shear force depends only on the magnitude of the load W. The magnitude of the bending moment depends both on the magnitude of the load W and the span L (Tables 4.1 and 4.2). Therefore the greater L is in relation to W, the greater the importance of the bending moment M in relation to the shear force V. The dimensions of machine parts are often determined by the shear force, and the bending moment is of secondary importance. In buildings the reverse is true, except for the foundation (Chapter 9). In the foundation the spans L are generally the same as for the superstructure, but the load W from all the floors above is transmitted to it, so that the magnitude of W for the foundation can be very much higher than for the superstructure. Under those circumstances the shear force may determine structural sizes.

For the superstructure, however, the structural sizes required for shear resistance are usually less than those required for bending resistance, and it is therefore the bending moment that determines the dimensions of a beam carrying a vertical load or of a column resisting a horizontal load. We therefore design for bending and then check the design to see if the shear resistance is adequate. If necessary, we reinforce the beam for shear.

Rectangular sections of timber, standard steel sections, and reinforced concrete slabs rarely require any modification for shear. However, steel plate girders and reinforced concrete beams generally need shear reinforcement.

Shear forces produce both tension and compression at 45° to the direction of the shear forces (Fig. 6.1.4). Thus diagonal tension and diagonal compression are both present simultaneously in all beams subject to shear. Diagonal tension does no harm to a steel plate, which has ample tensile strength, and diagonal compression does no harm to concrete beams that have adequate thickness to prevent buckling under a diagonal compressive force (Section 5.3). However, the diagonal compression may produce a shear failure in a steel plate if the plate is thin and consequently buckles, and the diagonal tension may cause failure in concrete that is weak in tension.

Thus, in steel plate girders, diagonal compression failure due to shear takes the form of buckling; the web crinkles at right angles to the compression. This can be prevented by welding stiffeners to the web of the plate girder at intervals.

Diagonal tension failure of a concrete beam takes the form of diagonal tension cracks at right angles to the tension. This is prevented by reinforcement (Fig. 6.1.5).

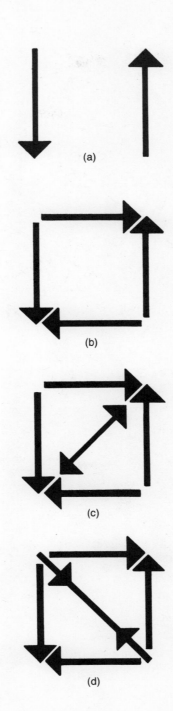

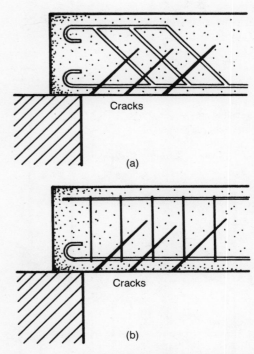

6.1.4. (a) A vertical shear force acting downwards is associated with another vertical shear force acting upwards; both forces are needed for the cutting effect of shear and also for vertical equilibrium.
(b) The pair of vertical shear forces is not in moment equilibrium (Section 2.6), because together they tend to rotate the beam. Another pair of horizontal shear forces is needed for equilibrium.
(c) The four shear forces can be combined to produce diagonal tension; each "pull" is produced by one pair of shear forces.
(d) The four shear forces can also be combined to produce diagonal compression.
The diagonal tension and the diagonal compression are just another way of representing the shear forces; both are acting at the same time.

6.1.5. Concrete has ample strength in shear and in compression, but a small diagonal tensile force suffices to produce a crack. Shear reinforcement can be provided by steel at right angles to the direction of a potential tension crack (a). In practice it is generally simpler to use vertical stirrups or hoops (b); the tensile force in these vertical bars has a 45° component that resists the diagonal tension due to shear. The diagonal tension differs from that due to torsion (Fig. 6.2.4.), because it points in the same direction on both faces of the beam.

6.2 Torsion

Torsion is very common in machine parts; for example, the shaft of every engine is in torsion.

In buildings torsion is also quite common, but it is mostly a secondary effect resulting from a nonuniform distribution of the load. Thus all spandrel (see Glossary) beams are subject to combined bending and torsion, because the load is entirely on one side of the beam. Any asymmetrical arrangement of the beams in a building or of the load on the beams produces secondary torsion. More substantial twisting moments occur in primary beams that support secondary beams on one side only (Fig. 6.2.1) and in balcony or corner beams (Fig. 6.2.2).

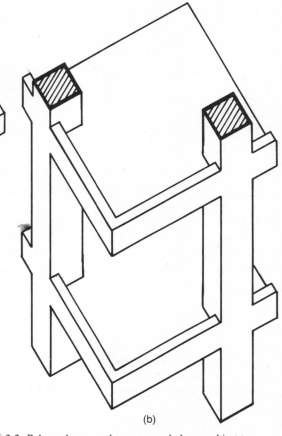

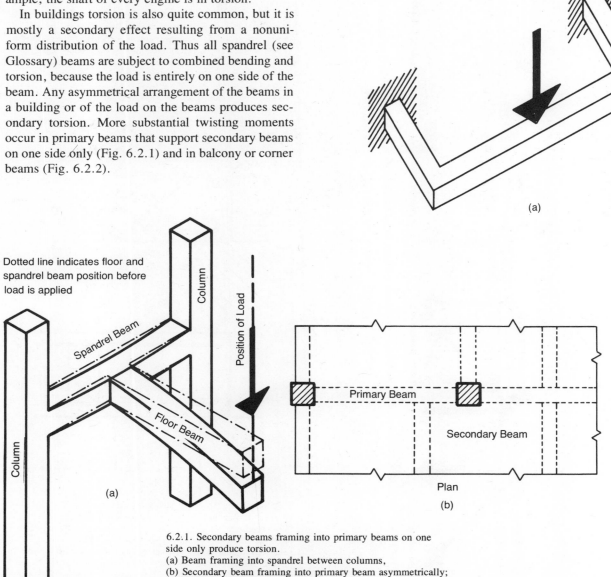

6.2.1. Secondary beams framing into primary beams on one side only produce torsion.
(a) Beam framing into spandrel between columns,
(b) Secondary beam framing into primary beam asymmetrically; if this occurs close to the column, the twisting moment can be very high.

6.2.2. Balcony beams and corner spandrels are subject to combined bending and torsion.
(a) Balcony girder cantilevered from facade.
(b) Spandrel on corner without a column.

In steel structures the requirements for torsional strength are in conflict with those for bending strength. For a rectangular section, the greatest bending strength is obtained with a narrow, deep section; in contrast, the square section has the greatest torsional strength (Fig. 6.2.3). Standard steel I-sections have excellent bending strength but poor torsional strength. The best torsional resistance is obtained by concentrating the material as far from the axis of torsion as possible. Thus a tubular section or two channel sections welded together (Fig. 5.3.3.d,e, and f) have excellent torsional resistance. These sections also have fairly good bending resistance. It is evidently necessary to compromise in the choice of the structural steel section when the layout of the beams or the distribution of the load produces significant twisting moments.

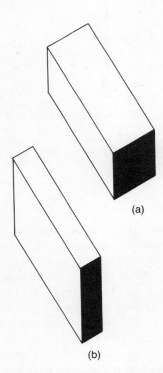

6.2.3. A square section (a), measuring d by d, and a rectangular section (b) measuring $d/\sqrt{2}$ by $\sqrt{2}/d$ (twice as deep as it is wide) have the same cross-sectional area. The first is stronger in torsion (T), the second is stronger in bending (M). For an elastic section (Ref. 6.1):

$$\frac{M_b}{M_a} = 1.41 \quad \text{and} \quad \frac{T_b}{T_a} = 0.84$$

Torsion is a form of shear, and in concrete torsion produces diagonal tensile stresses. This type of failure is easily demonstrated by twisting a bar of chalk of the type used for writing on a chalkboard. It fails with a 45° corkscrew fracture, and a concrete beam fails in precisely the same way. Thus we need shear reinforcement in the form of vertical closed hoops (Fig. 6.2.4) where the torsion occurs (Ref. 6.1). The same type of reinforcement is also suitable for resisting shear forces (Section 6.1). However, the reverse does not apply, because some shear reinforcement is not in the form of closed hoops.

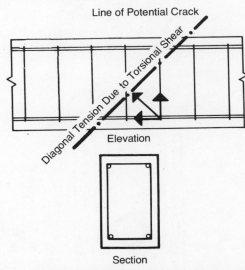

6.2.4. Torsional shear reinforcement normally takes the form of closed vertical hoops; the torsional shear stresses act on all four faces of the beam, and the diagonal tension points in opposite directions on opposite faces (as on a corkscrew). When torsion is present, all shear reinforcement must be in the form of closed hoops.

6.3 Stress Trajectories

We showed in Section 6.1 that every beam is subject not merely to a bending moment but also to a shear force. Many beams additionally have some torsion.

Most columns are subject to some bending, and in tall buildings all columns are subject to some bending. This places them in combined compression, bending, and shear.

The stress distribution in structural members is therefore quite complex. This is no problem for electronic computers, which can handle the necessary arithmetic. The combined stresses can also be measured on a structural model (Section 6.5).

By combining the stresses due to the various forces and moments at any point within the structure, we obtain the greatest tensile and compressive stresses at that point. These are called the *principal stresses*. Connecting the directions in which the principal stresses point produces a series of lines, called the *stress trajectories* or *isostatic lines* (Fig. 6.3.1). They indicate, for example, the direction in which reinforcement should ideally be arranged in concrete. Pier Luigi Nervi and Lev Zetlin have on several occasions made use of stress trajectories, not so much because of their structural economy as because of their interesting patterns (Fig. 6.3.2).

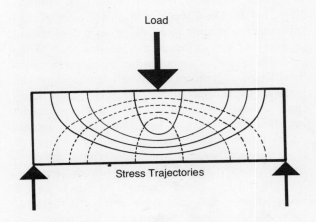

6.3.1. Stress trajectories (also called isostatic lines) in a beam carrying a concentrated load.

105

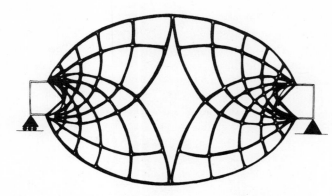

6.3.2. Drawing after beam designed by Lev Zetlin, based on the patterns of the stress trajectories.

6.4 Stress Concentrations

If the trajectories are drawn at equal increments of stress, a crowding of the lines indicates a stress concentration, just as the crowding of contours on a map of a mountainous region indicates a steep gradient.

Load concentrations, reentrant corners, notches, and holes invariably produce stress concentrations (Fig. 6.4.1). Local yielding in a small area of steel immediately adjacent to a hole or notch does not endanger the safety of the member, because the yielding is confined to a very small part of the structural member. Reentrant corners and load concentrations can produce larger areas of yielding, which make it appropriate to reinforce the structural steel member with a stiffener. However, as long as the steel is ductile, stress concentrations present no problems.

In very cold weather structural steel may turn brittle and crack without previous plastic deformation. Ductility and brittleness depend not merely on the composition of the material but also on the temperature. Metals become more ductile (and sometimes lose strength) at higher temperatures, and they become more brittle at lower temperatures. Thus at −30°C (−22°F) structural steel has a higher yield point but a lower ductility than at +20°C (+68°F). Special attention should therefore be given to stress concentrations when structural steel may be exposed to very low temperatures.

Because of its greater brittleness, stress concentrations also become important when high-carbon steel is used. This material has a higher strength but less ductility (Section 3.3).

Reentrant corners, which are unavoidable in buildings (for example, around door and window openings), must be suitably reinforced in reinforced concrete members, whether precast or site-cast (Fig. 6.4.2). Additional shear reinforcement should also be provided under load concentrations. Holes, which are unavoidable in steel and aluminum members, are rarely needed in concrete members, but when they are needed they require reinforcement.

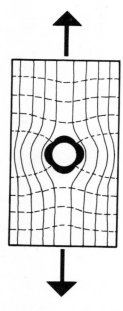

6.4.1. Stress trajectories in a steel plate in tension into which a hole has been cut for connecting it to another plate. Very high stresses are produced near the hole, sufficient to cause yielding of the steel. The black outline indicates that the steel has yielded. (*After Salvadori.*)

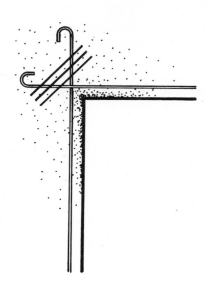

6.4.2. Reinforcing detail for reentrant corner in concrete. If the steel were bent around the reentrant corner, it would straighten out and break off the cover. Instead, we use two reinforcing bars, and each is anchored. It may also be necessary to place additional reinforcement diagonally to the corner, at right angles to the line of the potential crack.

6.5 Experimental Stress Analysis

Stresses cannot be measured directly, but strains can be measured by a number of mechanical, electrical, and optical devices. In an elastic material stress is proportional to strain (Section 3.1), so that strain gauges give a measure of stress.

The best technique for visualizing stress concentrations is photoelasticity, which is based on the property of certain glasses and plastics of breaking up incident light into two components polarized in the direction of the principal stresses (Ref. 6.2). The equipment needed is simple (Ref. 6.3), photoelastic materials are easily obtainable, and models are easily made from them; they include, for example, Plexiglas and araldite. When a model made from these materials is placed between two polarizing filters, oriented at right angles to one another and loaded, colored fringes (called isochromatics) appear; the closer the fringes are together, the higher the stresses. When monochromatic light (for example, from a sodium lamp) is used, the fringes are dark and light only (Fig. 6.5.1).

Photoelasticity is primarily useful for analyzing stress concentrations. More general complex stress systems can be explored, particularly for a preliminary design, by making a scale model from a suitable elastic material, such as Plexiglas, and measuring the strains with a suitable strain gauge. Electrical resistance strain gauges are most commonly used for this purpose. They consist of a length of zigzag wire, mounted on paper and attached to the structural model with a special glue. When the wire is subjected to tension, it becomes thinner and allows less current to pass; when it is compressed, it allows more current to pass. This change in electrical resistance is measured. With care, strains of the order of 10^{-6} (1 part in 1 million) can be ascertained. The technique of model analysis is described in several books on the subject (for example, Ref. 6.2).

6.5.1. Photoelastic patterns in a beam simply supported at its ends and carrying two concentrated loads. The stress concentrations under the load are shown by the closely spaced contours.

6.6 Problems for Chapter 6

If you have not yet read Section 5.9, you should do so before attempting the following problems.

In this short, mainly descriptive chapter, there are two problems. The first explains how to determine the glue-line stresses in laminated timber, and the second deals with shear reinforcement in concrete beams.

The size of diagonal braces to absorb the shear forces due to wind loads is determined in Problem 8.3 for a steel-framed building and in Problem 8.4 for a concrete-framed building.

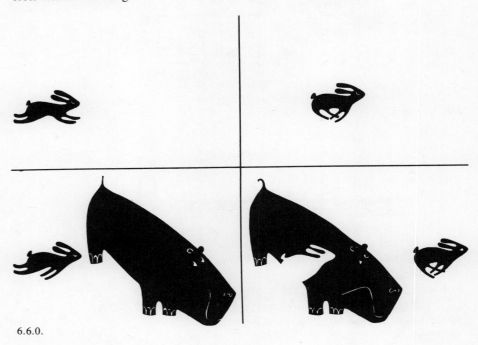

6.6.0.

Problem 6.1 The laminated timber beam shown in Fig. 6.6.1 carries a uniformly distributed load of 100 kN (*20 kips*) over a simply supported span of 3 m (*20 ft*). Determine the maximum stress in the glue.

The support reaction is $R = \frac{1}{2}W$. This reaction bends the beam, but it also has a tendency to shear or cut through it, which is resisted by the internal shear resistance of the beam (Fig. 6.6.2). The maximum shear force is immediately adjacent to the support, and it is

$$V = R = \frac{1}{2}W$$

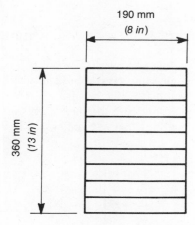

6.6.1. Dimensions of cross section of laminated timber beam in Problem 6.1.

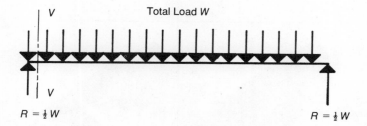

6.6.2. If a (hypothetical) cut is made along the dashed line immediately adjacent to the left-hand support of a simply supported beam, the shear force at that cut is obtained from the equations of equilibrium for the part of the beam to the left of the cut. For vertical equilibrium, $V = \frac{1}{2}W$ (Problem 6.1).

The shear force produces shear stresses within the beam that vary parabolically (Fig. 6.6.3); this is proved in more advanced textbooks (for example, Ref. 1.1, pp. 170–72). Therefore the maximum shear stress is

$$v = \frac{3V}{2bd} = \frac{3W}{4bd}$$

where W is the total load carried by the beam, b is its width, and d is its depth. The span of the beam does not enter into the calculations.

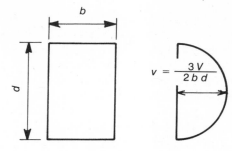

6.6.3. The shear stresses in a rectangular beam vary parabolically. Since the area under a parabola is equal to $\frac{2}{3}$ of the area of a rectangle of the same width and depth, the maximum shear stress is $\frac{3}{2}$ times the average shear stress (Problem 6.1).

We have a total load of 100 kN = 0.1 MN and beam dimensions of 0.36 m and 0.19 m. Therefore the maximum shear stress in the glued joints is

$$v = \frac{3 \times 0.1}{4 \times 0.36 \times 0.19} = 1.096 \text{ MPa}$$

We have a total load of 20 kips and beam dimensions of 13 in. and 8 in. Therefore the maximum shear stress is

$$v = \frac{3 \times 20}{4 \times 13 \times 8} = 0.144 \text{ ksi} = 144 \text{ psi}$$

The shear stress in the glue is within the admissible range (Ref. 5.1).

Problem 6.2 Determine the maximum shear stress in the reinforced concrete beams designed in Problem 5.10.

We determined the total load carried by these beams when we considered the identical loading problem for steel beams in Problem 5.6. Each beam carried a total load of 153.1 kN (*34.13 kips*).

From Fig. 6.6.2 the maximum shear force is

$$V = R = \tfrac{1}{2}W$$

This shear force produces shear stresses within the concrete beam. The nominal shear stress used in concrete codes (Refs. 5.10 and 5.11) is

$$v = \frac{V}{bd}$$

where b is the width of the beam and d is the effective depth, measured to the center of the reinforcement (Fig. 5.5.11). This formula is explained in specialized books on reinforced concrete design (for example, Ref. 5.7).

The width stated in Problem 5.10 is 300 mm (*12 in.*), and the overall depth is 600 mm (*24 in.*). From this overall depth we must deduct 50 mm (*2 in.*) for cover and bar radius (Fig. 5.5.11) to obtain the effective depth d. This gives $d = 550$ mm ($d = 22$ *in.*). The loads are the same as in Problem 5.6.

We have $V = \frac{1}{2}W = 76.55$ kN, $b = 0.3$ m, and $d = 0.55$ m. The nominal shear stress is

$$v = \frac{76.55}{0.3 \times 0.55} = 464 \text{ kPa}$$

We have $V = \frac{1}{2}W = 17.07$ *kips,* $b = 12$ *in., and* $d = 22$ *in. The nominal shear stress is*

$$v = \frac{17.07}{12 \times 22} = 0.0647 \text{ ksi} = 65 \text{ psi}$$

This is less than the maximum shear stress permissible without reinforcement, and only nominal shear reinforcement is required near the supports, as shown in Fig. 6.1.5.b.

References

6.1. REINFORCED AND PRESTRESSED CONCRETE IN TORSION by H.J. Cowan and I.M. Lyalin. St. Martin Press, New York, 1965. pp. 18−19.

6.2. MODELS IN ARCHITECTURE by H.J. Cowan, J.S. Gero, G.D. Ding, and R.W. Muncey. Elsevier, London, 1968. 228 pp.

6.3. BUILDING SCIENCE LABORATORY MANUAL by H.J. Cowan and F.J. Dixon. Applied Science, London, 1978. 156 pp.

Chapter 7 Structural Assemblies

Sir Christopher Wren
Said, "I am going to dine with some men.
If anyone calls,
Say I am designing St. Paul's."
(Edmund Clerihew Bentley)

We consider the distinction between mechanisms and structures and the distinction between structures that have just sufficient members and those that are more rigid than this minimum requirement. The rigidity of structures depends on the rigidity of the joints and on the number of joints relative to the number of members.

We then examine the special problems of roof trusses, of parallel-chord Pratt trusses and Vierendeel trusses, of portal frames, and of arches.

7.1 Pinned and Rigid Joints

The first structures to be designed by mechanical principles, in the late eighteenth century, were made from cast iron and wrought iron. Many of them were joined with pins. It was a simple matter to cast a hole into a cast iron member or to bend a wrought iron bar to form an eye. An iron pin could then be pushed through these holes to make a flexible joint. In comparison, bolts had to be cut individually and thus were expensive, and welding was difficult at the time.

Pins allow complete freedom of rotation; the joint cannot transmit a bending moment, and therefore the bending moment at a pin joint is always zero. However, a pin joint can transmit tension, compression, or shear (Fig. 7.1.1). Pin joints are still used for very long spans where it is important to ensure that at a particular joint the bending moment is, in fact, very close to zero; but they have now become an expensive type of joint, and in buildings they are rarely needed. Most really long spans occur in bridges.

A number of other, cheaper joints are today classed as "pin joints" because they transmit only a small fraction of the bending moment. Thus many steel structures with light bolted or welded connections may be regarded as pin joints (Figs. 7.1.2 and 7.1.3). Steel structures can also be made with rigid joints (Fig. 7.1.4).

In site-cast reinforced concrete, rigid joints are much cheaper than flexible joints. As the concrete is cast in one piece or suitably reinforced, it is automatically rigid (Fig. 7.1.5), and little is to be gained from producing a pin joint in site-cast concrete. Dry precast joints are normally pinned (Fig. 7.1.6). Precast units are sometimes made with projecting reinforcing bars, which are then cast into concrete on the site to produce a rigid joint.

Design calculations are greatly simplified by having joints that transmit either no bending moment at all *(pin joints)* or the entire bending moment *(rigid joints)*.

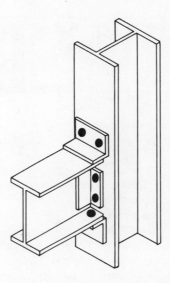

7.1.2. Light-bolted connection for a steel frame, which is considered a pin joint. The web angles are welded to the beam in the fabricating shop, and the single-line bolted connection to the column is made on the site.

A beam seat can be added for ease of construction. This is an angle welded to the column in the shop but connected to the beam with only one line of bolts on the site, so that it does not provide a rigid restraint. A small clip angle on top of the beam is also admissible in a pin joint. Only the site connections are shown.

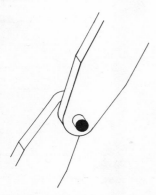

7.1.1. A pin joint can transmit tension, compression, and shear, but it cannot transmit any bending moment whatever.

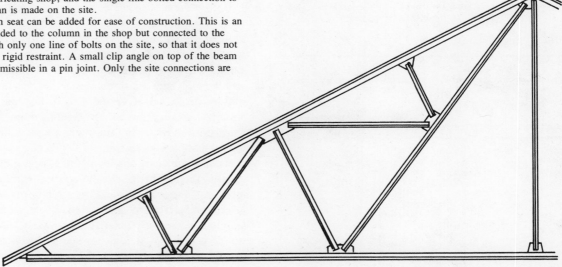

7.1.3. Light-welded connections for roof truss, which are considered pin joints. Each member is relatively long and flexible, and any moment transmitted to it is largely dissipated along its length.

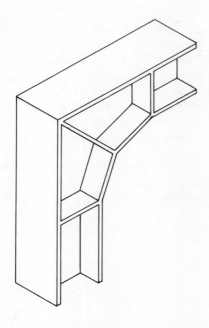

7.1.4. Rigid welded joints in steel portal frames, which may be considered to transmit the entire bending moment.

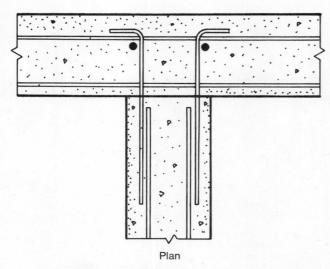

7.1.5. Rigid joints in site-cast concrete walls.

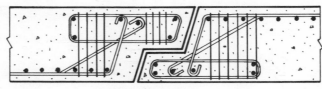

7.1.6. Pin joint in precast concrete.

7.2 Statically Determinate and Statically Indeterminate Structures

Computerized structural design is a relatively recent development. The first structural program was written at the University of Manchester in England in 1953, and until 1964 when the STRESS program became fully operational at the Massachusetts Institute of Technology, computerized structural design was not really convenient or economical for ordinary buildings (Section 8.7).

Before the advent of the computer it was a matter of great importance whether the structure could be worked out by statics alone (a *statically determinate structure*) or whether some additional, more complicated theory was required (a *statically indeterminate structure*). There are three equations of statical equilibrium (Eqs. 4.3, 4.4, and 4.5, in Section 4.4); in addition, at every pin joint the bending moment $M=0$, and therefore each pin joint gives an additional equation of statical equilibrium.

If the number of unknown forces and moments is equal to the number of statical equations, they can be found by statics alone, and the structure is statically determinate. If there are not enough equations, the structure is statically indeterminate; in addition to the statical equations, it is necessary to introduce equations derived from the elastic deformation of the structure. Computer programs for structural frames employ equations based on the elastic deformation of the structure for all problems, so that the distinction vanishes. However, statically determinate structures still retain their significance for preliminary design decisions.

Statically determinate structures are of special interest also for a study of the rigidity of structures. A structure can be regarded as a system of members and joints (Fig. 7.2.1). If there are insufficient members in relation to the number of joints, we have a mechanism, which collapses at the slightest push (Fig. 7.2.2). A mechanism can have one, two, three, or more degrees of freedom, depending on the number of members (Fig. 7.2.3). When we eliminate the last degree of freedom, we obtain a statically determinate structure. This structure has just sufficient members to prevent its collapse at the slightest push (Fig. 7.2.4). As we add further structural members, we obtain a statically indeterminate structure with one, two, three or more degrees of redundancy (Fig. 7.2.5).

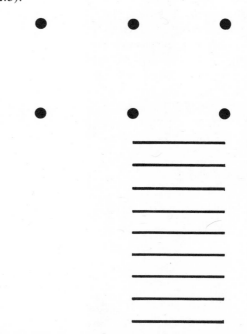

7.2.1. The geometry of a structural assembly is in the first instance determined by the geometry of the joints. Its rigidity depends on the number of members used to connect these joints.

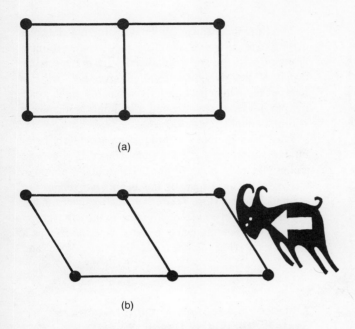

(a)

(b)

7.2.2. If all the joints are true pin joints, the number of members is only sufficient for a mechanism, which collapses at the slightest push.

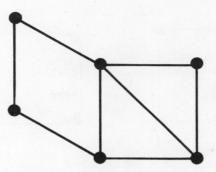

Mechanism with One Degree of Freedom

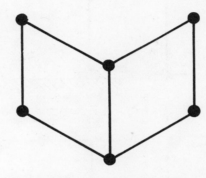

Mechanism with Two Degrees of Freedom

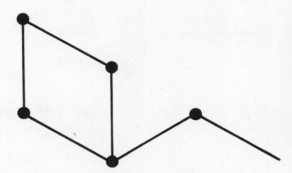

Mechanism with Three Degrees of Freedom

7.2.3. A mechanism can have 1, 2, 3, or more degrees of freedom, depending on how many members would be required to make it a statically determinate structure.

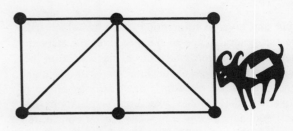

Statically Determinate Structure

7.2.4. A statically determinate structure has just sufficient members. Removal of one member turns it into a mechanism. Addition of a member makes it statically indeterminate. The statically determinate structure can collapse only if *one* of the members is broken or buckled.

One Redundancy

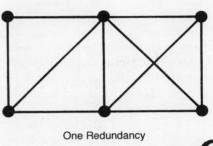

Two Redundancies

7.2.5. A statically indeterminate structure may have one, two, or more degrees of redundancy, depending on how many more members there are than in a statically determinate structure. A structure with two redundancies can collapse only if *three* of the members are broken or buckled.

113

For any given configuration of joints, which indicate the general shape of the structure, the statically determinate solution is therefore the borderline between a whole range of mechanisms (with more and more degrees of freedom) and a whole range of statically indeterminate structures (with more and more redundancies). Mechanisms are, of course, unsafe, and structures with many degrees of redundancy may be unnecessarily complex.

The statically determinate structure has a particular interest for a preliminary design. We can take a proposed structure and render it statically determinate by placing imaginary pin joints in places where we have reason to believe that the bending moment is approximately zero or (from past experience) that it could be made zero without serious consequences. We thus obtain a structure whose approximate maximum bending moments can be obtained quickly and easily without a computer: we can determine approximate structural sizes for a preliminary design and form an opinion whether the structure is likely to be suitable for our purpose. This process is described in more detail in Sections 7.6 and 7.7.

We can modify the rigidity of structural systems not merely by altering the number of members but also by introducing additional rigid joints or pin joints.

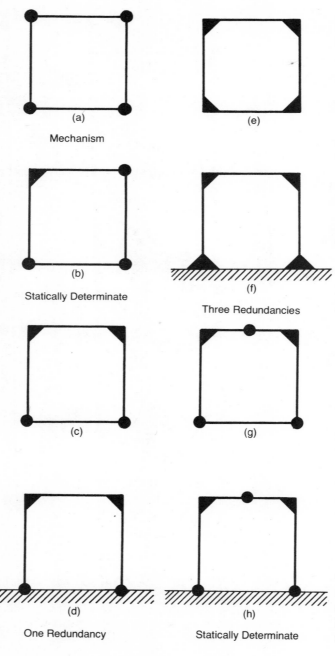

7.2.6. (a) Four pin joints joined by four members are a mechanism, not a structure.
(b) We can make this square panel into a statically determinate structure by adding one diagonal, or we can instead make one pin joint rigid. The panel is now statically determinate.
(c) If we make two joints rigid, we have a statically indeterminate structure with 1 redundancy. This is the prototype of an important structure (Section 7.6), the two-pin portal (d).
(e) If we make all four joints rigid, we have a statically indeterminate structure with three redundancies. This is a panel of a Vierendeel girder (Section 7.5). It is also the prototype of another important structure (Section 7.6), the rigid portal (f).
(g) We can restore the statically determinate conditions in the panel (c) by placing a pin joint in one of the members. This is another important structure (Section 7.6), the three-pin portal (h).

Thus any mechanism can be turned into a statically determinate structure or a statically indeterminate structure with several redundancies by making some or all of the pin joints rigid. Conversely, we can remove redundancies either by replacing rigid joints by pin joints or by placing pin joints somewhere within a member (Fig. 7.2.6).

This last-mentioned technique is particularly helpful for simplifying rigid frames (Fig. 7.2.6.h) and arches. By placing a pin joint at a strategic point, in this case the mid-span of the beam, we render the redundant frame (Fig. 7.2.6.d) statically determinate. This is not merely a useful structure in its own right; it also offers some guidance to the structural sizes of a rigid frame (Fig. 7.2.6.f) of the same span and height.

7.3 Roof Trusses: An Ancient Structural System Still Widely Used

The basis of the pin-jointed truss is triangulation. Three members connecting three pin joints make a statically determinate frame. For every additional joint we require two additional members if the truss is to remain statically determinate. Thus the truss is divided into triangles (Fig. 7.2.4).

The ancient Romans built triangulated trusses, but they were used only accidentally in medieval Europe, in China, and in Japan (Ref. 1.4, pp. 113–19). Andrea Palladio revived the Roman system of triangulation in the seventeenth century, and it was used extensively for industrial buildings in the nineteenth and early twentieth centuries. Today rigid frames are often preferred.

Triangulated trusses are particularly suitable for sloping roofs, and sloping roofs are particularly suitable for buildings of one to four stories; they keep the rain out at less cost than a flat roof. In a building tall enough to require an elevator, the requirements for its hoisting gear interfere with a sloping roof. A sloping roof is also visually unsatisfactory on a tall building.

If the roof forms part of the interior space (Fig. 7.3.1), as it does in many factories and some houses, rigid frames (Section 7.6) are more attractive in appearance, and they collect less dust. When a flat ceiling is used, a roof truss is an economical structure.

There are basically two structural types: the Pratt truss and the trussed rafter (Fig. 7.3.2). Either can be constructed from timber, steel, or aluminum, but reinforced concrete is not a good material for trusses, because concrete is not suitable for tension members. The extent of the subdivision into triangles depends on the span and on the need to keep the slenderness ratio of the compression members within an economic limit (Section 5.3).

Natural lighting can be admitted directly through the roof or through windows in the walls. North-light windows (south-light in the southern hemisphere) admit natural light without direct solar heat; there are two types: the vertical window for the subtropical zone and the sloping window for the temperate zone, where lighting levels are lower (Fig. 7.3.3).

The roof trusses support purlins at the panel points where the triangles meet, and these in turn support the roof sheets or the rafters that carry shingles or tiles.

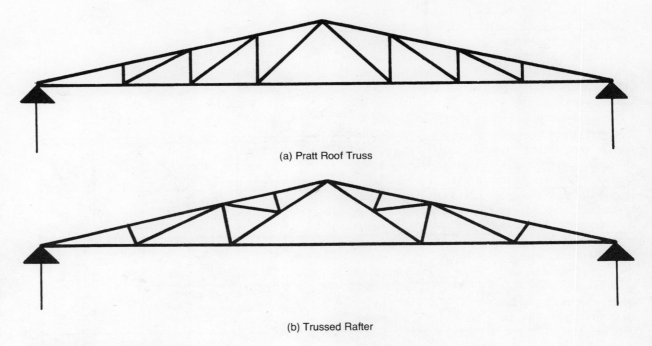

7.3.2. Roof trusses for sloping roofs.
(a) Pratt roof truss.
(b) Trussed rafter.

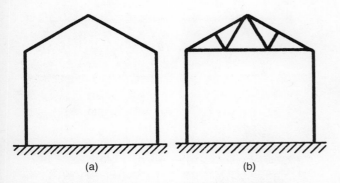

7.3.1. (a) Rigid frame for sloping roof. This is particularly suitable for a room without a flat ceiling.
(b) Roof truss for sloping roof. This is particularly suitable when there is a flat ceiling that can be attached to the underside of the truss.

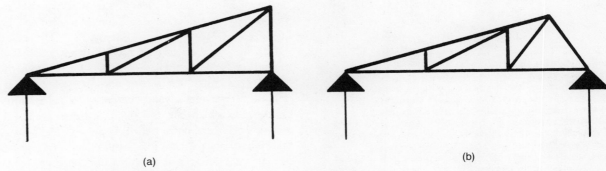

7.3.3. North-light roof (south-light roof in the southern hemisphere)
(a) for subtropical zone, (b) for temperate zone. In the tropics direct sunlight is received from both the north and the south, at different times of the year.

115

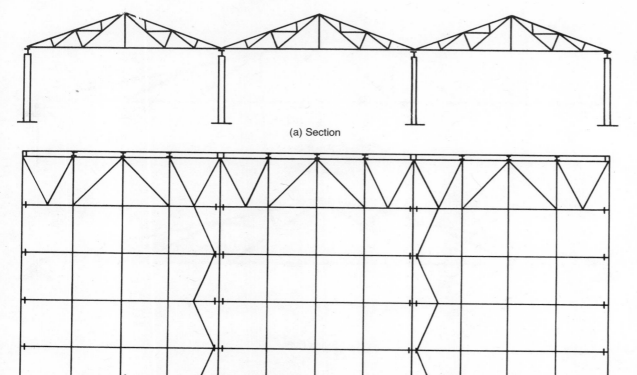

7.3.4. Wind and earthquake bracing for roof trusses, placed under the roof sheeting.

7.4 Parallel-Chord Trusses with Pin Joints: Beams with Big Holes or an Assembly of Tension and Compression Members

Parallel-chord trusses have already been considered, in Sections 4.7 and 5.1. If they are simply supported at both ends, as most pin-jointed trusses

The resistance to horizontal forces caused by wind or earthquake, at right angles to the roof trusses, is provided by triangulation placed under the roof covering (Fig. 7.3.4).

Half the members of each roof truss are in tension (including the horizontal members at the bottom of the truss), and half are in compression (including the rafters). There are several methods for calculating their sizes, which can be found in more advanced textbooks (for example, Ref. 1.1, pp. 81–99). For a preliminary design it is sufficient to know the size of the largest tension and compression members, and these can be found in a few minutes by taking moments about a suitable point in the truss. This is explained in Problem 7.1, near the end of this chapter.

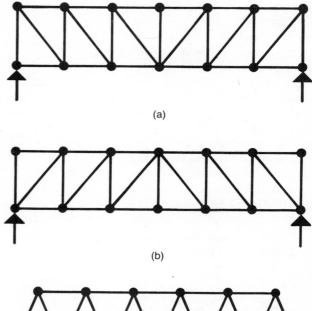

7.4.1. Pin-jointed parallel-chord trusses.
(a) Pratt truss with vertical members in compression and diagonal members in tension.
(b) Pratt truss with vertical members in tension and diagonal members in compression.
(c) Warren truss, with diagonal members only; half are in tension and half are in compression.

are, the upper chord is entirely in compression and the lower chord is entirely in tension. Half the bracing members are in tension, and half are in compression (Fig. 7.4.1); unless there are good reasons to the contrary, it is best to put the shorter bracing members in compression and the longer ones in tension, because they are not liable to buckle.

As we pointed out in Section 4.7, the chords resist the bending moment and the bracing members that hold the chords apart provide the lever arm for the moment. In addition, the bracing members resist the shear force (Section 6.1), and therefore the largest forces in the chords occur at mid-span and the largest forces in the bracing members occur at the supports.

The mechanics of roof trusses with sloping rafters is basically the same, but the argument is complicated by variation of the depth of the truss.

Parallel-chord trusses are suitable for flat roofs of medium or long span. If the depth of the truss is made equal to the height of an entire story—that is, 3m (10 ft) or more—it can carry a large load over a large span. This is useful where it is proposed to eliminate ground-floor columns (Fig. 7.4.2) or to hang the building from the roof (Fig. 5.2.3).

7.5 Parallel-Chord Trusses with Rigid Joints, and Shear Walls

The diagonals of braced trusses supporting a building frame whose ground-floor columns have been partly eliminated are liable to interfere with the fenestration (Fig. 7.4.2). In practice this limits the use of braced trusses to service floors that do not require windows and to windowless walls.

The difficulty can be overcome by replacing the pin joints with rigid joints, so that the diagonal members become superfluous. The resulting structure is called a *Vierendeel truss* (Fig. 7.5.1). The truss is statically indeterminate, and the members are therefore subjected to moments as well as to tension or compression; thus the truss requires more steel than a pin-jointed truss of the same depth and span. If both trusses are designed to use as little material as possible, the Vierendeel truss has the greater deflection because of the absence of diagonal tension members. (Fig. 7.5.2).

A rigid-jointed truss is also suitable for a reinforced concrete structure, but as concrete can be used to form the external wall into which the windows are placed, it is more appropriate to think of this structure as a shear wall than as a Vierendeel truss. Shear walls are very effective for resisting shear distortion, and the solution in Fig. 7.5.3.c is likely to be more economical than those in Figs. 7.5.3.a or b.

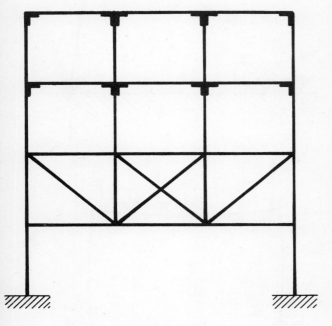

7.4.2. Truss of full-floor depth used to eliminate ground floor columns. The diagonals are liable to interfere with the fenestration.

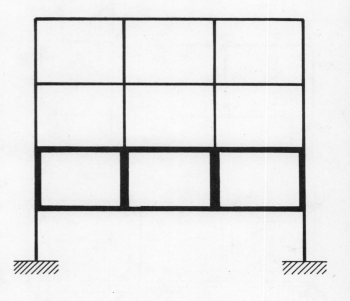

7.5.1. Vierendeel truss used to eliminate ground-floor columns in multistory building. If the building is only a few stories high, the upper stories can be built with conventional flexible joints.

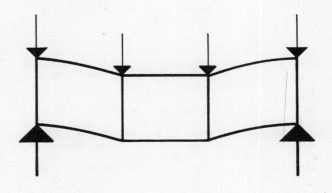

7.5.2. Because of the absence of diagonal tension members, Vierendeel trusses may have an appreciable deflection.

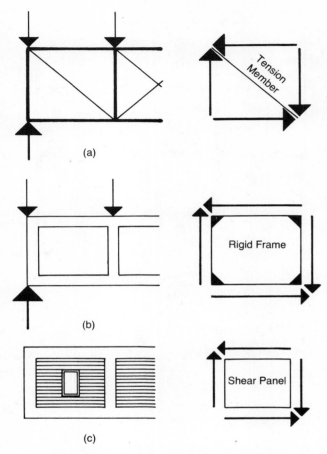

7.6 Portal Frames: Simple Frames with Three Pins and More Complicated Rigid Frames

If we join a beam to its two supporting columns with rigid joints, we get a new type of structure, called a *portal frame* (Fig. 7.6.1). When we load a beam, it deforms elastically. This slight deformation is readily accommodated on simple supports (Fig. 4.6.6.c), but when we join the beam to columns we create a practical problem. Evidently the columns would spread (Fig. 7.6.1.b), and it would be necessary to put them on rollers so that they could move back and forth every time the load on the beam

7.6.1. A rectangular portal frame, simply supported on the ground (a), is statically determinate, just as a simply supported beam is statically determinate. Unlike a simply supported beam, it is not a practical structure, because the deflection of the beam under the vertical load would cause the columns to spread (b). It would be possible to fit one of the columns with a roller bearing to allow the movement to take place, but the cost of doing so would be out of all proportion to the advantages gained. It is more realistic to pin the columns to the ground (c); when the portal frame is loaded, the distance between the pin joints is kept constant by horizontal reactions R_H (d). The structure is therefore statically indeterminate.

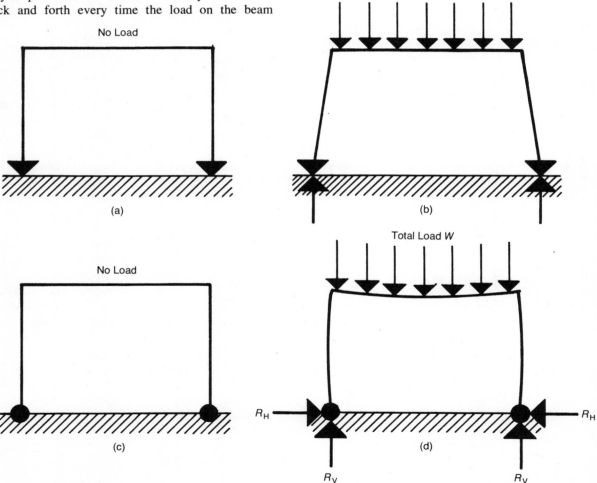

7.5.3. Comparison of Pratt truss, Vierendeel truss, and shear wall used to eliminate ground floor columns. The load of the upper part of the building causes a shear distortion of the deep girder at the second-floor level. This vertical shear produces a horizontal shear (Fig. 6.1.3)
(a) In the Pratt truss (Fig. 7.4.2) this shear distortion is resisted by a relatively small tension member.
(b) The Vierendeel truss (Fig. 7.5.1) is not well suited to resist this shear distortion, and this accounts for the high deflection. The performance of the girder is greatly improved if the opening is filled with a nonloadbearing panel of bricks or blocks, which resists the shear distortion.
(c) A reinforced concrete shear panel is effective in resisting the shear distortion, provided that it is suitably reinforced with horizontal and vertical steel bars to prevent cracking of the concrete. A small window does not detract from the effectiveness of the shear panel, provided that the reentrant corners are suitably reinforced (Fig. 6.4.2).

changed. It is preferable to fix the portal to the floor (Fig. 7.6.1.c) and provide the horizontal reactions R_H necessary to prevent the spread of the columns (Fig. 7.6.1.d).

The portal frame now has more reactions than we can determine with the equations of equilibrium, Eqs. (4.3) to (4.5). Equation (4.3) tells us that the two horizontal reactions R_H are equal, but it does not tell us what they are. Equation (4.4) tells us that the load on the beam is equal to the sum of the two vertical reactions, and we can conclude from the last equation, (4.5), that the two vertical reactions are equal, by taking moments about one of the pin joints at the supports. There is therefore one unknown reaction, namely, R_H, and the frame is statically indeterminate with one redundancy.

We can make it statically determinate by adding another pin joint in a place through which the reaction, R_H, does not pass. The most convenient place is the middle of the beam (Fig. 7.6.2.b).

We can also make all the joints in the portal frame rigid (Fig. 7.6.2.c), and such a frame is easier to construct, particularly if we use reinforced concrete. The frame then has three redundant reactions, namely, two support moments and the horizontal reactions R_H.

Portal frames can be constructed with a flat roof, with a sloping roof, with a north (or south) light, or with monitors that hold high-light windows. They can be constructed as single-bay or multibay frames (Figs. 7.6.3 and 7.6.4). Multistory rigid frames are considered in Chapter 8.

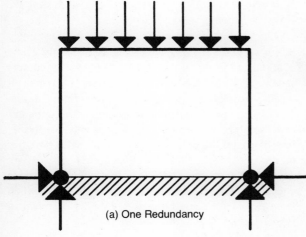

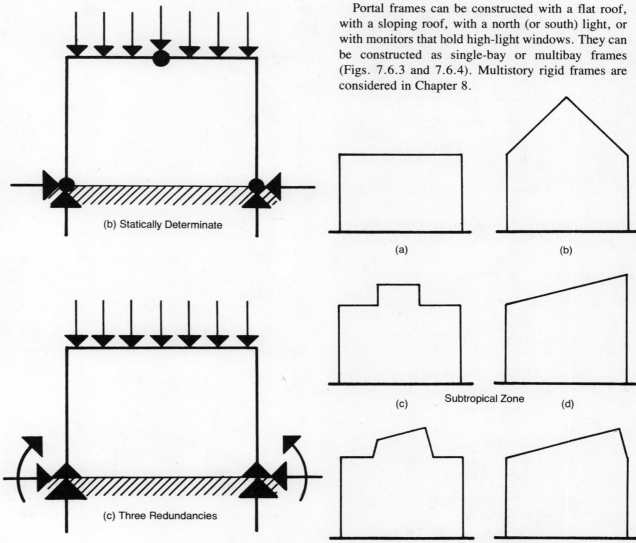

7.6.2. The portal frame with two pin joints has one redundancy (a). We can make it statically determinate by adding a third pin; the most convenient place for this pin is at the mid-span of the beam (b). For most materials the cheapest portal frame has rigid joints only, and consequently three redundancies (c).

7.6.3. Rigid frames for single-bay, single-story buildings. (a) flat roof; (b) gabled roof; (c) monitor roof for the subtropics; (d) north-light (south-light) roof for the subtropics; (e) monitor roof for the temperate zone; (f) north-light roof for the temperate zone.

Portal frames are rigid only within their plane. If the frames are connected merely with light purlins that support the roofing material, then diagonal bracing is needed to provide resistance to horizontal forces at right angles to the rigid frames (Fig. 7.3.4).

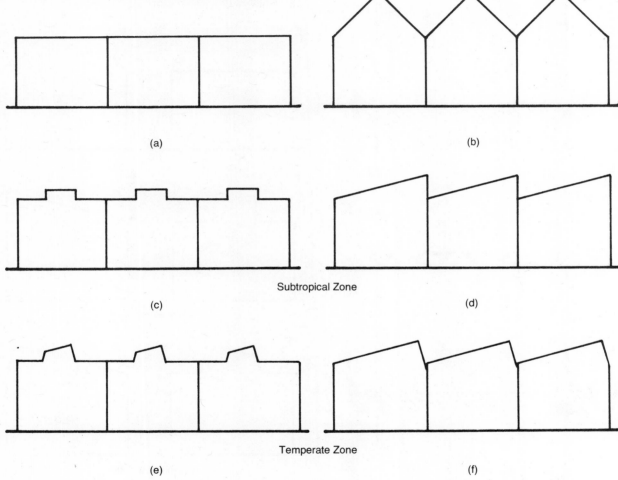

7.6.4. Rigid frames for multibay, single-story buildings: (a) flat roof; (b) gabled roof; (c) monitor roof for the subtropics; (d) north-light (south-light) roof for the subtropics; (e) monitor roof for the temperate zone; (f) north-light roof for the temperate zone.

Structural steel, laminated timber, and reinforced and prestressed concrete are all suitable materials for portal frames. Each has its own advantages and disadvantages (see Table 3.1 and Section 3.7).

The three-pin portal frame is easily analyzed (Fig. 7.6.5). It is convenient to take it apart at the center pin and replace the force exerted by the other half of the frame by another force (whose magnitude we do not need to know). From an inspection of the half portal (Fig. 7.6.5.b) it is obvious that the vertical reaction is

$$R_V = \tfrac{1}{2} W \qquad (7.1)$$

We obtain the horizontal reaction by taking moments about the top pin:

$$R_H \cdot H = \tfrac{1}{2} W \cdot \tfrac{1}{2} L - \tfrac{1}{2} W \cdot \tfrac{1}{4} L = \tfrac{1}{8} WL$$

That is,

$$R_H = \frac{WL}{8H} \qquad (7.2)$$

The whole purpose of the top pin is to render the frame statically determinate, by providing a point that is not in line with the force R_H and at which the bending moment is known to be zero.

The bending moment is zero at the bottom pin; it builds up gradually as we move up the column. There is no other horizontal force, and therefore the moment increases uniformly as we get further from R_H. It reaches its maximum, $R_H H$, at the top of the column. The top of the column is joined rigidly to the beam, and therefore the entire bending moment is transferred around the corner into the beam. The horizontal force R_H remains at the distance H as we pass along the beam, and therefore the bending moment due to it remains constant and equal to $R_H H$, which from Eq. (7.2) is equal to $-\tfrac{1}{8} WL$. This is shown in Fig. 7.6.5.c.

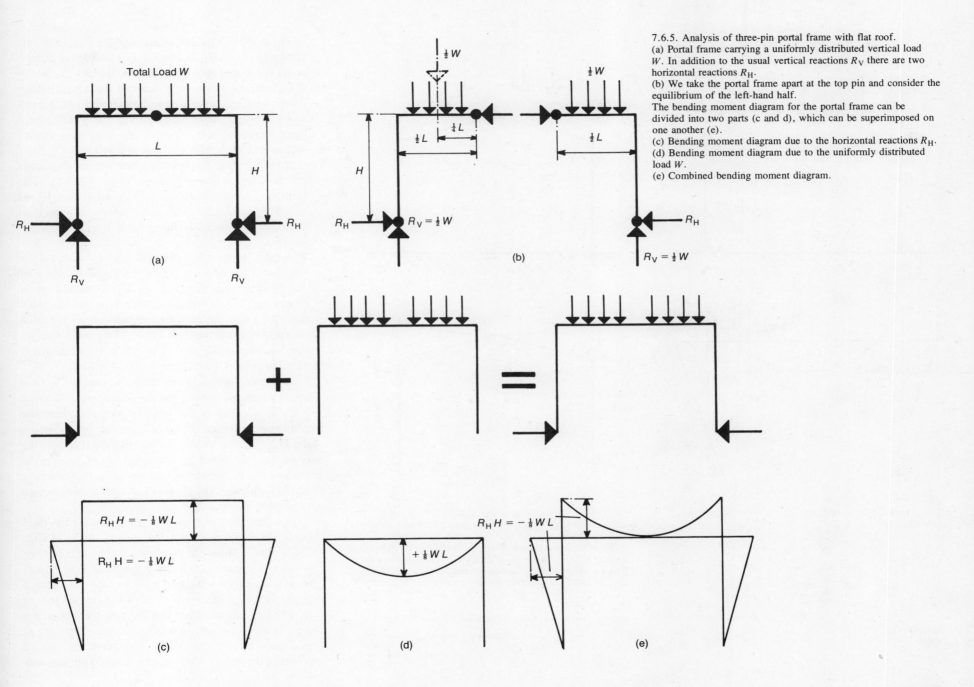

7.6.5. Analysis of three-pin portal frame with flat roof.
(a) Portal frame carrying a uniformly distributed vertical load W. In addition to the usual vertical reactions R_V there are two horizontal reactions R_H.
(b) We take the portal frame apart at the top pin and consider the equilibrium of the left-hand half.
The bending moment diagram for the portal frame can be divided into two parts (c and d), which can be superimposed on one another (e).
(c) Bending moment diagram due to the horizontal reactions R_H.
(d) Bending moment diagram due to the uniformly distributed load W.
(e) Combined bending moment diagram.

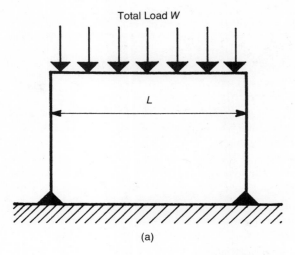

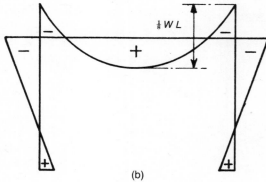

7.6.6. Bending moment diagram (b) for a rigid portal frame with a flat roof (a).

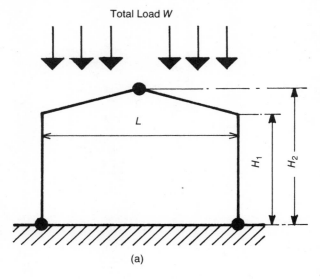

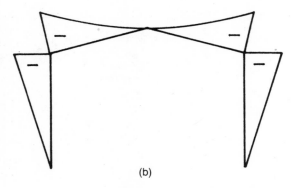

7.6.7. Bending moment diagram (b) for a three-pin portal frame with a moderately sloping roof, carrying a uniformly distributed vertical load (a).

If the roof slope is not too steep, the bending moment builds up linearly to its maximum (negative) value at the knee and then declines parabolically to zero at the top pin. The horizontal reaction, obtained by taking moments about the top pin, is

$$R_H = \frac{WL}{8H_2} \quad (7.3)$$

and the maximum bending moment is

$$M = R_H H_1 = \frac{WL}{8H_2}H_1 \quad (7.4)$$

The other component of the total bending moment, shown in Fig. 7.6.5.d, consists of a uniformly distributed load W and two reactions of $\frac{1}{2}W$ each. This is the standard case of simply supported uniform loading, which we have previously encountered (Figs. 4.6.4.f and 4.7.6 and Table 4.1, line 9). The bending moment diagram for this part of the load is parabolic, with a maximum of $+\frac{1}{8}WL$ at mid-span.

If we combine the two components, we obtain the net bending moment diagram (Fig. 7.6.5.e). The bending moment increases uniformly from 0 to $-\frac{1}{8}WL$, turns the corner at the rigid joint, and then decreases parabolically from $-\frac{1}{8}WL$ to 0 at the pin joint.

In fact, it was not really necessary to work out the horizontal reaction R_H. Once the general pattern of the bending moment diagram is established, it is evident that the bending moment must be zero at all three pin joints (because they are pin joints) and must be a maximum at the rigid joints. Therefore it follows that this maximum bending moment, for a uniformly distributed load W and span L, is $-\frac{1}{8}WL$.

Knowing the maximum bending moment, we can select a suitable section for the portal frame (Section 5.8). In addition to the bending moment, there are compressive forces acting both on the beam and on the column; the column carries a load R_V, and the beam is compressed by the force R_H. However, the stresses due to these forces are much smaller and they can be neglected for purposes of preliminary design (Problems 7.3 to 7.5).

Determination of the bending moment for the rigid portal takes much longer; but some features of Fig. 7.6.5 also apply to the rigid portal (Fig. 7.6.6). The portal still carries a statically determinate load system W as part of its total load system, and therefore a parabolic curve of overall depth $\frac{1}{8}WL$ is still part of the bending moment diagram. Because there is no pin joint in the beam, the bending moment is zero not at mid-span but at two other symmetrically spaced points in the beam, whose location we do not know at present. Because there are no pin joints at the base, the points of zero bending moment occur some distance up the column; again we do not know exactly where these points are. However, we can draw the general shape of the bending moment diagram (Fig. 7.6.6).

Unlike the three-pin portal, whose bending moment is entirely negative, the bending moment in the rigid portal varies from positive to negative and back to positive. It is important to bear this in mind when reinforced concrete is used, because it determines the location of the tension reinforcement.

The maximum bending moment is evidently less than $\frac{1}{8}WL$. It must also be more than $\frac{1}{16}WL$, because the maximum positive and negative bending moments must add up to $\frac{1}{8}WL$. From the solution of the three-pin portal we can therefore determine approximately the magnitude of the bending moment, and we know approximately whether it is positive or negative at any one section.

A three-pin portal frame with a sloping roof is solved almost as easily as one with a flat roof (Figs. 7.6.7 and 7.6.8).

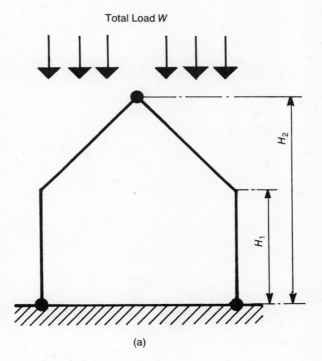
(a)

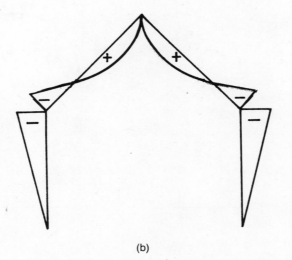
(b)

7.6.8. In a three-pin portal frame with a steeply sloping roof (a), there is a reversal of bending moment in the rafter. If the roof is sufficiently steep, the maximum positive moment in the rafter could be larger than the maximum negative moment at the knee of the frame. For determining approximate sizes for a preliminary design, it is usually sufficient to take the bending moment at the knee,

$$M = \frac{WL}{8H_2}H_1 \qquad (7.5)$$

The only exceptions are buildings in which the roof slope is made particularly steep for visual reasons, for example, in a church to create a feeling of height.

7.7 Arches

The mechanical principle of the arch is precisely the same as that of the portal frame. The straight pieces of material joined by sharp bends are smoothed out into a continuous curve. This increases the cost of construction but greatly reduces the stresses.

The geometry of the curve further affects the cost of construction and the stresses. The circular arch is easiest to construct; the catenary arch is most efficient. Like portals, arches can be three-pinned, two-pinned, or rigid.

We will examine first a three-pinned semicircular arch. As we noted in Fig. 7.6.4, the bending moment can be divided into two parts, one representing the effect of the two horizontal reactions and the other the uniformly distributed load and the two vertical reactions. We will do the same for the arch.

Let us consider a point on the arch, which is x to the right of the left-hand support. The height of the arch at that point is y (Fig. 7.7.1). Therefore the bending moment due to the horizontal reaction at the left-hand support at a distance x along the span is

$$R_H y$$

This is true for any number of points a distance x along the span and y above the springings (see Glossary) of the arch. Consequently the shape of the bending moment diagram due to the horizontal reactions R_H looks exactly like the shape of the arch, in this case a semicircle.

As for the portal frame, the bending moment due to a uniformly distributed load W simply supported over a span L is a parabola of depth $+\frac{1}{8}WL$. The combined bending moment diagram is drawn in Fig. 7.7.2. It can be shown (Ref. 1.1, pp. 144–47) that for a semicircular arch carrying a uniformly distributed load the maximum bending moment is

$$M = -\tfrac{1}{32}WL \qquad (7.9)$$

Evidently if the two component parts of this bending moment were identical, the bending moment would vanish altogether. For a load uniformly distributed in plan, this is so when the arch is parabolic.

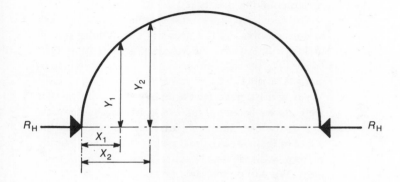

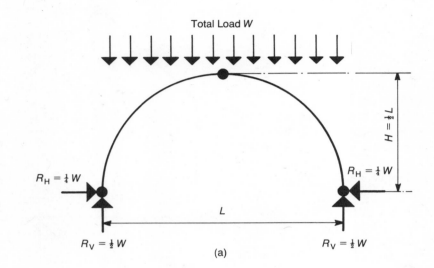

7.7.1. At any point x along the span the height of the arch is y. Therefore the moment of the horizontal force R_H about that point is $R_H y$, and the shape of the bending moment diagram due to R_H is identical with the shape of the arch. We therefore obtain the net bending moment diagram by superimposing the shape of the arch on the load diagram (Fig. 7.7.2).

7.7.2. Analysis of three-pin semicircular arch carrying a total load W uniformly distributed in plan.
(a) Because the arch is semicircular, its height is $H = \frac{1}{2}L$. By symmetry
$$R_V = \tfrac{1}{2}W \tag{7.6}$$
Taking moments about the pin at the crown
$$R_H \cdot \tfrac{1}{2}L = \tfrac{1}{2}W \cdot \tfrac{1}{2}L - \tfrac{1}{2}W \cdot \tfrac{1}{4}L = \tfrac{1}{8}WL$$
This gives
$$R_H = \tfrac{1}{4}W \tag{7.7}$$
(b) The bending moment diagram consists of two parts, a parabolic positive bending moment, whose maximum is $\tfrac{1}{8}WL$, and a semicircular negative bending moment. Because the net bending moment at the pin at the crown must be zero, the maximum negative bending moment of the first component part must also be $\tfrac{1}{8}WL$.

The combined bending moment diagram is therefore the difference between a semicircle and a parabola of the same height $\tfrac{1}{8}WL$, and it is negative. It can be shown that the maximum bending moment is
$$M = -\tfrac{1}{32}WL \tag{7.8}$$

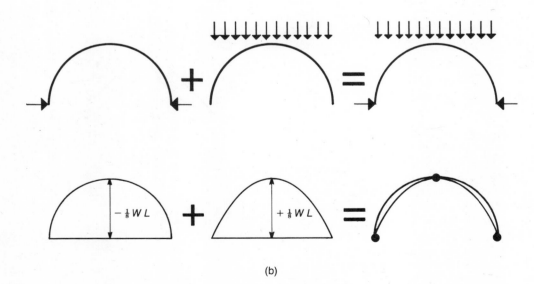

124

The arch carries mainly its own weight, and because of its curvature its weight per unit length of span is greatest near the springings. When this is taken into account, it is found that the moment-free arch is a catenary. This is the curve assumed by a cable hanging under its own weight, and its equation is

$$y = c \cosh \frac{x}{c} = \frac{1}{2} c \ (e^{x/c} + e^{-x/c})$$

where x and y are the horizontal and vertical coordinates and c is a constant. (The hyperbolic cosine [cosh] is a function listed in mathematical tables. Many digital calculators have a key for e^x.)

We reached the same conclusion, using a different line of argument, in Section 4.7 and Fig. 4.7.4.

The arch has a long and distinguished architectural tradition. Until the price of iron dropped in the eighteenth century to a level where it could be used as a building material, there were basically two groups of structural material: those that burned and those that had no tensile strength (Fig. 7.7.3).

It is possible that a sophisticated ancient civilization could have erected buildings of quite long span built entirely from timber, bamboo, or reeds, which have vanished without a trace. There were many stone buildings with timber roofs (for example, most of the Ancient Greek temples), and the timber structures disappeared long ago. As early as the first century B.C., Vitruvius remarked that temples and other important buildings should have roofs arched to protect them against fire.

Horizontal slabs of very strong stone, such as granite or diorite, can be used over moderate spans (Section 5.5). However, this is an expensive form of construction which only very wealthy governments, such as those of ancient Egypt and of Athens at the peak of its power, could afford.

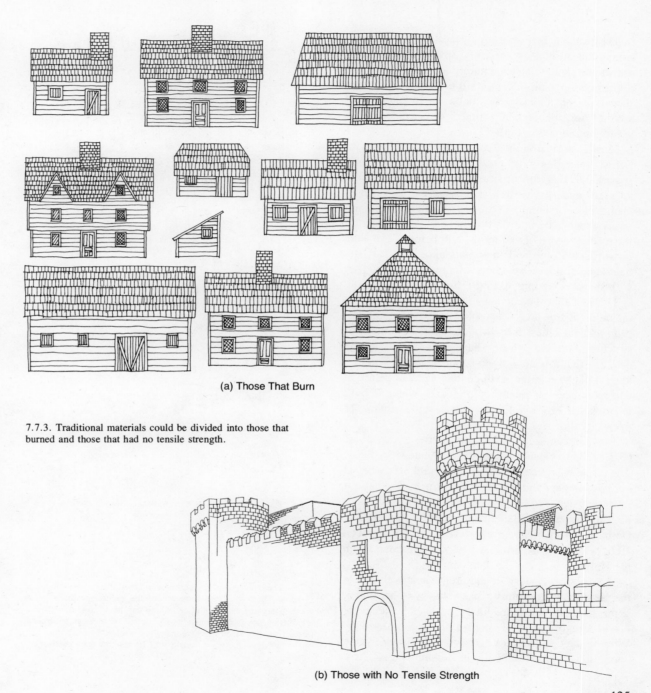

(a) Those That Burn

7.7.3. Traditional materials could be divided into those that burned and those that had no tensile strength.

(b) Those with No Tensile Strength

The arch has therefore been used since time immemorial because it is the only structural form that allows the stone to be used purely in compression; the dome and the vault (Chapter 10) are variations on the theme of the arch. In any book on the history of architecture the arched (domed or vaulted) roofs greatly outnumber all other kinds.

Most historical arched roofs were not, in modern terms, long-span. The great Gothic cathedrals were remarkable feats of structural daring, and some are very tall even by modern standards, but the span of the nave was generally of the order of 14 m (45 ft) and rarely exceeded 16 m (51 ft). The only long spans, in the modern sense of that term, built before 1850 were in imperial Rome and in Renaissance and post-Renaissance Europe; we know of fewer than a dozen with spans exceeding 30 m (98 ft 6 in.).

The fact that the catenary is the correct shape for the moment-free arch was probably known empirically in ancient times in the Middle East. A number of structures survive whose shape is so close to the catenary that it is unlikely to be an accident. The best known of these is the sixth-century Persian palace at Ctesiphon, now a suburb of Baghdad, which is built of brick 7 m (23 ft) thick at the base and spanning 25 m (84 ft) (Fig. 7.7.4).

The Saracen arch, consisting of two circular arcs (Fig. 7.7.5), closely resembles the catenary shape. The circle was the only curve that could easily be set out with ancient and medieval instruments, because it has a constant radius of curvature; a string tied to a nail at one end can be used to scribe a circle. The Gothic arch, which closely resembles the Saracen arch, was probably brought back by Crusaders from the Holy Land in the eleventh century.

The catenary theory of the arch was developed in the seventeenth century, possibly by Christopher Wren himself, but more likely by Robert Hooke or David Gregory (Ref. 1.4, p. 179). It was used by Wren in the design of the brick cone that supports the dome of St. Paul's Cathedral in London (Fig. 7.7.6).

7.7.4. Ruins of the Great Hall of the Palace of Ctesiphon (near Baghdad), built for a sixth-century Persian king.

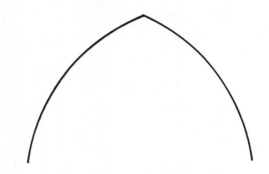

7.7.5. The Saracen or Gothic arch closely resembles the catenary arch.

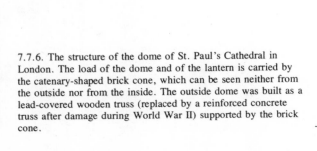

7.7.6. The structure of the dome of St. Paul's Cathedral in London. The load of the dome and of the lantern is carried by the catenary-shaped brick cone, which can be seen neither from the outside nor from the inside. The outside dome was built as a lead-covered wooden truss (replaced by a reinforced concrete truss after damage during World War II) supported by the brick cone.

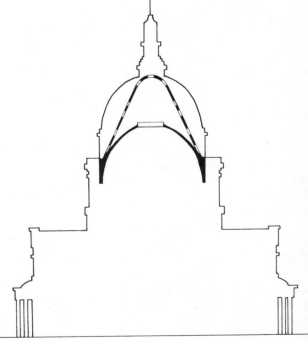

In the eighteenth century the correct line of the thrust of an arch or dome was on several occasions determined experimentally by means of a string of beads, each bead being made proportional to the weight of the stone used at that level. Antonio Gaudi still used models employing weighted strings in the early twentieth century.

In 1842 St. Isaac's Cathedral in St. Petersburg (now Leningrad) became the first building to use an iron structure for its dome. The building had a classical appearance, but the structure of the dome consisted of arch ribs of iron, designed by a new elastic theory. This proved more economical than masonry. The dome of the U.S. Capitol in Washington, D.C., originally conceived as a masonry dome, was redesigned and built in iron. At the end of the nineteenth century iron was gradually replaced by steel, and reinforced concrete arches made their appearance in the early twentieth century.

The failure of masonry arches (built from unreinforced concrete, bricks, or blocks of stone) differs from that of arches built from a material with tensile strength.

The design of arches made from structural steel, laminated timber, or reinforced concrete is generally based on their elastic deformation. The maximum stresses due to the thrust (compressive force) and the bending moment must not exceed the maximum permissible stresses for the material. When the arch is overloaded, "pin joints" are formed by the failure of the material. If the arch is rigid, it is statically indeterminate with three redundancies (Fig. 7.6.2), and it cannot actually collapse until it has formed four such "pin joints" by local failure of the material.

In a masonry arch, failure occurs by the opening of four joints (Fig. 7.7.7). Each open joint acts as a pin, allowing rotation at that point of the arch. Thus three open joints render the rigid masonry arch statically determinate, and the fourth turns it into a mechanism, so that it collapses. The joints begin to open when the line of thrust falls outside the arch. The thickness of the arch is therefore determined by the need to keep the line of thrust within it (Fig. 7.7.8). The blocks of stone forming a masonry arch are not necessarily damaged by the failure of the arch. If they could be caught in a net, the arch could be reassembled from them. Therefore the strength of the material is not critical for the strength of a masonry arch, unless it is very weak in compression. Failure occurs because the joints between the blocks have negligible tensile strength.

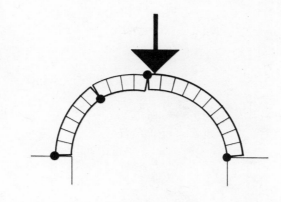

7.7.7. Failure of masonry arch.

Nevertheless, the masonry arch is quite an efficient structure, and it is not particularly difficult to design tension-free arches, vaults, and domes. Arched, domed, and vaulted masonry structures vanished in countries such as the United States and Great Britain because of the high labor cost involved in building them. The old skills (Fig. 7.7.9) survive in many parts of the Middle East and Latin America, and it may be appropriate to revive them for buildings of small to medium span. Masonry construction (including construction in brick and mud brick) does not require steel and portland cement, which in many of these countries have to be imported. It is labor-intensive, but this may not be a disadvantage if there is high unemployment and wages are low. It may preserve the continuity of an architectural tradition, and it may produce better buildings in countries that presently lack the experience to produce reinforced concrete structures of adequate quality outside metropolitan areas. However, it should be pointed out that masonry structures have poor resistance to earthquakes.

The thrust in a rigid arch is not very different from that of a three-pin arch, and the bending moment is likely to be smaller. Useful preliminary design decisions on rigid arches can therefore be made by treating them as three-pin arches.

Vaults and domes are considered in Chapter 10.

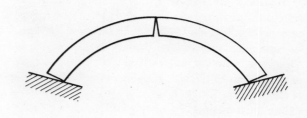

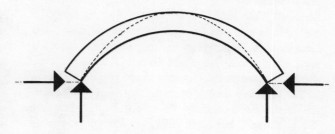

7.7.8. The line of thrust must remain within the thickness of a masonry arch.

7.7.9. Traditional masonry construction in the Middle East.

7.8 Problems for Chapter 7

If you have not yet read Section 5.9, please do so before attempting the following problems. The next two sections contain two problems on trusses, three on portal frames, and three on arches.

7.9 Problems on Trusses

Roof trusses with sloping rafters behave like beams with large holes, as do parallel-chord trusses. In simply supported parallel-chord trusses, the largest forces invariably occur at mid-span. The depth of roof trusses with sloping rafters varies from zero at the supports to a maximum at mid-span, and consequently the largest forces in the members usually occur at the supports.

For a preliminary design these forces are easily determined by resolving horizontally and vertically at the supports. Once the forces in the members are known, suitable steel or timber sizes can be determined from section tables, as in Section 5.13.

To obtain preliminary sizes for a parallel-chord truss, we can assume a suitable size for the chords (which could be of steel, timber, or aluminum) and calculate the minimum depth. Alternatively, we can fix the depth and determine the cross-sectional area of the chords.

7.9.0.

In Vierendeel trusses the rigid joints and the absence of diagonal members produce large secondary bending moments in the parallel chords of the truss. Thus the members must be much larger than the simple method used in Problem 7.2 would suggest. An approximate solution can be obtained by assuming pin joints at the centers of all members of the truss, as in Fig. 8.4.3.

Problem 7.1 Determine the forces in the members AB and AC of the roof truss shown in Fig. 7.9.1.

In a parallel-chord truss the members that carry the largest forces occur at mid-span, where the bending moment is a maximum. In trusses with sloping rafters the greatest forces usually, though not invariably, occur at the supports, where the truss has a small depth.

The vertical forces set up a bending moment, and this produces compression in the top members (or rafters) and tension in the bottom members (or ties), as it does in a beam. Thus the force T_{AB} in the member AB is the largest compressive force, and the force T_{AC} in the member AC is the largest tensile force.

We noted in Problem 5.1 that the internal resistance forces in the members are equal and opposite to the external forces. The internal forces T_{AB} and T_{AC} and the reaction R keep one another in equilibrium (Fig. 7.9.2).

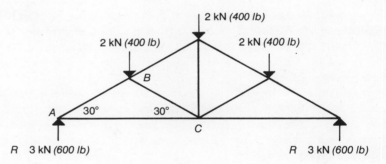

7.9.1. Forces acting on the roof truss in Problem 7.1.

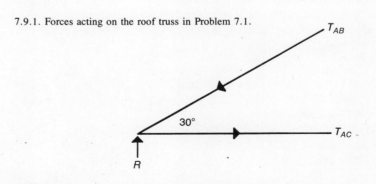

7.9.2. Conditions for equilibrium at the left-hand support of the roof truss in Problem 7.1.

Resolving vertically, we have

$$T_{AB} \sin 30° = R$$

Resolving horizontally, we have

$$T_{AB} \cos 30° = T_{AC}$$

The reactions are 4 kN each, and therefore

$$T_{AB} = \frac{3 \text{ kN}}{\sin 30°} = 6 \text{ kN}$$

$$T_{AC} = 6 \text{ kN} \cos 30° = 5.196 \text{ kN}$$

The reactions are 600 lb each, and therefore

$$T_{AB} = \frac{600 \text{ lb}}{\sin 30°} = 1200 \text{ lb}$$

$$T_{AC} = 1200 \text{ lb} \cos 30° = 1040 \text{ lb}$$

If the members of a steel truss are joined by bolting, an allowance must be made in the tension members for loss of area due to the bolt holes. Compression members press on the bolts, and no deduction need therefore be made. If the members of the truss are joined by welding, there are no holes in the tension members and consequently no loss of area.

If the truss is made from steel, the members are usually angle sections, used singly or in pairs. If the truss is made from timber, the members are usually rectangular. They can be joined by nails, glue, bolts, or timber connectors. Nails are not considered to cause a loss of cross-section in tension members, but bolt holes must be allowed for.

The full permissible stress can be used for tension members, but the stress in compression members must be reduced in accordance with their slenderness ratio.

We have already considered the design of tension and compression members for a given force, in Problems 5.11 to 5.13.

Problem 7.2 Steel trusses of the type shown in Fig. 7.9.3 are to be used to carry a flat roof over a span of 15 m (*50 ft*). Each truss carries a total load of 240 kN (*50 kips*). Determine the overall depth.

The truss consists of pairs of angles for the top and bottom chords, welded to diagonal members. Although the individual members of the truss are either in tension or in compression, the truss as a whole resists the load as a simply supported girder. From Table 4.1, line 9, the bending moment is

$$M = \tfrac{1}{8} W L$$

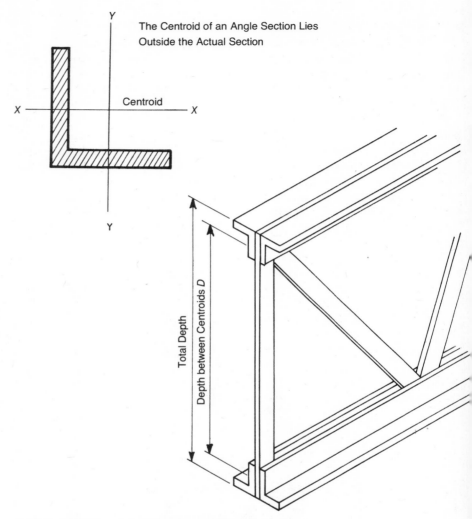

7.9.3 Overall depth and depth between centroids of chords (D) of the parallel-chord truss in Problem 7.2.

The maximum permissible compressive stress is lower than the maximum permissible tensile stress. If C is the compressive force in the upper chord and the (equal) tensile force in the lower chord at mid-span, the resistance moment provided by the truss at mid-span is

$$M = C D$$

The depth D is the distance between the centroids of the steel angles (Fig. 7.9.3); the overall depth of the truss is a little larger.

Let us try 75 mm by 75 mm by 8 mm angles, that is, angles in which each leg is 75 mm long and 8 mm thick. From steel section tables (not reproduced in this book), the cross-sectional area of each of these angles is 1140 mm². We reduce the maximum permissible compressive steel stress to 112 MPa, as explained in Problem 5.12.

The bending moment is

$$M = \tfrac{1}{8} \times 240 \times 15 = 450 \text{ kN m} = 0.45 \text{ MN m}$$

and the resistance moment is

$$M = 112 \text{ MPa} \times 2 \times 1140 \times 10^{-6} \text{ m}^2 \times D$$

The depth is

$$D = \frac{0.45}{112 \times 2 \times 1140 \times 10^{-6}} = 1.762 \text{ m}$$

The overall depth required is slightly greater, say, 1.85 m.

Let us try 3 in. by 3 in. by $\tfrac{5}{16}$ in. angles, that is, angles in which each leg is 3 in. long and $\tfrac{5}{16}$ in. thick. From steel section tables (not reproduced in this book), the cross-sectional area of each of these angles is 1.78 in.² We reduce the maximum permissible compressive stress to 16 ksi, as explained in Problem 5.12.

The bending moment is

$$M = \tfrac{1}{8} \times 50 \times 50 = 312.5 \text{ kip ft} = 3750 \text{ kip in.}$$

and the resistance moment is

$$M = 16 \text{ ksi} \times 2 \times 1.78 \text{ in.}^2 \times D$$

The depth is

$$D = \frac{3750}{16 \times 2 \times 1.78} = 65.84 \text{ in.}$$

The overall depth is slightly greater, say, 5 ft 8 in.

This is a reasonable depth for the truss in the light of its span. If the depth determined proved unsuitable, we would have to choose angles of a different size.

7.10 Problems on Portal Frames and Arches

The following three problems deal with portal frames. We determine the preliminary sizes for a rectangular portal frame, a portal frame with a sloping roof, and a portal frame with monitors for side lighting.

We then determine the preliminary sizes of a parabolic arch and a semicircular arch, and we calculate the size of the ties required to absorb the horizontal reactions within the concrete floor.

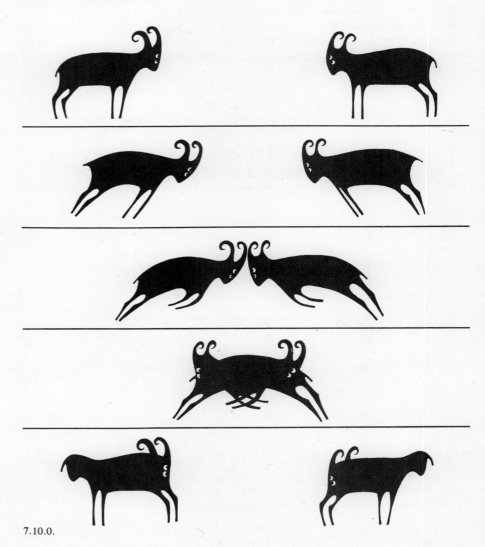

7.10.0.

Problem 7.3 Determine the size of the structural steel section required for the portal frame in Fig. 7.10.1.

The distribution of the bending moment is shown in Fig. 7.6.4. The maximum occurs at the rigid joint, and it is

$$M = R_H H = \frac{WL}{8}$$

Both the beam and the columns of the portal are also subject to compressive forces, but the stresses due to them are so much smaller than the flexural stresses that they can be neglected (see also Problem 7.7). The steel section is therefore determined by the section modulus calculated from the bending moment. From Eq. (5.10)

$$S = \frac{M}{f}$$

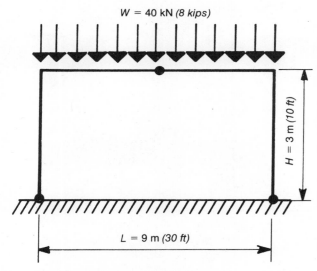

7.10.1. Dimensions of the flat-roofed portal frame in Problem 7.3.

At the rigid joint the bending moment is

$$M = \frac{40 \text{ kN} \times 9 \text{ m}}{8} = 45 \text{ kN m} = 45 \times 10^{-3} \text{ MN m}$$

The maximum permissible flexural stress is $f = 165$ MPa, and the section modulus required is

$$S = \frac{45 \times 10^{-3}}{165} = 272.7 \times 10^{-6} \text{ m}^3 = 272\,700 \text{ mm}^3$$

From Table 13.1 we choose a wide flange section 200 mm deep with a mass of 31 kg/m, which has $S = 299\,000$ mm^3.

At the rigid joint the bending moment is

$$M = \frac{8 \text{ kips} \times 30 \text{ ft}}{8} = 30 \text{ kip ft} = 360 \text{ kip in.}$$

The maximum permissible flexural stress for A 36 steel is 24 ksi, and the section modulus required is

$$S = \frac{360}{24} = 15.00 \text{ in.}^3.$$

From Table 13.2 we choose a wide flange section 8 in. deep with a mass of 18 lb/ft, which has $S = 15.2$ in.^{3}.

Problem 7.4 Determine the size of the structural steel section required for the portal frame in Fig. 7.10.2.

From Fig. 7.6.6 and Eq. (7.4), the maximum bending moment is

$$M = R_H H_1 = \frac{W L H_1}{8 H_2}$$

$$M = \frac{40 \times 9 \times 3}{8 \times 5} = 27 \text{ kN m} = 27 \times 10^{-3} \text{ MN m}$$

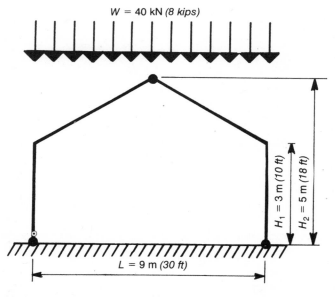

7.10.2. Dimensions of the portal frame with a sloping roof in Problem 7.4

The section modulus required is

$$S = \frac{27 \times 10^{-3}}{165} = 163.6 \times 10^{-6} \text{ m}^3 = 163\,600 \text{ mm}^3$$

From Table 13.1 we choose a wide flange section 200 mm deep with a mass of 22 kg/m, which has $S = 194\,000 \text{ mm}^3$.

$$M = \frac{8 \times 30 \times 10}{8 \times 16} = 18.75 \text{ kip ft} = 225 \text{ kip in.}$$

The section modulus required

$$S = \frac{225}{24} = 9.38 \text{ in.}^3$$

From Table 13.2 we choose a wide flange section 8 in. deep with a mass of 13 lb/ft, which has $S = 9.91$ in.^{3}.

Problem 7.5 Determine the size of the structural steel section required for the portal frame in Fig. 7.10.3.

The maximum bending moment occurs at the rigid joint above the support, and it follows from Eq. (7.4), which also applies to this portal frame.

$$M = R_H H_1 = \frac{W L H_1}{8 H_2}$$

$$M = \frac{40 \times 9 \times 3}{8 \times 4} = 33.75 \text{ kN m} = 33.75 \times 10^{-3} \text{ MN m}$$

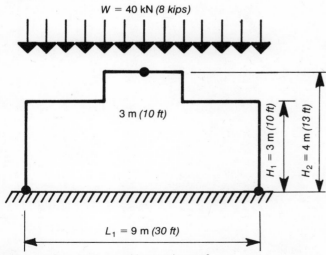

7.10.3. Dimensions of the portal frame with a monitor roof in Problem 7.5.

The section modulus required is

$$S = \frac{33.75 \times 10^{-3}}{165} = 204.5 \times 10^{-6} \text{ m}^3 = 204\,500 \text{ mm}^3$$

From Table 13.1 we choose a wide flange section 200 mm deep with a mass of 27 kg/m, which has $S = 249\,000 \text{ mm}^3$.

$$M = \frac{8 \times 30 \times 10}{8 \times 13} = 23.08 \text{ kip ft} = 277.0 \text{ kip in.}$$

The section modulus required is

$$S = \frac{276.9}{24} = 11.54 \text{ in.}^3$$

From Table 13.2 we choose a wide flange section 8 in. deep with a mass of 15 lb/ft, which has $S = 11.8$ in.^{3}.

Problem 7.6 Determine a suitable size for the laminated timber arch shown in Fig. 7.10.4.

Parabolic arches are difficult to form from steel. They can be cast in concrete, but the formwork is expensive. They are relatively easily made in laminated timber, because the individual laminations are flexible and can be bent to the correct shape and glued in position.

As we noted in Section 7.7, parabolic arches are in pure compression under the action of a uniformly distributed load, and the arch is therefore quite thin in comparison to the structures we have considered earlier in this section. The arch can be designed as a pure compression member only if two precautions are observed:

1. The lower part of the building must be sheltered from wind by surrounding buildings, because a horizontal wind pressure would create bending stresses. This is possible if the building is in a built-up area. (The effect of wind on the upper part of the building is an uplift (Fig. 2.3.4), which reduces the compressive force.)
2. The arches must be interconnected at intervals by the purlins (which carry the roof sheeting) to prevent the buckling of the arches. Without this restraint the arches would be excessively long compression members. Because buckling is the main problem, it is best to use a square cross section. There must be enough of these purlins to ensure that the permissible compressive stress does not fall too low, and we therefore keep the ratio L/d below 30, where L is the spacing of the purlins and d is the depth of the square section.

The maximum permissible compressive stress of glued laminated Douglas fir with a ratio $L/d = 30$ is (Ref. 5.1) $f = 4$ MPa *(0.6 ksi)*.

From Fig. 7.10.4, the vertical reactions are

$$R_V = \tfrac{1}{2} W$$

Taking moments about the top pin, we have

$$R_H \cdot H = R_V \cdot \tfrac{1}{2} L - \tfrac{1}{2} W \cdot \tfrac{1}{4} L = \tfrac{1}{2} W (\tfrac{1}{2} L - \tfrac{1}{4} L) = \tfrac{1}{8} WL$$

so that $R_H = WL/8H$.

The thrust (or compressive force) at the top pin is R_H, and it increases until it reaches its maximum at the bottom pins, where it is (Fig. 7.10.5)

$$R = \sqrt{R_H{}^2 + R_V{}^2}$$

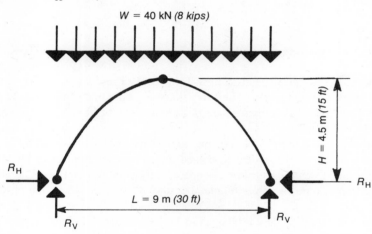

7.10.4. Dimensions of the parabolic arch in Problem 7.6

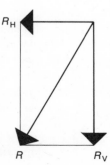

7.10.5. Component horizontal and vertical reactions and resultant reaction for the arch in Problem 7.6.

The horizontal and vertical reactions are

$$R_V = \tfrac{1}{2} \times 40 = 20 \text{ kN}$$

and

$$R_H = \frac{40 \times 9}{8 \times 4.5} = 10 \text{ kN}$$

The resultant is $R = \sqrt{20^2 + 10^2} = 22.36$ kN. This is also the maximum thrust in the arch.

The cross-sectional area required is

$$A = \frac{R}{f} = \frac{22.36 \times 10^{-3}}{4} = 5.59 \times 10^{-3} \text{ m}^2 = 5590 \text{ mm}^2$$

A section 75 mm square has $A = 5625$ mm². For $L/d = 30$, $L = 75 \times 30 = 2250$ mm = 2.25 m. The parabolic arch has a length of approximately 15.5 m*, so that it must be subdivided into at least seven parts to prevent a buckling failure. This requires six purlins (Fig. 7.10.6).

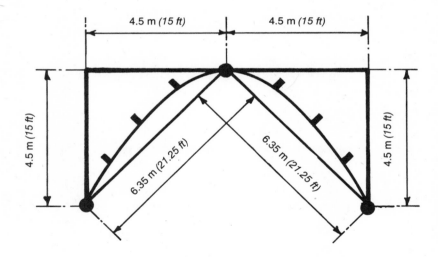

7.10.6. Calculation of the length of the parabolic arch in Problem 7.6, and arrangement of the six purlins.

---

*The length of the parabola is not required accurately. The length of a square portal of 9 m span and 4.5 m height is 18 m. The length of a triangular arch of the same dimensions is 12.7m. The parabolic arch is intermediate (Fig. 7.10.6). In order to determine the minimum number of purlins, it is sufficient to assume 15.5 m.

The horizontal and vertical reactions are

$$R_V = \frac{1}{2} \times 8 = 4 \text{ kips}$$

and

$$R_H = \frac{8 \times 30}{8 \times 15} = 2 \text{ kips}$$

The resultant is $R = \sqrt{4^2 + 2^2} = 4.472$ kips. This is also the maximum thrust in the arch.

The cross-sectional area required is

$$A = \frac{R}{f} = \frac{4.472}{0.6} = 7.453 \text{ in.}^2$$

A section 3 in. square has $A = 9$ in.². For $L/d = 30$, $L = 3 \times 30 = 90$ in. $= 7$ ft 6 in. The parabolic arch has a length of approximately 52 ft*, so that it must be divided into at least seven parts to prevent a buckling failure. This requires six purlins (Fig. 7.10.6).

Problem 7.7 Determine a suitable size for the laminated timber semicircular arch shown in Fig. 7.10.7.

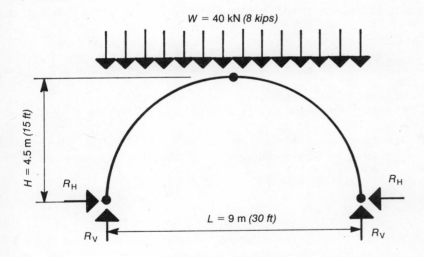

7.10.7. Dimensions of the semicircular arch in Problem 7.7.

---

*The length of the parabola is not required accurately. The length of a square portal of *30 ft* span and *15 ft* height is *60 ft*. The length of a triangular arch of the same dimensions is *42.5 ft*. The parabolic arch is intermediate (Fig. 7.10.6). In order to determine the minimum number of purlins, it is sufficient to assume *52 ft*.

We noted in Fig. 7.7.2 that a three-pin arch has its maximum bending moment approximately midway between the lower pin joints and the top pin joint. Its maximum value (Eq. 7.7) is

$$M = -\frac{WL}{32} \tag{7.7}$$

The minus sign has no significance for a timber arch. In a concrete arch the reinforcement is required on the outer face of the arch.

The thrust in a three-pin arch is the resultant of the components of R_H and R_V in line with the direction of the arch (Fig. 7.10.8). It increases from a minimum at the crown to a maximum at the springings. Since $R_H = \frac{1}{4}W$ and $R_V = \frac{1}{2}W$, the thrust ranges from $\frac{1}{4}W$ to $\frac{1}{2}W$, and the maximum thrust is

$$P = R_V = \tfrac{1}{2}W \tag{7.5}$$

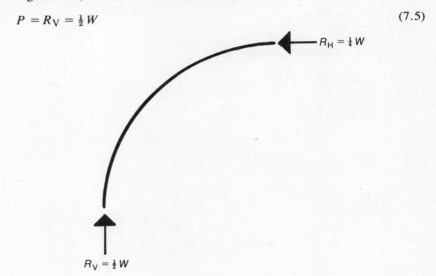

7.10.8. The thrust in a semicircular arch varies from $R_H = \frac{1}{4}W$ at the crown to $R_V = \frac{1}{2}W$ at the springings.

Although the bending moment is small compared to that in a portal frame, it dominates the design. For a rectangular section, from Eq. (5.6), the section modulus is $S = bd^2/6$. Because the section is also subject to compression, we will use a square section, so that

$$S = \frac{d^3}{6}$$

where d is the side of the square.

The maximum bending moment is

$$M = \frac{WL}{32} = \frac{40 \times 9}{32} = 11.25 \text{ kN m}$$

The maximum permissible flexural stress for glued laminated Douglas fir is $f_b = 15$ MPa, and the section modulus required is

$$S = \frac{11.25 \times 10^{-3}}{15} = 0.75 \times 10^{-3} \text{ m}^3 = 750\,000 \text{ mm}^3$$

The square section requires a side length of

$$d = \sqrt[3]{6S} = 165.1 \text{ mm}$$

Because there is also a thrust to be considered, we will use a section 200 mm square.
The maximum thrust is

$$P = \tfrac{1}{2}W = \tfrac{1}{2} \times 40 = 20 \text{ kN} = 0.02 \text{ MN}$$

and the maximum compressive stress due to the thrust is

$$\frac{0.02 \text{ MN}}{0.2 \text{ m} \times 0.2 \text{ m}} = 0.5 \text{ MPa}$$

which is less than 5% of the 15 MPa stress due to the bending moment. It must be added to the stress due to the bending moment, but as we have chosen a section much larger than is needed to resist bending alone, there is an adequate margin of area to resist the additional compressive stress.

The maximum bending moment is

$$M = \frac{WL}{32} = \frac{8 \times 30}{32} = 7.5 \text{ kip ft} = 90 \text{ kip in.}$$

The maximum permissible flexural stress for glued laminated Douglas fir is $f_b = 2.2$ ksi, and the section modulus required is

$$S = \frac{90}{2.2} = 40.91 \text{ in.}^3$$

The square section has a side length of

$$d = \sqrt[3]{6S} = 6.262 \text{ in.}$$

Because there is also a thrust to be considered, we will use a section 8 in. square.

The maximum thrust is

$$P = \tfrac{1}{2}W = 4 \text{ kips}$$

and the maximum compressive stress due to this thrust is

$$\frac{4}{8 \times 8} = 0.0625 \text{ ksi} = 62.5 \text{ psi}$$

which is less than 5% of the 2.2 ksi stress due to the bending moment. It must be added to the stress due to the bending moment, but as we have chosen a section much larger than is needed to resist bending alone, there is an adequate margin of area to resist the additional compressive stress.

Problem 7.8 Determine the size of the tie bars required to absorb the horizontal reactions of the arches in Problems 7.6 and 7.7.

We can either use abutments to prevent the arches from spreading or, alternatively, tie them together with steel reinforcing bars buried in the concrete floor. The second method is simpler and cheaper. The horizontal reactions are, in fact, quite small.

The smallest reinforcing bars that are reasonable for this purpose are:

1. a Canadian metric No. 10 bar, with a diameter of 11.3 mm and a cross-sectional area $A_S = 100$ mm^2
2. a British or Australian 12 mm diameter bar, with $A_S = \tfrac{1}{4}\pi \times 12^2 = 113$ mm^2
3. *a No. 4 bar, which has a diameter of $\tfrac{1}{2}$ in., with $A_S = \tfrac{1}{4}\pi \times (\tfrac{1}{2})^2 = 0.196$ in.^{2}*

The maximum permissible stress for A 36 steel is 150 MPa (*22 ksi*). The tensile force provided by the bar is therefore $P = f A_S$, which has the value

1. $150 \times 10^3 \times 100 \times 10^{-6} = 15.0$ kN
2. $150 \times 10^3 \times 113 \times 10^{-6} = 16.95$ kN
3. *$22 \times 0.196 = 4.312$ kips*

In Problem 7.6 we calculated that the horizontal reactions of each arch are 10 kN (*2 kips*), so that the one small bar is adequate.

Chapter 8 Multistory Buildings and Tall Buildings

Don't clap too hard—it's a very old building.
(John Osborne)

The majority of buildings more than three stories high have steel or reinforced concrete skeleton frames, although loadbearing walls are making a comeback for buildings up to twenty stories high. Simple design procedures exist for low-rise buildings with steel or reinforced concrete skeleton frames and for buildings with loadbearing walls, provided that earthquake loads need not be considered.

Computer-based methods are invariably used for the structural design of high-rise buildings. Economies in material can be obtained by the use of new structural systems, utilizing the structure of the service core and of service floors.

As buildings get taller, it may be necessary to use energy dissipation systems to an increasing extent, particularly in earthquake zones. Bridges between tall buildings could produce further structural economies.

8.1 A Brief Historical Note

The development of tall buildings depended on two inventions: the skeleton frame and the passenger elevator. The elevator was the more important, because people were unwilling to live or work in buildings more than a few stories high as long as they had to walk up the stairs. Elevators had been in use in mines since ancient Roman times, but there had been many accidents. In 1854 Elisha Graves Otis invented an automatic safety device that prevented the elevator from falling if the hoisting gear failed or the hoisting rope broke. The first Otis elevator was installed in New York in 1857 and in Chicago in 1870.

In 1871 the entire city center of Chicago was destroyed by fire. When the city was rebuilt, many of the new buildings were taller than those they replaced, and many had elevators.

Iron was used extensively because it was considered fireproof. Iron had already been in use for a century, frequently in factory construction, and on a few occasions in architect-designed buildings, such as the British Museum in London, the Prince Regent's Pavilion in Brighton, and the Théâtre Français in Paris. But this was the first time that an entire city center was built with iron as a major structural material. In the earlier postfire Chicago buildings, the interior columns were generally of cast iron and the beams of wrought iron, but the exterior walls were loadbearing masonry. In 1885 the first building was erected in which the entire vertical load was carried by iron or steel beams and iron columns; it was also the first use of steel, as distinct from iron, in the structure of a building.

The *height* of the tallest building increased rapidly in the half-century from 1880 to 1930, and there has been only a modest increase in the half century since then. On the other hand, the *number* of tall buildings increased sharply after 1950 (Section 3.8). Before that time tall buildings had been limited to a few American cities, notably New York and Chicago. In many European countries and in Australia the maximum height of buildings had been limited, and tall buildings were not allowed. This was partly due to a desire to prevent the erection of buildings that were out of scale with the historical city center and partly due to fire regulations (Section 2.4), which restricted building heights to about 50 m (150 ft).

Few cities today restrict the height of buildings by fire regulations, although the precautions required increase sharply if the building is designed to be taller than the fire brigade's ladders (Section 2.4). The esthetic conflict between modern tall buildings and older buildings remains a problem, although many American and Australian cities have acquired through their tall buildings a character they previously lacked.

8.2 The Design of Low-Rise Steel Frames for Vertical Loads

Although the invention of the skeleton frame, almost a century ago, was due to a desire to build taller, it was soon used as well for buildings only a few stories in height. Today the low-rise, skeleton-framed buildings outnumber the high-rise buildings by a factor of at least one hundred. Consequently the typical skeleton-frame building is a low-rise building. It is a matter of opinion whether the widespread replacement of loadbearing walls by skeleton frames has been a step in the right direction (Section 8.9).

The simple theory of the steel skeleton frame devised in the late nineteenth century is still valid for the smaller multistory buildings, although the final calculations for multistory buildings are now usually made by computer, with the use of a more sophisticated theory (Section 8.7). The simple theory assumes that the beams have relatively flexible (or "pin-jointed") connections to the columns, as in Fig. 7.1.2.

In analyzing the structural behavior of the building under vertical loading, we can conceive it as a series of columns, cantilevering from the ground, carrying simply supported beams (Fig. 8.2.1). Thus each beam may be considered separately as a simply supported beam carrying the load at that particular floor. The reactions of these beams are then transmitted to the columns.

Thus the bending moment in each beam (Table 4.1, line 9) is

$$M = +\tfrac{1}{8}WL \qquad (8.1)$$

where W is the load carried by the beam and L is the spacing of the columns. The compressive force transmitted to the columns is

$$P = nW \qquad (8.2)$$

where n is the number of stories above the column section and W is the load on each floor carried by the column.

The properties of the structural steel sections required follow from Eqs. (5.9) and (5.10), in Section 5.8.

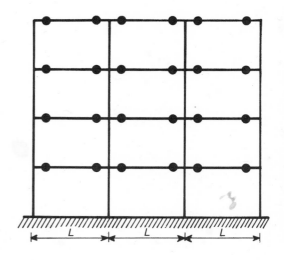

8.2.1. The simple steel skeleton frame, carrying vertical loads. We assume that the beams are simply supported on the columns.

8.3 The Design of Low-Rise Reinforced Concrete Frames for Vertical Loads

Site-cast reinforced concrete is monolithic (see Glossary), and therefore the beams and slabs (Section 5.5) are continuous over the columns (Fig. 8.3.1).

It is most important in reinforced concrete structures to place the reinforcement on the face of the beam or slab on which the tension occurs. This is shown in Figs. 4.6.6 and 4.6.8 (Section 4.6). If there is no tension reinforcement on the tension face, the strength is only that of an unreinforced concrete beam or slab and is therefore negligibly small.

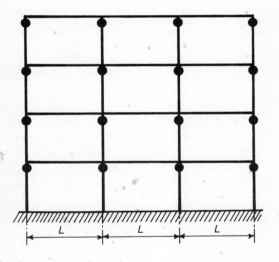

8.3.1. The simple reinforced concrete frame. We assume that the reinforced concrete beams or slabs are continuous over the columns and that the columns apply only a negligible moment restraint to the floor structure.

Structural concrete codes, such as the American ACI Code (Ref. 8.1), give bending moment coefficients for the positive moment at mid-span (reinforcement on bottom face) and the negative moment over the supports (reinforcement at top face) for various parts of the floor structure. Alternatively, the bending moments can be worked out on the assumption that the beams or slabs are continuous over the columns (Fig. 8.3.1), using the bending moment coefficients in Table 4.2. Having determined the bending moment

$$M = (\text{coefficient})\, W L \qquad (8.3)$$

we can calculate the amount of reinforcement required from Eq. (5.11), in Section 5.8.

Because the floor is rigidly joined to the columns, some bending moment is transmitted to the columns; this is normally expressed as an eccentricity of loading, ϵ (Fig. 8.3.2). The American ACI Code (Ref. 8.1) requires a minimum eccentricity of $0.10\, h$, where h is the width of the column.

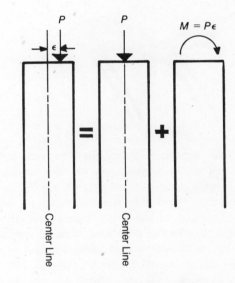

8.3.2. A column loaded eccentrically at a distance ϵ from its center line is subject to combined bending and compression; the bending moment in the column is
$$M = P\, \epsilon \qquad (8.4)$$
Conversely, a column subject to a bending moment M can be regarded as a column loaded with an eccentricity ϵ.

The formulas for the design of reinforced concrete columns with eccentric loading (or to put it another way, of reinforced concrete sections subject to combined bending and compression) are very complicated, and the solution is invariably obtained with the aid of tables or charts. A set of these is issued in conjunction with each national concrete code (for example, Refs. 8.2, 8.3, 8.4), and readers who wish to obtain numerical answers for reinforced concrete column problems should make use of these. They are also available as computer programs.

As already mentioned in Section 8.2, the columns must carry the load of all the floors above them, so that the column section must be designed for

$$P = n W \qquad (8.2)$$

and in addition for the eccentricity ϵ (this assumes that there are no horizontal forces acting on the reinforced concrete frame).

8.4 The Effect of Horizontal Loads on Simple Frames

Many multistory frames do not need to be designed for horizontal loads. If the building is not in an earthquake zone and is sheltered from wind by surrounding buildings of equal or greater height, there is no need to consider horizontal forces.

If the building is subject to wind or earthquake loads (Section 2.3) but is only of moderate height, simple frame theory can still be used. However, the pin joints assumed in Figs. 8.2.1 and 8.3.1 must be modified. Evidently the slightest horizontal force (Fig. 8.4.1) would push the frames sideways. This does not actually happen because we have, for the purpose of determining the vertical loads, simplified the frames. The beam-column joint does, in fact, have appreciable stiffness, which is not brought into play by a vertical load but resists distortion of a frame by a horizontal load.

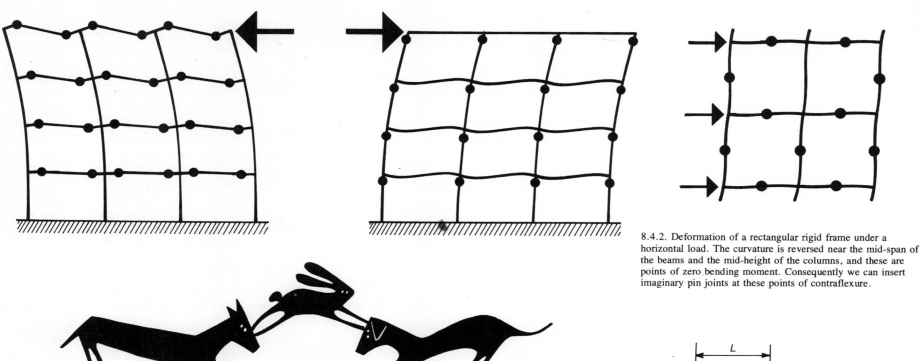

8.4.2. Deformation of a rectangular rigid frame under a horizontal load. The curvature is reversed near the mid-span of the beams and the mid-height of the columns, and these are points of zero bending moment. Consequently we can insert imaginary pin joints at these points of contraflexure.

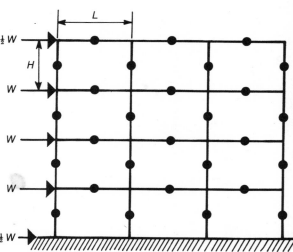

8.4.1. The assumptions made in Figs. 8.2.1 and 8.3.1 are satisfactory for vertical loads, however small. Evidently a small horizontal force can push either frame sideways.

For the purpose of determining the bending moments due to horizontal forces we now assume the beam-column joints to be completely rigid. The rectangular rigid frame then deforms as shown in Fig. 8.4.2. Near the middle of the beams and near the middle of the columns there is a reversal of curvature, and at this point of contraflexure the bending moment is zero; this is proved in more advanced textbooks (for example, Ref. 1.1, p. 242). We can therefore insert an imaginary pin joint at each point of contraflexure (Figs. 8.4.2 and 8.4.3). This makes the frame statically determinate, and we can calculate the bending moments due to the horizontal forces W (Fig. 8.4.4).

8.4.3. The pinned-frame concept for small horizontal loads. Since the bending moment is zero at the points of contraflexure (Fig. 8.4.2), we can insert imaginary pin joints, which render the frame statically determinate.

The maximum bending moments in the columns and in the floor structure occur near the floor-column junction, and these moments must be superposed on the moments and forces due to the vertical loads (Figs. 8.4.5 and 8.4.6).

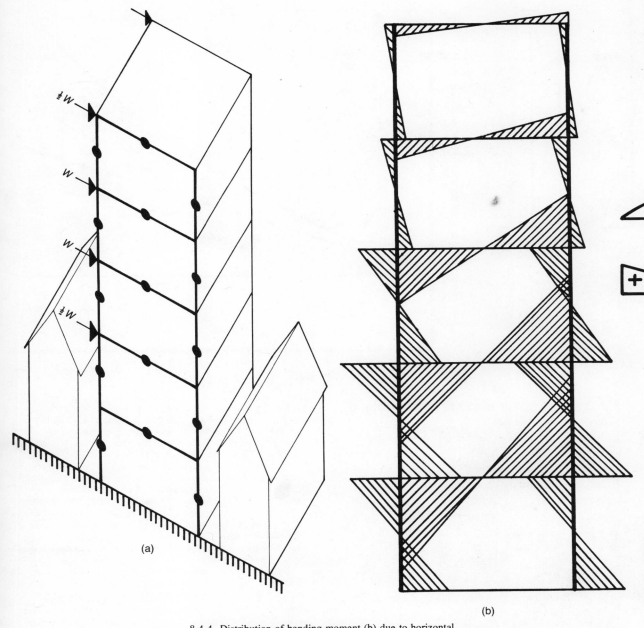

8.4.4. Distribution of bending moment (b) due to horizontal forces acting on multistory frame (a).

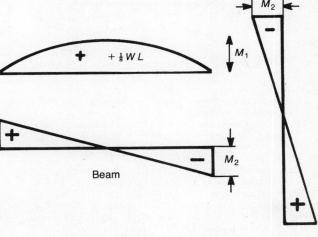

8.4.5. In a simple steel frame the bending moment in the beams consists of:

1. a positive bending moment, M_1, due to the vertical loads, which has a maximum at mid-span and reduces to zero at the columns (Section 8.2); and
2. a uniformly varying bending moment due to the horizontal loads, M_2, which is positive at one end, negative at the other, and zero at mid-span (Fig. 8.4.4.b).

If M_2 is smaller than M_1, the maximum bending moment calculated for vertical loads remains unaltered, and there is no need to enlarge the beams.

The bending moment in the columns, previously zero, now varies from $+M_2$ to $-M_2$. The maximum permissible compressive stress, f, must not exceed the sum of the compressive stresses due to the column load, P, and the bending moment, M_2:

$$f = \frac{P}{A} + \frac{M_2}{S} \qquad (8.5)$$

A new, larger column may be necessary to satisfy this equation.

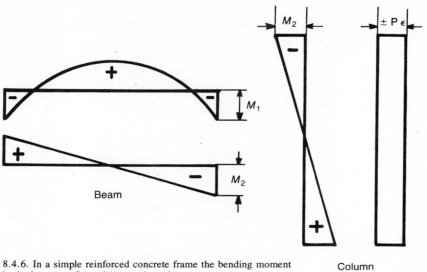

Beam

Column

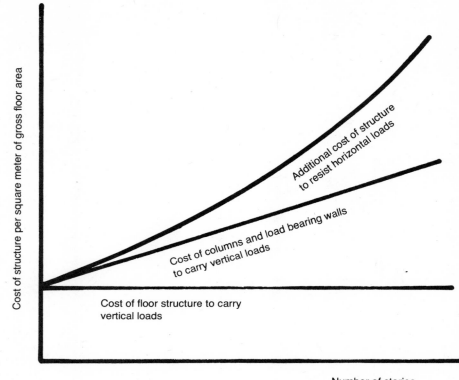

Cost of floor structure to carry vertical loads

Number of stories

8.4.6. In a simple reinforced concrete frame the bending moment in the beams or floor slabs consists of:
1. a bending moment due to the vertical loads, which varies from a maximum positive moment at mid-span to a maximum negative moment $-M_1$ at the columns (Section 8.3); and
2. a uniformly varying bending moment M_2, which is negative at one end and positive at the other (Section 8.4).

The reinforcement must be checked and may have to be increased for the greater negative bending moment, $-(M_1 + M_2)$; if M_2 is large, it will also be necessary to increase the depth of the beam or slab.

The bending moment in the column due to the horizontal loads, M_2, must be added to the bending moment resulting from the eccentricity of loading (Fig. 8.3.2), and the column checked (Section 8.3) and possibly enlarged, for the greater bending moment,

$$M = P\epsilon + M_2 \qquad (8.6)$$

8.5 The Vertical Structure in High-Rise and Low-Rise Buildings

As buildings increase in height, the cost per square meter of floor area increases (Fig. 8.5.1). The cost of the floor structure required to support the vertical loads remains almost constant, but the size of the columns required to support the vertical loads increases with the number of stories. As the building gets taller, additional structural material is required to resist the horizontal loads. This depends to some extent on the structural system chosen and the skill of the designer, but it is also greatly influenced by the magnitude of the horizontal loads; some buildings have shelter from surrounding buildings, and some regions are not prone to high wind or earthquake loads.

8.5.1. The cost of the structure increases with the number of stories. If we neglect the additional cost due to the building services and the additional space taken up by the vertical transportation system, the cost of the floor structure to carry the vertical loads remains constant as the number of stories increases. The cost of the columns to carry the vertical loads only increases in proportion to the number of stories (Eq. (8.2), Sections 8.2 and 8.3). The additional cost due to horizontal forces is often zero for low-rise buildings, because the building is sheltered or because the additional bending moments (Section 8.4) can be absorbed by the structure required in any case for the vertical loads. A great deal of structural material can be saved by the skillful use of diagonal ties or shear panels (Figs. 4.8.1.a and b and 5.2.1).

The bending moment due to horizontal loads acting on a vertical cantilever (Fig. 4.8.2) increases rapidly with the height of the building, and it dominates the structural design of very tall buildings.

To this additional structural cost must be added the additional cost of the building services, notably the vertical transportation system. The cost of usable floor area is further increased by the need for more and more space for the vertical transportation system, which increases with the height of the building, and by the limitations on usable floor space imposed by building authorities, which become more restrictive as buildings become taller.

The cost per unit floor area must therefore increase appreciably with building height. On the other hand, certain advantages accrue both to the public and to the building owner. From the public point of view, there is a saving in horizontal transportation and service lines to buildings. From the owner's point of view, tall buildings use less land per unit floor area, and land values are invariably high in districts where tall buildings are erected (although it is sometimes difficult to say whether the tall buildings or the high land values came first). Space in tall buildings in fact commands high rentals, and most tall buildings used for offices or high-quality apartments have proved good investments for their owners; low-cost apartments have been less successful. (The relative social advantages and disadvantages of high-rise and low-rise buildings and of high-density and low-density cities are beyond the scope of this book; one could write a book on that subject alone.)

Horizontal forces were neglected in the earliest skeleton-framed buildings erected in the 1880s. All had external walls of solid masonry which, in any case, acted as shear panels (Fig. 4.8.1.b). In the early 1890s, when buildings began to exceed 20 stories in height, crossed diagonals of wrought iron (Fig. 4.8.1.a) were added to the skeleton frame.

In 1893 the Old Colony Building in Chicago became the first to use a structural frame with rigid joints (Figs. 8.5.2 and 8.5.3). The frame was designed as a double series of vertically stacked portal frames (Section 7.6). The portal-frame method of design was widely used until the 1930s. The Empire State Building, which is only 14% lower than the tallest building in existence today, was designed by it.

In the 1890s the inadequacies of the existing structural theory imposed several limitations on the structural design of high-rise buildings. By 1930 the rigid-frame theory had been solved in principle, but the numerical evaluation of tall frames still presented practical problems. These have since been solved by the invention of the electronic digital computer (Section 8.7).

Evidently the strength of the structure is increased by treating it as a series of rigid frames (Fig. 8.5.4) both in theory and in the construction of the frame.

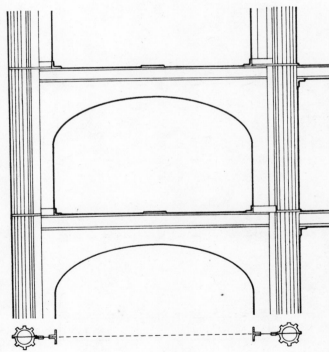

8.5.2. Riveted steel frame for the Old Colony Building, erected in Chicago in 1893 and 1894. The wind is resisted by the portal frames on the left. The interior bay has simply supported beams.

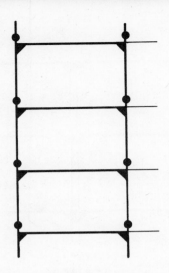

8.5.3. Structural principle of the steel and wrought iron structure shown in Fig. 8.5.2. The outer bays resist the horizontal forces by virtue of the rigid joints at the top of each portal.

The resistance of these rigid frames to horizontal forces is supplemented by any walls in the building (which need not be loadbearing) acting as shear walls (Fig. 4.8.1.b) and by the service cores (Figs. 8.5.5 and 8.5.6).

The service core becomes more important as the height of the building increases. The greater need for vertical transportation increases its size, and therefore the contribution it makes to the structure; indeed, it is possible to design a building to be supported entirely by its service core (Fig. 5.2.2).

The size of the columns also increases greatly with height. An exterior column does not necessarily waste space inside the building (Fig. 8.5.7), but an interior column inevitably does. As the building becomes taller, it therefore becomes advantageous to eliminate all interior columns and use a floor structure capable of spanning from the service core to the exterior columns. This stronger floor structure can also make an appropriate contribution to resisting the horizontal forces.

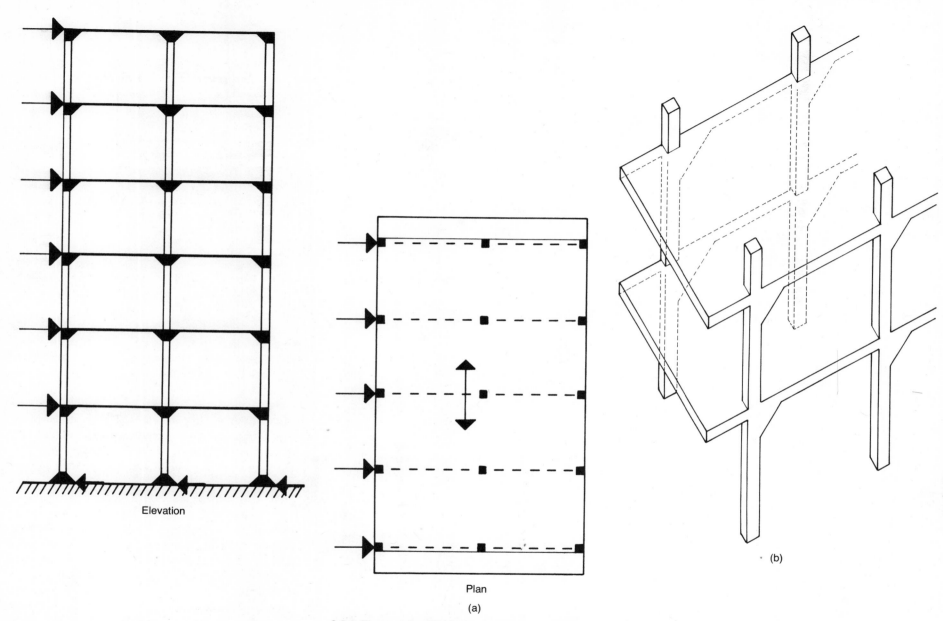

8.5.4. The strength of the structure is increased if all pin joints are eliminated and it is designed and constructed as a rigid frame.

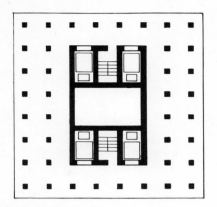

8.5.5. Single, central service core.

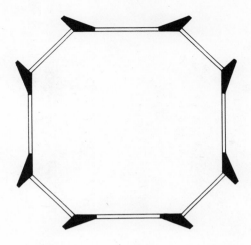

8.5.7. Most building regulations now limit the ratio of floor area to ground area instead of requiring setbacks at higher levels. The facade of the building is therefore generally set back from the property line, and columns can be external to the facade without infringing on neighboring property. In these circumstances the size of the exterior columns is not critical, because they do not waste valuable rentable interior floor area.

Every tall building requires one or more service floors. These are usually higher than normal floors because of the space required for large items of equipment; they do not require normal fenestration and sometimes need no windows at all. It is therefore possible to place a deep truss or reinforced concrete wall around these floors, and this greatly stiffens the interconnection between the inner tube of the service core and the outer tube of the facade through the floor structure (Fig. 8.5.10). Diagonals placed in the outer tube also increase its strength (Fig. 5.2.1) and therefore reduce the amount of structural material required.

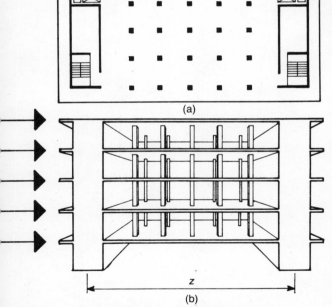

8.5.6. The functional requirements of the building should always determine the location of the elevators; however, if it is convenient to arrange them in two widely separated service cores (a), their resistance moment to horizontal forces is greatly increased by the length of the moment arm z between the two cores (b).

The service core is frequently of reinforced concrete to protect the elevators against fire. It then effectively forms a vertical tube of reinforced concrete, pierced by a few holes to provide access to the elevators and other services. However, even a steel core with lightweight fireproofing is effective as a vertical tube.

The outer columns and the spandrel beams, whether of steel or of reinforced concrete, form another, outer tube, even though the holes in that tube are much larger. The resistance to the horizontal forces is therefore provided by a vertical cantilever whose cross section is a tube within a tube (Fig. 8.5.8). Each tube contains a great deal of material, and it has great depth; in recent tube-in-tube structures this has ranged from 40 to 60 m (130 to 200 ft). The structural efficiency of the system is shown in Fig. 8.5.9.

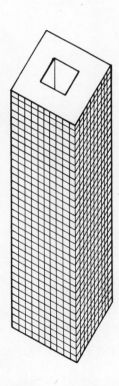

8.5.8. Provided that the exterior columns and the spandrel beams are suitably designed, they can be considered to form an outer vertical "tube with holes in it," which surrounds the inner vertical tube of the service core.

It is possible to subdivide the building into a number of tubes joined together with common walls (Fig. 8.5.11). This gives greater freedom in determining the shape of the building, and it overcomes the image of the "upturned cigarette carton" conveyed by so many buildings of the 1950s and 1960s.

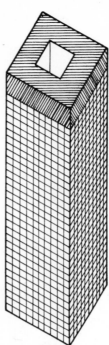

8.5.10. "Top hat" or stiffener placed in the service floor at the top of a tube-in-tube structure to interconnect the inner and outer tubes. It can take the form of a reinforced concrete shear wall or, in a steel frame, a truss running the full depth of the service floor, rigidly connected to the structure of the floors above and below.

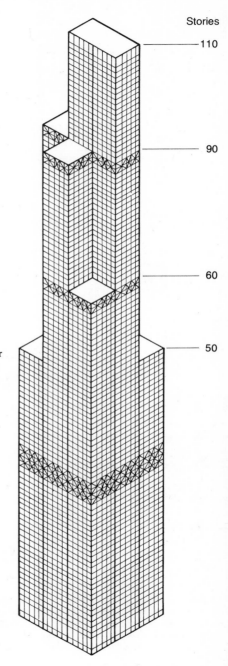

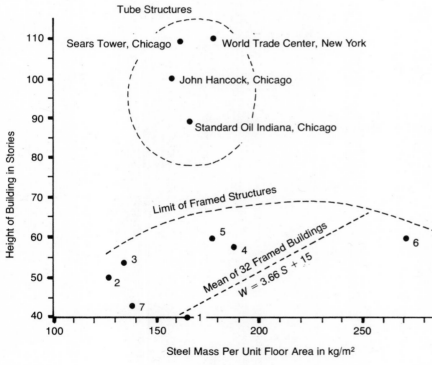

1 Union Guardian, New York
2 General Motors, New York
3 U.S. Steel, Pittsburgh
4 Pan American, New York
5 First National Bank, Chicago
6 Chase Manhattan Bank, New York
7 B.H.P. House, Melbourne

8.5.9. The tube-in-tube structure is a more efficient system for resisting horizontal forces than the conventional framed system, because of the large amount of structural material in the two tubes and their great depth. (*Drawing after an A.I.S.C. chart; by courtesy of the Australian Institute of Steel Construction.*)

8.5.11. The structure of the Sears Tower in Chicago, completed in 1974, at the time of writing the tallest building (442 m or 1450 ft). It consists of a number of tubes joined by common walls. Each of the nine tubes can be terminated at a different level.

8.6 The Floor Structure

The floor structure is essentially the same for high-rise and for low-rise buildings, except to the extent that the column spacing may differ. There is no advantage in having long spans in low-rise buildings, unless the activities planned for the building require open spaces. In high-rise buildings, however, the space wasted by large interior columns at the lower levels alters the economic considerations (Section 8.5).

The reinforced concrete structures considered in Section 5.5 are all suitable for multistory buildings; one-way slabs (solid, ribbed, or hollow tile), two-way slabs (solid or ribbed), flat slabs (solid or ribbed), and flat plates.

Structural steel frames are often constructed with reinforced concrete floors. Alternatively, steel plate or steel troughing with a concrete topping can be used.

Structural steel beams must be protected against fire, either by an insulating material such as spray vermiculite or by concrete. The concrete fire protection can be considered to contribute to the strength of the structure (Fig. 8.6.1); this is called *composite construction*.

In buildings with permanent partitions or walls, the size of the open spaces can be doubled by the use of staggered trusses (Fig. 8.6.2).

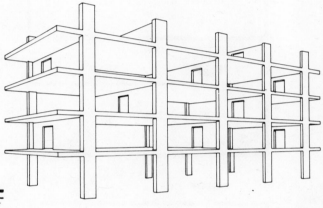

8.6.2. In steel-framed buildings with permanent partitions or walls, trusses that run the full height of the floor can be used. The floor structure is alternately supported on top of the trusses and hung from them. Because the trusses are deep, they use structural material efficiently.

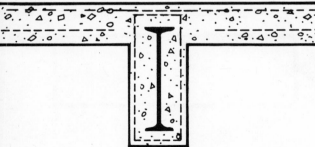

8.6.1. The concrete fire protection of structural steel can be considered part of the structure, provided that the bond stresses between the steel and the concrete and the shear stresses in the concrete are checked (Ref. 8.5). Shear reinforcement is frequently required.

8.7 The Rigid Frame Theory

In structural terminology, elastic *design* means that the overall geometry of the structure (spans, heights), the forces acting on it, and the maximum permissible stresses for the materials are specified and the structural designer determines the sizes of the structural members. Elastic *analysis* means that the geometry of the structure, the sizes of the structural members, and the maximum permissible stresses are specified and the structural designer determines the forces.

All statically determinate structures (Section 7.2) can be either designed or analyzed. Most statically indeterminate structures cannot be designed, only analyzed. Thus rigid frames are generally analyzed. The structural designer specifies the sizes of the structural members either by approximate calculations or by guesswork based on previous knowledge of structures of a similar type. Alternatively, a computer can be programmed to produce a preliminary design. If the analysis shows that the members are too small or that they are much larger than necessary, the sizes are revised and the analysis repeated.

Before structural computer programs were devised, the analysis of rigid frames was laborious. The most popular technique was the *moment distribution method*, which is still used today.

If a manual (that is, noncomputer) method of analysis is employed, it is too laborious to analyze the entire frame at once. The rigid frame in Fig. 8.5.4 is therefore broken up into a series of bents (Fig. 8.7.1), one for each story, and these are analyzed in turn. Initially all joints in the first bent are thought of as clamped to some large imaginary backboard. Thus the bent consists of a series of built-in beams or slabs; from Fig. 4.6.6.f we can see that each beam has at its supports a bending moment $-WL/12$. The columns are at that stage moment-free.

The joints are then released in turn, and in each case the bending moment at the end of the beam is distributed to the other members rigidly joined to it in proportion to their stiffnesses. Because each member deforms elastically, this distribution generates bending moments at the far ends of the members, and these "carried-over" bending moments upset the previous distribution, which must therefore be repeated.

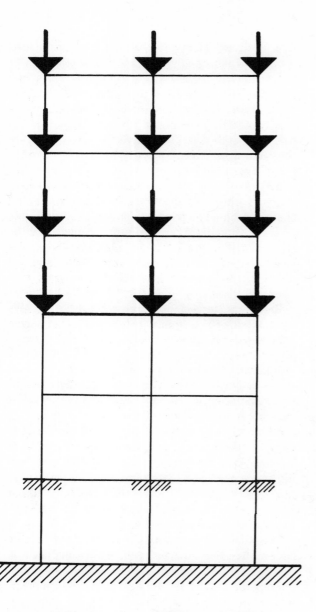

8.7.1. To simplify the analysis of rigid frames under vertical loading by the moment-distribution method when undertaken without a computer, the rigid frame of Fig. 8.5.4 is broken up into a series of bents, one for each story. The columns above and below the floor under consideration are assumed to be rigidly restrained at their far ends.

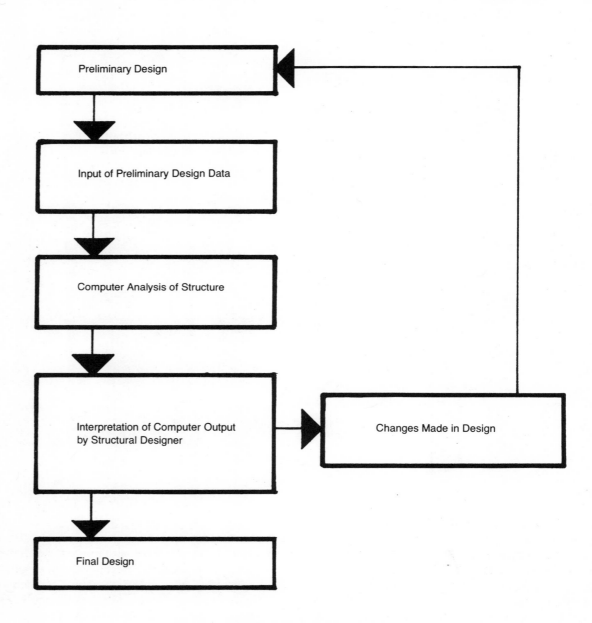

8.7.2. Flow chart for computer analysis of rigid frame.

The error decreases with each distribution, and for most problems three distributions suffice. The method is described in detail in more advanced books (for example, Ref. 1.1, pp. 244–51).

Analysis by the moment distribution method can be carried out with a pencil and paper only, with a digital pocket calculator, or with an electronic digital computer. However, if a computer is to be used (and that is now the normal practice for the analysis of rigid frames), it is better to employ a program based on a matrix algebra, which is particularly suited to the computer's capabilities (Fig. 8.7.2).

The two principal methods are the *matrix-force method*, or flexibility method, and the *matrix-displacement method*, or stiffness method. The latter is more convenient for most frames.

Using a computer does not alter the structural problem. We have a rigid frame that is statically indeterminate. To solve it we must determine the redundancies (Section 7.2), and we need one equation for each redundancy. In the matrix-force method the equations for the redundant forces are written in matrix form and then solved by inverting the matrix. In matrix-displacement methods the displacements (or elastic deformations) are the unknown quantities, but the process is otherwise the same (Ref. 8.6, pp. 188–225, and Ref. 8.7).

So far we have assumed that all forces acting on the structure are static. This is not sufficient for high-rise steel frames, which may vibrate due to wind forces, and for both medium-rise and high-rise frames in earthquake zones. A dynamic analysis is then required (Ref. 8.8).

8.8 Energy Dissipation Systems for Tall Buildings

The energy generated by wind or earthquake forces can be dissipated in several ways:

1. By the natural damping of structural materials. Most steel structures include some bolted connections, and energy can be dissipated by minute slipping of the bolts (Section 5.7). Since horizontal forces do not always come from the same direction, the process is reversible.
2. By elastic deformation (Fig. 3.1.1). As the structure deforms elastically, it stores strain energy, which is released when it moves back to and beyond its original position. However, the vibration caused by this elastic pendulum action can be distressing for the occupants of the building if it occurs at too low a wind velocity, and it may have to be damped. Concrete structures generally have less elastic deformation than steel structures.
3. By plastic deformation (Fig. 3.2.2.a). Far more energy can be absorbed by plastic than by elastic deformation, and far more by a ductile than a brittle material, as shown by the hatched area under the curve in Fig 3.2.2.a. Thus steel structures can absorb more energy than concrete structures; however, the reinforcement in concrete is also capable of some plastic deformation without excessive cracking of the concrete.
4. By viscous damping. It is possible to install shock absorbers in buildings, as is normal practice in automobiles. This has been done in a few very tall structures (see also Section 8.10).

Steel structures are generally capable of dissipating more energy than concrete structures, but they are also liable to greater vibrations. The relative significance of these two factors depends on the time interval between the recurrence of horizontal forces of various magnitudes. We can identify three categories:

1. Those that occur frequently, say, more often than once every two years. The building should not vibrate to an extent that would distress the occupants; indeed, for certain occupancies (for example, a theater or a restaurant), any perception of motion due to wind would probably be unacceptable. If natural damping is insufficient, viscous vibration absorbers may need to be installed. This problem arises only in very tall steel-framed buildings.
2. Those that occur no more frequently than, say, every ten years. In regions subject to earthquakes or hurricanes some minor damage (for example, deformation sufficient to break a few windows) must be accepted from time to time.
3. Those that are not expected to occur more frequently than once every hundred years. In a major earthquake significant damage must be accepted, provided that there is no loss of life. Since the energy generated by these rare catastrophic events is high, steel-framed buildings are preferable in earthquake regions, because the plastic deformation of the steel frame is capable of absorbing more energy than that of a reinforced concrete frame. However, large plastic deformation of a frame can cause appreciable damage to the building fabric.

8.9 Back to the Loadbearing Wall

In the first half of this century the invention of the skeleton frame was hailed as one of the most important innovations made in the design of buildings. In retrospect it is by no means certain that this judgment was correct.

The nineteenth-century method of designing loadbearing walls treated each wall as acting in isolation. The flexural resistance of a single unreinforced masonry wall to horizontal forces is low, because the material has a low tensile strength. However, if we design the walls at right angles to one another and the reinforced concrete floors spanning between them, as a structural unit, the resistance to horizontal forces is greatly increased (Fig. 8.9.1). If this is taken into account, loadbearing walls need not be confined to low-rise structures.

In the 1890s the 16-story Monadnock Building was erected in Chicago with loadbearing walls 1.83 m (6 ft 0 in.) thick; it is still standing. The design was by no means extravagant according to the structural knowledge of the time. The new design concepts that allow for the contribution of the reinforced concrete floors and the masonry cross walls were introduced in the 1950s, and the 21-story Liberty Park Towers in Pittsburgh have walls of brick and concrete masonry only 0.33 m (1 ft 3 in.) thick, which is 21% of the thickness of the walls in the lower part of the Monadnock Building. In many countries the tallest building is not much higher than 21 stories. Loadbearing walls are therefore a definite design alternative for a large number of buildings.

Masonry walls have good thermal insulation and a high thermal inertia. In view of the current interest in energy conservation, there are therefore nonstructural reasons for their use (Chapter 11). They are also an economical solution for buildings where permanent partitions or internal walls are required.

The structural principle is similar to that of the tube concept considered in Section 8.5. By taking the entire wall system as a unit we have a large depth, and therefore a large section modulus, S, to resist the bending moment due to the wind load, M. This still leaves us with an appreciable tensile stress, M/S, in the masonry, which it, unlike the steel or reinforced concrete tube of Section 8.5, is incapable of resisting.

However, we also have a dead load due to the weight of the building, P, which is carried by the masonry walls (with cross-sectional area A) as a vertical load, producing a compressive stress P/A (Fig. 8.9.2). Thus tension does not develop if P/A is greater than M/S.

If this criterion cannot be satisfied, the design is not necessarily unacceptable. The dead load acts all the time, the maximum wind load only for a short period.

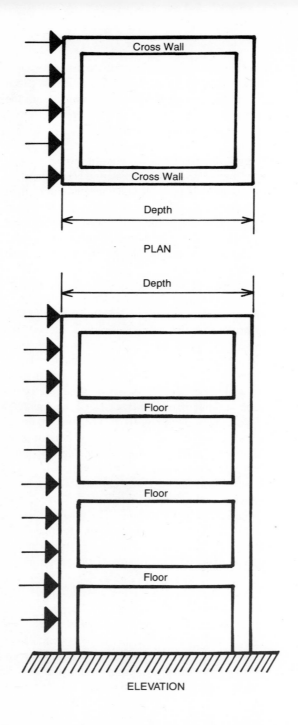

8.9.1. The loadbearing wall is part of a three-dimensional structure. In plan, the structure gains depth from the cross walls, and in elevation it gains depth from the reinforced concrete floors.

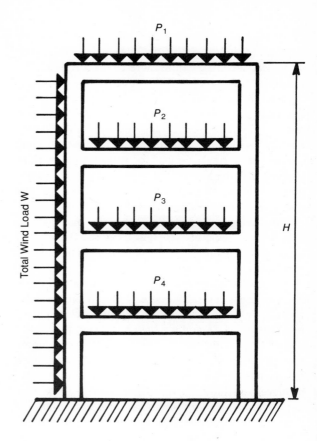

8.9.2. The masonry walls carry the dead weights P_1 to P_4 transmitted to them by the reinforced concrete floors, and in addition they resist the bending moment M due to the wind load W. For a building height H,
$$M = \tfrac{1}{2} W H$$
The total vertical load is
$$P = P_1 + P_2 + P_3 + P_4$$

If the masonry joints opened up a little for a short time, the structure would not be unsafe, provided the uplift was small, say, not extending over more than one fifth of the length of the wall (Fig. 8.9.3). Gothic structures have survived under those conditions for centuries.

Loadbearing wall and floor panels of precast concrete have been used intermittently since the early years of this century. They were particularly popular in Eastern Europe in the 1950s and 1960s, and early 1970s, but their use has declined since then. The structural principles are precisely the same as those of other loadbearing walls. The collapse of a block of apartments at Ronan Point in London, following a gas explosion in a kitchen, has drawn attention to the importance of proper connections between the units (Section 5.7).

The problems associated with the use of large precast wall units are mainly economic. The manufacture of the units needs a large factory, and their transport and handling need heavy equipment. This is justified only if there is a demand for a large number of almost identical units within a reasonable distance from the factory.

Complete rooms have also been prefabricated and then assembled as box units (Fig. 8.9.4). One noteworthy example is the Habitat apartments designed by Moshe Safdie, built in Montreal in 1966. The concept of arranging the apartments as a cluster with an individual open space for each unit is an entirely separate problem from their prefabrication. Precisely the same units could have been built, faster and at a lower cost, with conventional brick or block walls and site-cast concrete floors.

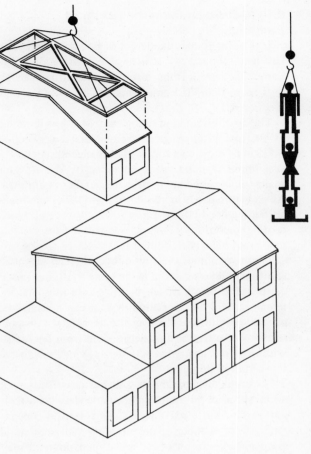

8.9.4. Erection of a block of apartments from precast concrete box units. These have not proved economical because of the high handling costs.

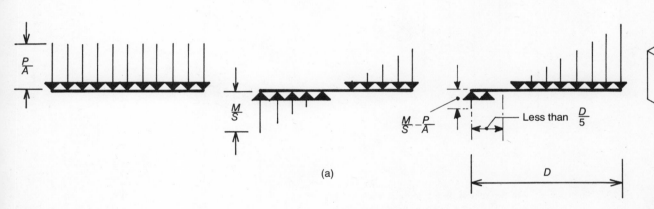

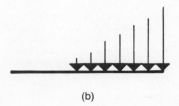

8.9.3. (a) The vertical load produces a uniform compressive stress P/A in the masonry walls. The wind load produces a stress that varies from a maximum tensile stress M/S to a maximum compressive stress M/S. If M/S is greater than P/A, uplift develops in a high wind. Some specifications permit this, provided that the net uplift is less than $D/5$. Since the masonry cannot resist any tensile stresses, the diagram at the right of (a) must be adjusted so that the compressive stresses (b) balance both the vertical force P and the moment M.

151

8.10 Factors Limiting the Height of Buildings

It is unlikely that the height of buildings is or ever has been limited by structural considerations. We know that ancient Rome already had apartment buildings more than 70 Roman feet (68 ft or 21 m) high (Section 4.1).

The height limitation set by staircases was removed by the invention of the passenger elevator (Section 8.1), but before the end of the nineteenth century it was found that the space taken up by the elevator shafts set an economic limit to the height of buildings. As elevator speeds increased, this limit became higher, but the height of the Empire State Building, which held the record for 42 years, was still not the greatest that could have been built at that time.

The invention of the sky lobby, whereby elevators at different levels can run in the same shaft, has raised the possible height of tall buildings. In most the internal vertical transportation system can now handle people more efficiently than the horizontal transportation system that brings them to and from the buildings. It is the latter that sets the present limit to the size of tall buildings.

Government departments and business firms seek office space in tall buildings because they wish their staff to be near to one another. This need to concentrate a large number of people in a small area could disappear as a result of technical innovation. It may become cheaper to obtain a computer printout from a central data bank than to carry a file from one office to another. It may become simpler to have a business conference through television consoles than to congregate in one room. If business and government become dispersed as a result, the need for tall office buildings will vanish.

However, for the last century buildings have steadily increased in size, and there is no indication so far that this trend is being reversed. If buildings are to grow taller, new structural systems will be needed to control the cost that arises from greater height (Fig. 8.5.1). We have seen how the tube concept and the top-hat concept reduced the cost of the structure (Figs. 8.5.8 to 8.5.10). This principle can also be applied to groups of buildings. The idea of putting two tall buildings close together is not new; it was done in the World Trade Center in New York, completed in 1973. It remains to interconnect them above ground level (Fig. 8.10.1).

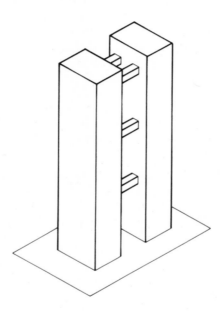

8.10.1. Interconnection of two tube structures at three levels to increase their strength and stiffness in relation to wind loads and to improve communications between the two buildings.

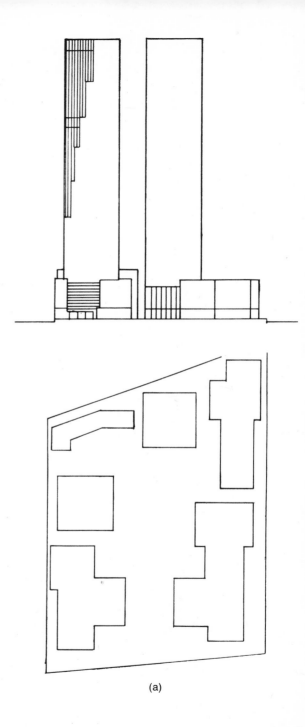

8.10.2. Plan and elevation of (a) the World Trade Center (1973) and (b) the Empire State Building (1931), both in New York. The enlarged base of the Empire State Building is needed to provide space for 50 elevators and the people using them, but it also increases the strength and stiffness of the vertical cantilever. The World Trade Center has a greater area of usable floor space, but fewer elevators because of the use of sky lobbies. (*By courtesy of P. R. Smith, University of Sydney.*)

(a)

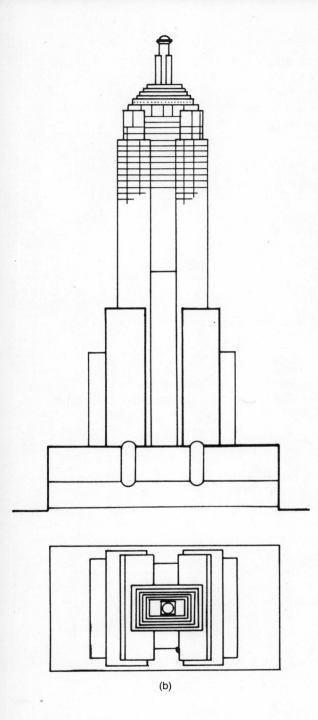

The individual towers of the World Trade Center have a height-to-width ratio of 7:1, much greater than in the Empire State Building, which is almost the same height (Fig. 8.10.2). As a result it was necessary to use 20 000 viscoelastic dampers (Section 8.8). Interconnection increases both the strength and the stiffness of the structure in relation to wind loads (at present it would be unwise to build to the same height in earthquake zones). The structural analysis of interconnected buildings poses some unsolved (but not insoluble) problems.

Interconnection would greatly improve communications between buildings and reduce the size of the vertical transportation system. The concept could easily be extended to four or more buildings (Fig. 8.10.3). This would be a very large project, but perhaps no bigger than was the building of Rockefeller Center in New York at the time it was undertaken.

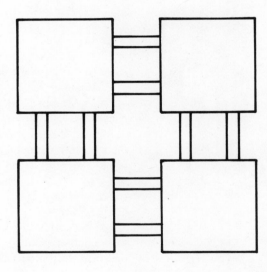

8.10.3. Interconnection of four tube structures at three levels.

(b)

153

8.11 Problems for Chapter 8

If you have not yet read Section 5.9, please do so before attempting the following problems.

The design of the floor structure for vertical loading has already been considered, in Chapters 5 and 6. Problems 5.8 and 5.9 dealt with the size of reinforced concrete floor slabs for multistory buildings with either steel or concrete frames. In Problems 5.4 and 5.6 we designed the floor beams for a steel frame, and in Problem 5.10 we considered the size of the floor beams for a reinforced concrete frame. Problem 5.12 was concerned with column sizes for a steel frame. In Problem 5.11 we calculated the size of the hanger for the unusual case when the floors are hung from the roof.

Problems 8.1 and 8.2 consider in more detail the sizing of columns for an eight-story building, one in steel and the other in reinforced concrete, assuming that there is no wind or earthquake load.

8.11.0.

Problems 8.3 and 8.4 give a simple solution for absorbing the wind load with a single set of diagonals on each facade, one for a steel-framed building, and the other for a concrete-framed building.

Problem 8.5 deals with an approximate method of determining the bending moment in a rigid frame due to wind loads by assuming pin joints at the mid-span of the beams and the mid-height of the columns; the same method can be used for earthquake loads. In Problem 8.6 the perimeter columns are assumed to resist the entire wind load by tube action.

All the problems in this chapter involve approximations.

Problem 8.1 Determine the approximate size of the columns required for an eight-story steel-framed building with the floor structure shown in Fig. 5.11.4.

All columns are subject to some bending moment in addition to the compressive force, P. However, if the columns are spaced symmetrically and there are no horizontal wind or earthquake loads, the flexural stresses are so much smaller than the compressive stresses due to the load P that they can be neglected in a preliminary design.

From Eq. (8.2) the column at the lowest floor carries a load of

$$P = nW$$

where $n = 8$ and W is the load of a floor panel measuring 6.5 m × 6.5 m (*21 ft × 21 ft*). This applies to an interior column; an exterior column carries a smaller share of the vertical load. The load (Problem 5.6) is 7 kPa (*150 psf*).

We assume that the steel frame is encased in concrete for fire protection, providing a 50 mm (*2 in.*) cover, and that the combined weight of the steel and the concrete is 24 kN/m³ (*150 lb/ft³*).

From the calculations in Problem 5.6 we note that the nominal dimensions of the steel beam are 460 mm deep weighing 60 kg/m (*18 in. deep weighing 40 lb/ft*). From Tables 13.1 and 13.2 we note that actual dimensions of the steel beam are 455 mm × 153 mm (*17.90 in. × 6.015 in.*). Taking the nearest modular dimensions, we find that the concrete-encased steel beam measures 575 mm × 275 mm (*22 in. × 10½ in.*).

For the purpose of assessing the dead load we assume that the concrete-encased steel column is 0.6 m (*2 ft*) square, and each floor is 3 m (*10 ft*) high. For lightweight fireproofing the weight would be less.

We noted in Problem 5.12 that the compressive stress in a column is reduced to allow for the effect of the slenderness ratio, but this ratio is unknown until a column section has been chosen. We therefore take 80% of the value applicable to a zero slenderness ratio. For A 36 steel this gives maximum permissible stresses of 112 MPa (*16 ksi*) (Problem 5.12).

The load per floor is

$$7 \text{ kPa} \times 6.5 \text{ m} \times 6.5 \text{ m} = 295.8 \text{ kN}$$

To this we add for the beams and their concrete encasement

$$24 \text{ kN/m}^3 \times 0.575 \text{ m} \times 0.275 \text{ m} \times 2 \times 6.5 \text{ m} = 49.4 \text{ kN}$$

For eight floors

$$nW = 8(295.8 + 49.4) = 2761.6 \text{ kN}$$

The weight of the upper-floor columns is

$$7 \times 24 \text{ kN/m}^3 \times 0.6 \text{ m} \times 0.6 \text{ m} \times 3 \text{ m} = 181.5 \text{ kN}$$

The total column load is

$$P = 2762 + 182 = 2944 \text{ kN} = 2.95 \text{ MN}$$

The cross-sectional area of the column is

$$A = \frac{2.95}{112} = 26.34 \times 10^{-3} \text{ m}^2 = 26\,330 \text{ mm}^2$$

From Table 13.1 we choose a 310 mm wide flange shape with a mass of 226 kg/m, which has $A = 28\,900$ mm².

The load per floor is

$$150 \text{ psf} \times 10^{-3} \times 21 \text{ ft} \times 21 \text{ ft} = 66.15 \text{ kips}$$

To this we add for the beams and their concrete encasement

$$\frac{150 \text{ lb/ft}^3 \times 10^{-3} \times 22 \text{ in.} \times 10.5 \text{ in.} \times 2 \times 21 \text{ ft}}{144} = 10.11 \text{ kips}$$

For eight floors

$$nW = 8(66.15 + 10.11) = 610 \text{ kips}$$

The weight of the upper floor columns is

$$7 \times 150 \text{ lb/ft}^3 \times 10^{-3} \times 2 \text{ ft} \times 2 \text{ ft} \times 10 \text{ ft} = 42 \text{ kips}$$

The total column load is

$$P = 610 + 42 = 652 \text{ kips}$$

The cross-sectional area of the column is

$$A = \frac{652}{16} = 40.75 \text{ in.}^2$$

From Table 13.2 we choose a 12 in. wide flange shape, with a mass of 152 lb/ft, which has $A = 44.7$ in.².

Problem 8.2 Determine the approximate size of the columns required for an eight-story reinforced-concrete framed building with the floor structure shown in Fig. 5.11.4.

The problem is basically the same as Problem 8.1. However, the determination of the size of reinforced concrete columns requires us to use tables, charts, or a computer program, or else to employ drastic approximations. Suitable tables may be found in Refs. 8.2 and 8.3.

In this problem we will obtain a quick answer by means of a rough approximation. Concentrically loaded steel columns are designed with a reduced compressive stress, which depends on the slenderness ratio. As we noted in Section 8.3, concentrically loaded reinforced concrete columns are designed with an eccentricity of loading, which also depends on the geometry of the section. This increases the size of the column. The reinforcement contributes from 20% to 60% of the total load-bearing capacity, and it can be varied by the designer. The more reinforcement, the smaller the column. We will omit both the eccentricity and the contribution of the reinforcement. The resulting (unreinforced) cross section is roughly the same size as a reinforced concrete column designed for the minimum eccentricity specified by the concrete code.

Reinforced concrete columns in multistory buildings are not normally so slender that an allowance need be made for buckling.

Omitting the reinforcement and the eccentricity, we find that the *ultimate* load for a concrete column is

$$P' = 0.85 f'_c A$$

where A is the cross-sectional area of the concrete column and f'_c is the cylinder crushing strength of the concrete*.

The ultimate column load (see Glossary) is obtained by multiplying the service load (see Glossary) by a factor of safety. The American (ACI) concrete code (Ref. 5.10) and other national codes apply safety factors to the dead and live load and also to the design equation. For a preliminary design it is sufficient to use one factor of safety. We will use a high value, namely, 2.5.

---

*In Great Britain concrete is tested by crushing cubes instead of cylinders. These have a higher crushing strength, and the British concrete code (Ref. 5.11) gives a different formula. We hope British readers will pardon us if we confine ourselves to the use of one formula.

The concrete beams chosen in Problem 5.10 are 600 mm × 300 mm (*24 in. × 12 in.*). These beams are only slightly larger than those in Problem 8.1, and the weight of the reinforced concrete is approximately the same as that of concrete-encased steel. The calculations are similar to those in Problem 8.1: the service load is $P = 3.00$ MN (*672 kips*).

We will use concrete with a cylinder crushing strength of 30 MPa (*4 ksi*).

The ultimate load is

$$P' = 2.5 \times 3.00 = 7.50 \text{ MN}$$

The cross-sectional area required is

$$A = \frac{P'}{0.85 f'_c} = \frac{7.50}{0.85 \times 30} = 0.2941 \text{ m}^2$$

A 600 mm × 600 mm square section has an area of 0.36 m².

The ultimate load is

$$P' = 2.5 \times 672 = 1680 \text{ kips}$$

The cross-sectional area required is

$$A = \frac{P'}{0.85 f'_c} = \frac{1680}{0.85 \times 4} = 494.1 \text{ in.}^2$$

A 24 in. × 24 in. column has a cross-sectional area of 576 in.².

It should be emphasized that this is only a very approximate calculation, based on the assumption that there are no bending moments or eccentricities of loading in the columns, apart from the minimum eccentricity specified by the concrete code. The actual column needs between 1% and 8% of reinforcement. As this is in compression, it must be tied at intervals of about 300 mm (*12 in.*) to prevent buckling of the reinforcing bars (Fig. 10.14.7).

Problem 8.3 Fig. 8.11.1 shows an eight-story, four-bay, steel-framed building, with the floor structure shown in Fig. 5.11.4. The horizontal wind loads are to be resisted by one pair of crossed diagonals on each face of the building. Make a preliminary determination of the size of the diagonals.

One diagonal is subjected to tension when the wind blows from the right and the other when it blows from the left.

We will assume that the combined effect of wind pressure and wind suction is 1 kPa (*20 psf*). The wind on the lowest half story is resisted directly by the ground; that is, the maximum shear force produced by the wind is

$$V = 7\tfrac{1}{2} W$$

where W is the pressure acting on an area one story high and 13 m (*42 ft*) wide. The wind pressure on the other 13 m (*42 ft*) of the building is resisted by another set of diagonals on the far side of the building. Thus the tensile force in the diagonal is (Fig. 8.11.2)

$$T = \frac{V}{\cos \theta}$$

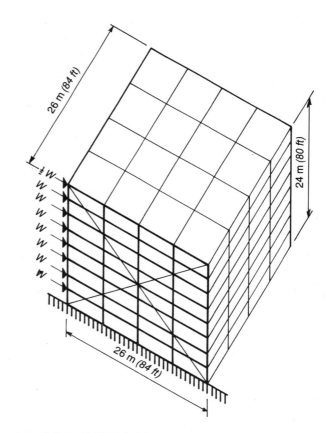

8.11.1. Dimensions of diagonally braced eight-story building; the floor plan is shown in Fig. 5.11.4. (Problems 8.3 and 8.4.)

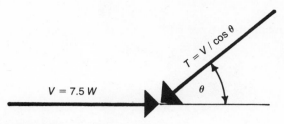

8.11.2. The tensile strength of the diagonal, T, must be sufficient to transmit the horizontal shear force V. (Problems 8.3 and 8.4.)

The length of the diagonal and the value of $\cos \theta$ are obtained from Fig. 8.11.3.

From Fig. 8.11.3

$$L = \sqrt{24^2 + 26^2} = 35.38 \text{ m}$$

$$\cos \theta = \frac{26}{35.38} = 0.735$$

The wind load is $W = 1 \text{ kPa} \times 3 \text{ m} \times 13 \text{ m} = 39 \text{ kN}$, and the maximum shear force is $V = 7.5 W = 292.5 \text{ kN}$. The tensile force in the diagonal is

$$T = \frac{292.5}{0.735} = 398.1 \text{ kN}$$

Using a maximum permissible tensile stress for A 36 steel of 150 MPa, we find that the cross-sectional area required is

$$A = \frac{398.1 \times 10^{-3}}{150} = 2.654 \times 10^{-3} \text{ m}^2 = 2654 \text{ mm}^2$$

From Fig. 8.11.3

$$L = \sqrt{80^2 + 84^2} = 116 \text{ ft}$$

$$\cos \theta = \frac{84}{116} = 0.724$$

The wind load is $W = 20 \times 10^{-3} \times 10 \times 42 = 8.4$ kips, and the maximum shear force is $V = 7.5 W = 63$ kips. The tensile force in the diagonal is

$$T = \frac{63}{0.724} = 87.0 \text{ kips}$$

The maximum permissible tensile stress for A 36 steel is 22 ksi (Ref. 5.9), and the cross-sectional area required is

$$A = \frac{87.0}{22} = 3.955 \text{ in.}^2$$

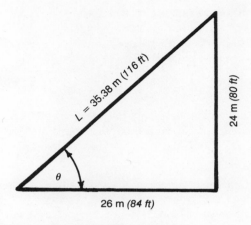

8.11.3. Length of diagonal bracing member and value of $\cos \theta$. (Problems 8.3 and 8.4.)

The type of steel section used depends on whether the tie is to be covered by the facade or exposed. For an exposed tie it is appropriate to use a steel section especially fabricated for satisfactory appearance. The cross-sectional area required is quite small and easily accommodated.

Problem 8.4 Fig. 8.11.1 shows an eight-story, four-bay, concrete-framed building, with the floor structure shown in Fig. 5.11.4. The horizontal wind loads are to be resisted by one pair of crossed diagonals on each face of the building. Make a preliminary determination of the size of the diagonals.

It is difficult to produce tension members from reinforced concrete, because concrete has negligible tensile strength, and the steel must resist the tensile forces. If the tension in the steel is too high, it produces large cracks in the concrete, which are unsightly and may cause corrosion. We shall therefore prestress the tie but design the remainder of the building in reinforced concrete.

The magnitude of the diagonal tensile force is the same as for Problem 8.3, namely, 398.1 kN (*87.0 kips*).

We considered the stresses to be used in prestressed ties in Problem 5.11. Allowing for temperature movement, we found that a residual prestress of 560 MPa (*80 ksi*) was appropriate. As these diagonal tension members apply a permanent force to a reinforced concrete structure, we make a further allowance of 15% for the contraction of the concrete that results from moisture movement under a permanently applied force (Section 5.6). This reduces the effective prestress to 476 MPa (*68 ksi*).

The cross-sectional area of steel is therefore

$$\frac{398.1 \times 10^{-3}}{476} = 0.836 \times 10^{-3} \text{ m}^2 = 836 \text{ mm}^2$$

$$\frac{87.0}{68} = 1.280 \text{ in.}^2$$

As we noted in Problem 8.3, the arrangement of the prestressing steel is mainly governed by esthetic considerations, but the area is quite small and easily accommodated.

Problem 8.5 Determine the effect of wind loading on the members of the rigid frame shown in Fig. 8.11.4.

The frame consists of bays measuring 6.5 m *(21 ft)* in each direction. These dimensions are the same as in Problem 5.6; we determined that the maximum bending moment in the beams due to the vertical loads is 165.9 kN m *(1434 kip in.)*, and we selected a steel beam 460 mm *(18 in.)* deep with a mass of 60 kg/m *(40 lb/ft)*.

The calculations for the size of the columns due to the vertical load are similar to those in Problem 8.1. However, this frame has only five stories, and furthermore each column is an exterior column and thus carries only half the load transmitted by the floors. Using a procedure similar to that followed in Problem 8.1, we find that the column service load is

$P = 1.50$ MN *(330 kips)*

We will assume that the combined effect of wind pressure and wind suction may be taken as equivalent to a pressure on the windward face of 1 kPa *(20 psf)*. This pressure acts on a width of the building of 6.5 m *(21 ft)* and a story height of 3 m *(10 ft)*, so that the wind load per story is

$W = 1 \times 6.5 \times 3 = 19.5$ kN

$W = 20 \times 21 \times 10 = 4200$ lb $= 4.2$ kips

Since the two lowest stories are sheltered by neighboring small houses, the wind loading, as shown in Fig. 8.11.4, has a total value of

$3W = 58.5$ kN *(12.6 kips)*

This horizontal load produces shear forces in the vertical structure of the building. Their effect is perhaps visualized more readily if we think of the building as a horizontal beam acted on by vertical loads (Fig. 8.11.5).

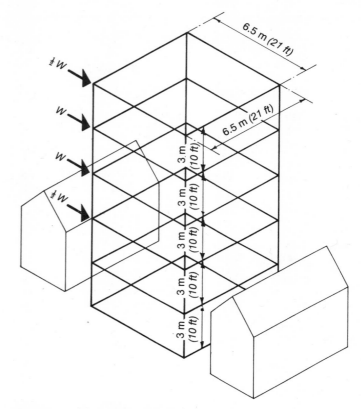

8.11.4. Overall dimensions of the frame of the building considered in Problem 8.5. The frame is one bay wide but extends to several bays in the other direction at right angles. The building is five stories high, but the lowest two floors are protected by adjoining buildings from the effects of wind. The wind forces therefore act only on the three upper floors. They are shown in this figure as wind pressure acting on one side of the building, but in fact the wind presses on one face of the building (the windward side) and sucks on the other face (the leeward side). Although wind suction must be considered in the detailed design of the facade, its effect on the overall stability is the same as that of the wind pressure: It is a lateral force on the building that complements the wind pressure.

We now insert imaginary pin joints in the frame, as in Fig. 8.4.4, and divide the frame into a number of bents (Fig. 8.11.6.a). Considering the uppermost of these (i), we can deduce the bending moment due to the wind load (Fig. 8.11.6.b).

The shear force $\tfrac{1}{2}W$, considered to be acting at the pin joint halfway along the upper-story column, produces a bending moment that reaches a maximum of

$$\tfrac{1}{2}W \cdot \tfrac{1}{2}H = \tfrac{1}{4}WH$$

at the rigid joint between the beam and the column at roof level and then decreases to zero at the pin joint halfway along the beam.

The maximum bending moments occur in bent (v), shown in Fig. 8.11.6.c, and in the identical bent (x).

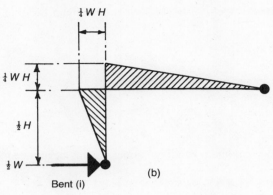

Bent (i) (b)

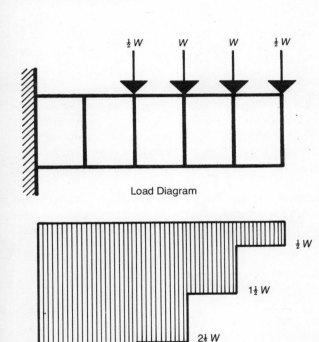

Load Diagram

Shear Force

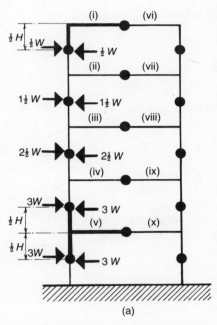

(a)

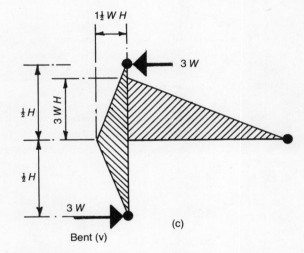

Bent (v) (c)

8.11.5. The effect of the wind pressure (and the wind suction) on the overall stability of the building is best visualized if we think of the building as a cantilever turned on its side. The wind forces, now shown in vertical positions, thus induce shear forces in the rigid frame. The shear force is $\tfrac{1}{2}W$ in the upper story, and it increases to $3W$ at the level of the third floor. As the lower part of the frame is protected from the wind, the shear force then remains constant until it is absorbed by the foundation.

8.11.6. We noted in Section 8.4 that in a rigid frame distorted by a lateral force the curvature changes from convex to concave near the mid-span of the beams and near the mid-span of the columns. At these points of contraflexure the bending moment is zero, and we can therefore insert imaginary pin joints at these points. The solution of the problem is only approximate, because we assume that these pin joints are exactly (instead of approximately) at the mid-span of the beams and at the mid-height of the columns.

At the "pin joints" the shear forces have the magnitude shown in Fig. 8.11.5. The shear force acts in the wind direction just above the pin, and its reaction acts in the opposite direction just below the pin.

We can divide the frame into ten bents, as shown in (a). We first analyze bent (i), shown in (b). This is similar to one half of a three-pinned portal frame (*Problem 7.3*). The shear force produces a bending moment in the column, which increases from zero at the "pin joint" to a maximum $\tfrac{1}{2}W \cdot \tfrac{1}{2}H = \tfrac{1}{4}WH$ at the rigid joint. The rigid joint transmits the full bending moment to the beam, and it decreases uniformly to zero at the "pin joint" in the beam supporting the flat roof.

The maximum bending moment occurs in bent (v), shown in (c), and in the bent (x). The bending moment in the half columns above and below the beam increases from zero at the "pin joints" to a maximum $3W \cdot \tfrac{1}{2}H = 1\tfrac{1}{2}WH$ at the right joint. Each of these two moments is transmitted by the rigid frame to the beam, so that the bending moment in the beam at the right joint is $1\tfrac{1}{2}WH + 1\tfrac{1}{2}WH = 3WH$. This bending moment then decreases from $3WH$ to zero at the "pin joint" in the beam.

The shear force of $3W$ produces a maximum bending moment in the columns of

$$3W \cdot \tfrac{1}{2}H = 1\tfrac{1}{2}WH$$

which is transmitted by each column to the beam, so that the maximum bending moment in the beam is

$$2 \times 1\tfrac{1}{2}WH = 3WH$$

Thus the maximum bending moment in any beam due to wind load is

$$M = 3WH = 3 \times 19.5 \text{ kN} \times 3 \text{ m} = 175.5 \text{ kN m}$$

$$M = 3WH = 3 \times 4.2 \text{ kips} \times 10 \text{ ft} = 126 \text{ kip ft} = 1512 \text{ kip in.}$$

This is slightly higher than the bending moment due to the dead load (Problem 5.6, page 94: kN m or 1434, kip in.). However, the American (AISC) steel code states (Ref. 5.9, Part 5, Section 1.5.6) that "Allowable stresses may be increased one third above the values provided in Sect. 1.5.1, 1.5.2, 1.5.3, 1.5.4, and 1.5.5 when produced by wind or earthquake loading."

The wind-load moments, M_2, and the moments due to vertical loading, M_1, in a simply supported beam are not additive but are alternative design conditions, as Fig. 8.4.5 shows. Since the bending moments due to wind load are only 6% more than those due to the vertical loads, and we are allowed an increase in the maximum permissible stresses of 33%, the wind load is evidently not critical for the design of the beams of this particular steel frame.

However, the wind load influences the design of the columns. The vertical load on each column, as stated above, is

$$P = 1.50 \text{ MN } (330 \text{ kips})$$

and the maximum bending moment, as stated above, is

$$M = 1\tfrac{1}{2}WH = 87.8 \text{ kN m } (756 \text{ kip in.})$$

At the section of the column where the bending moment M attains its maximum value, the column load also has its full value. Therefore, for the purpose of this preliminary design, we will halve the maximum permissible stresses, so that one half is available for the load P and one half is available for the bending moment M.

In Problem 8.1 we assumed a maximum permissible compressive stress for a column of 112 MPa (*16 ksi*); half of that is 56 MPa (*8 ksi*).

The maximum permissible flexural stress of 165 MPa (*24 ksi*) is increased for wind loads by 33% to 220 MPa (*32 ksi*); half of that is 110 MPa (*16 ksi*).

We therefore require a column section with a cross-sectional area of

$$A = \frac{1.50 \text{ MN}}{56 \text{ MPa}} = 26.800 \text{ mm}^2$$

$$A = \frac{330 \text{ kips}}{8 \text{ ksi}} = 41.25 \text{ in.}^2$$

and a section modulus of

$$S = \frac{87.8 \text{ kN m}}{110 \text{ MPa}} = 798\,200 \text{ mm}^3$$

$$S = \frac{756 \text{ kip in.}}{16 \text{ ksi}} = 47.25 \text{ in.}^3$$

From Tables 13.1 and 13.2 we require a wide flange shape with a nominal depth of 310 mm (*12 in.*). The cross-sectional area A requires a section with a mass of 226 kg/m (*152 lb/ft*), whereas the section modulus S requires only a section with a mass of 60 kg/m (*40 lb/ft*). Evidently our choice in the division of stresses favored the bending moment M due to the wind load more than the column force P due to the vertical load.

We next allocate three quarters of the maximum permissible stress to the column load P and one quarter to the bending moment M. This gives maximum permissible stresses of 84 MPa (*12 ksi*) for the column load P, and 55 MPa (*8 ksi*) for the bending moment M.

The cross-sectional area required is then

$$A = \frac{1.50 \text{ MN}}{84 \text{ MPa}} = 17\,900 \text{ mm}^2$$

$$A = \frac{330 \text{ kips}}{12 \text{ ksi}} = 27.5 \text{ in.}^2$$

and the section modulus required is

$$S = \frac{87.8 \text{ kN m}}{55 \text{ MPa}} = 1\,596\,000 \text{ mm}^3$$

$$S = \frac{756 \text{ kip in.}}{8 \text{ ksi}} = 94.5 \text{ in.}^3$$

These two requirements are sufficiently close to make a preliminary choice of section: a wide flange shape with a nominal depth of 310 mm (*12 in.*) and a mass of 143 kg/m (*96 lb/ft*).

Problem 8.6 Make a preliminary selection of the size of the perimeter columns of the building shown in Fig. 8.11.7.

We will assume that the columns are closely spaced at 1 m (*3 ft 4 in., or 40 in.*) centers. For visual reasons the corners of the building require special attention, but any contribution of the corner details to the strength of the structure is ignored at this stage. A wider column spacing is required at ground level; this is also ignored at present.

We assume that the spandrel beams and the floor structure at each level provide adequate restraint against buckling, and we will take the maximum permissible compressive stress for A 36 steel, as in Problem 8.1, as 112 MPa (*16 ksi*). There are 29 columns on each face, so that the outer tube is formed by 116 columns.

The load carried by the columns is estimated by the method already described in Problem 8.1. The total vertical (dead and live) load is 320 MN (*74 000 kips*) for the entire 40-story building. Let us assume half that load is carried by the inner service core and the other half (160 MN or *37 000 kips*) by the perimeter columns.

The vertical load per column is then

$$\frac{160}{116} = 1.38 \text{ MN}$$

$$\frac{37\,000}{116} = 319 \text{ kips}$$

The cross-sectional area required for each column is

$$A = \frac{1.38}{112} = 12.32 \times 10^{-3} \text{ m}^2 = 12\,320 \text{ mm}^2$$

$$A = \frac{319}{16} = 20.0 \text{ in.}^2$$

From Tables 13.1 and 13.2 we select wide flange shapes of nominal depth 310 mm (*12 in.*) weighing 107 kg/m (*72 lb/ft*), which have a cross-sectional area of 13 600 mm² (*21.1 in.²*).

For the purpose of assessing the resistance of the outer steel frame to wind loads we transform the 116 columns into an equivalent steel tube that has the same area as the columns. The total cross-sectional area of the 116 columns is

$$116 \times 13\,600 = 1\,577\,600 \text{ mm}^2$$

$$116 \times 21.1 = 2\,447.6 \text{ in.}^2$$

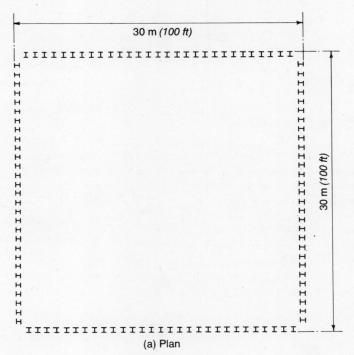

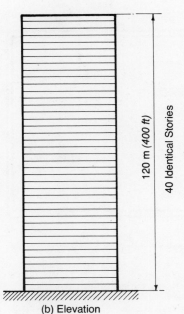

8.11.7. Plan (a) and elevation (b) of a 40-story building with perimeter columns spaced at 1 m (*40 in.*) centers. The columns are assumed to carry half the vertical load and the entire wind load.

We will now determine the thickness t of a steel tube measuring 30 m by 30 m (*100 ft by 100 ft*) that has the same cross-sectional area (Fig. 8.11.8):

$$t = \frac{1\,577\,600}{4 \times 30\,000} = 13.15 \text{ mm}$$

$$t = \frac{2447.6}{4 \times 100 \times 12} = 0.510 \text{ in.}$$

The section modulus of this tube is the difference between the section modulus of a solid square measuring B by D and another square measuring $B - 2t$ by $D - 2t$, representing the empty space inside the tube (Fig. 8.11.8). Expressed in algebraic terms

$$S = \tfrac{1}{6}[BD^2 - (B - 2t) \cdot (D - 2t)^2]$$

Since $B = D$ for the square tube, $S = \tfrac{1}{6}(D^3 - (D - 2t)^3)$.

$$S = \tfrac{1}{6}(30^3 - 29.9737^3) = \tfrac{1}{6} \times 70.948 = 11.82 \text{ m}^3$$

$$S = \tfrac{1}{6}(1200^3 - 1198.98^3) = \tfrac{1}{6} \times 4\,402\,800 = 733\,800 \text{ in.}^3$$

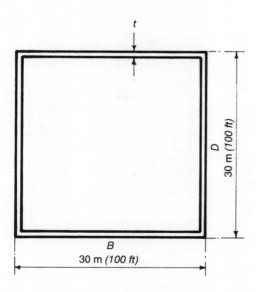

8.11.8. For the purpose of assessing the approximate resistance of the perimeter columns to horizontal loads we replace them by an imaginary square tube of thickness t.

The steel tube acts as a cantilever under the action of a wind load, which produces pressure on the windward face and suction on the leeward face (Fig. 8.11.9). If the combined effect of the wind pressure and wind suction amounts to a total uniformly distributed horizontal load W, the cantilever moment (Table 4.1, line 5) is

$$M = \tfrac{1}{2}WH$$

This is equal to the resistance moment of the tube,

$$M = fS$$

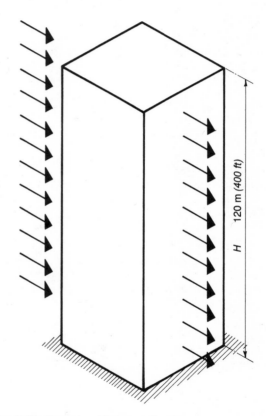

8.11.9. The dynamic wind loads can, as a first approximation, be thought of as static horizontal forces. The wind pressure on the windward side of the building and the wind suction on the leeward side are both horizontal forces acting in the same direction, so that the tube formed by the perimeter columns may be considered a very tall cantilever.

Most steel codes allow an increase in the maximum permissible stresses due to wind load. The American (AISC) steel code (Ref. 5.9, Part 5, Section 1.5.6) permits an increase of one third above the value allowable for vertical live loads.

One third of the maximum permissible flexural stress (which for A 36 steel is 165 MPa *or 24 ksi*) is therefore available for resisting the wind load when columns have been designed for vertical load only, before it becomes necessary to increase the column size in order to resist loads. This gives a resistance moment of $M = fS$.

$$M = \tfrac{1}{3} \times 165 \times 11.82 = 650.1 \text{ MN m}$$

$$M = \tfrac{1}{3} \times 24 \times 733\,800 = 5.870 \times 10^6 \text{ kip in.}$$

The total permissible wind load is $W = 2M/H$.

$$W = \frac{2 \times 650}{120} = 10.83 \text{ MN}$$

$$W = \frac{2 \times 5.87 \times 10^6}{400 \times 12} = 2446 \text{ kips}$$

The surface area of one facade of the building is 30 m by 120 m (*100 ft by 400 ft*), so that the permissible wind load per unit area is

$$w = \frac{10.83 \text{ MN}}{30 \text{ m} \times 120 \text{ m}} = 3.01 \text{ kPa}$$

$$w = \frac{2446 \text{ kips}}{100 \text{ ft} \times 400 \text{ ft}} = 61 \text{ psf}$$

The load per unit area w includes both the wind pressure on the windward facade of the building and the wind suction on the leeward facade. A wind load of 3 kPa (*61 psf*) is produced by a wind with a mean velocity of approximately 70 meters per second (*155 miles per hour*).

Evidently perimeter columns carrying half the vertical load can be designed to resist the entire wind load.

References

8.1 BUILDING CODE REQUIREMENTS FOR REINFORCED CONCRETE (ACI 318–77). American Concrete Institute, Detroit, 1977. 103 pp.

8.2 DESIGN HANDBOOK IN ACCORDANCE WITH THE STRENGTH DESIGN METHOD OF ACI 381–71 by the American Concrete Institute, Publication SP–17(73), *Volume 1*. American Concrete Institute, Detroit, 1973. 403 pp.

8.3 CODE OF PRACTICE FOR THE STRUCTURAL USE OF CONCRETE, *Part 2, Design Charts for Singly Reinforced Beams, Doubly Reinforced Beams and Rectangular Columns,* CP 110: Part 2: 1972. British Standards Institution, London, 1972. 90 pp.

8.4 AUSTRALIAN REINFORCED CONCRETE DESIGN HANDBOOK—ULTIMATE STRENGTH, S.I. UNITS. Cement and Concrete Association of Australia, Sydney, 1976. 168 pp.

8.5 STRUCTURAL STEEL DESIGN, Second Edition, edited by Lambert Tall. Ronald Press, New York, 1974. pp. 469–479.

8.6 DESIGN OF BUILDING FRAMES by J.S. Gero and H.J. Cowan. Applied Science Publishers, London, and Halstead Press, New York, 1976. 498 pp.

8.7 COMPUTER METHODS IN STRUCTURAL ANALYSIS by H.B. Harrison. Prentice-Hall, Englewood Cliffs, New Jersey, 1973. 337 pp.

8.8 DYNAMICS OF STRUCTURES by R.W. Clough and J. Penzien. McGraw-Hill, New York, 1975. 634 pp.

Chapter 9 Foundations

I like work: it fascinates me. I can sit and look at it for hours. I love to keep it by me: the idea of getting rid of it nearly breaks my heart. (Jerome K. Jerome)

We examine the properties of the principal foundation materials; for large buildings it is important that these be ascertained by a site investigation. Individual footings are used for walls and columns in small buildings, but in large buildings these merge into combined footings and raft foundations. Some types of soil make it necessary to use piles in addition.

Raft foundations in deep basements give buoyancy to buildings, which increases their loadbearing capacity. However, this produces horizontal forces in the walls of the basement, for which additional reinforcement must be provided.

Finally, we consider some special foundation problems for tall buildings and for long-span buildings.

9.0.1.

9.1 Foundation Materials

The designer whose building is founded on rock is fortunate; unless the rock is badly weathered or fissured, the architect has nothing to worry about. However, if the foundation material is a cohesive soil, its bearing capacity and consolidation characteristics require careful consideration.

Soils containing organic material should always be removed, because the organic material may decompose and cause subsidence. Such soils are usually confined to a surface layer.

The chemical composition of soil particles is not otherwise significant. The difference between soils is mainly due to their particle size. We usually divide them into seven groups:

Boulders:	greater than 60 mm	
Gravel:	60 mm to 2 mm	
Coarse sand:	2 mm to 0.6 mm	granular soils
Medium sand:	0.6 mm to 0.2 mm	
Fine sand:	0.2 mm to 60 μm	
Silt:	60 μm to 2 μm	cohesive soils
Clay:	less than 2 μm	

Fine sand and silt are intermediate in their properties between granular and cohesive soils, although classified, respectively, as a granular soil and a cohesive soil.

All soils contain voids occupied by air and water. The voids in clay can amount to more than half its total volume. In sand and gravel the voids are smaller because the particles are more tightly packed.

All soils contain some water. This may be stationary, or it may flow slowly through the soil across the site; the latter does no harm. Lowering of the water table can cause settlement in buildings, particularly if the foundations are inadequate. In Europe some medieval buildings were damaged during the late nineteenth and early twentieth centuries, after having stood safely for centuries. This was caused by excessive pumping of water from the ground, which resulted from the increased demand for water after the industrial revolution.

Settlement can also occur without any lowering of the water table, because water is squeezed out of the pores of the soil by the weight of the building. This is a particular problem in clay. In granular soils the particles usually touch one another, so that only a limited amount of compression occurs. Furthermore, the water moves with relative freedom through the large pores in sand, so that the compression occurs almost as soon as the load is placed on the soil.

The pores in clay are much finer, and they behave like capillary tubes (see Glossary) through which water can only flow slowly. Thus the *consolidation* (see Glossary) of clays may take many years. Furthermore, the particles of clay do not always touch one another; because they are so fine, they are sometimes kept apart by electrostatic forces. In such cases the amount of consolidation that eventually occurs is much larger. The compressibility of clays can be determined by a consolidation test, carried out on a sample of soil taken during a site investigation.

Uniform settlement of the entire building is not necessarily harmful, although it obviously should be avoided. The Monadnock Building, built in 1891 in Chicago, whose city center is founded on clay, was expected to settle because of its heavy masonry. It was therefore built with its ground floor 200 mm (8 in.) above street level. In 1905 the *Journal of the Western Society of Engineers* reported that "it had settled that and several inches more." The total settlement is by now about 0.6 m (2 ft); the building is still in use. However uneven settlement often leads to the destruction of the building (Fig 2.5.3).

Settlement due to consolidation occurs only under heavy buildings. However, moisture movement of clay can also be troublesome for quite small buildings, such as single-story houses. Certain clays expand appreciably in wet weather and shrink again in hot, dry weather, thus leaving the foundation partly unsupported. Where such clays occur it is necessary to place the foundations on piers or bored piles that penetrate the layer of clay liable to this moisture movement.

Ice occupies a larger volume than the same mass of water, and frost may therefore cause the soil to heave. It is then necessary to take the foundation below the level to which the soil is liable to freeze. This is why basements are common in small buildings in the northeastern United States and in northeastern Europe but rare in Great Britain and Australia.

In parts of northern Canada and Alaska the frozen soil does not melt completely during the summer. Not many buildings are erected in permafrost regions, but their foundations present special problems (Ref. 9.1).

9.2 Soil Pressure

Quite dramatic failures of soil can occur in embankments or dams by a slip down-hill (Fig. 9.6.5). However, a foundation material can only heave upwards (Fig. 9.2.1), and this type of failure can be prevented by keeping the soil pressure at a safe level.

A weakness in the superstructure is often detected in time and can be repaired. Deficiencies in the foundation are less easily noticed, and they are difficult to remedy (Fig. 9.2.2). Foundations are therefore designed with a substantial margin of safety.

For large buildings thorough site investigations are essential. The properties of the foundation material, including the strength of the soil, can then be determined from samples in the laboratory. A particularly careful investigation is needed if the foundation contains clay.

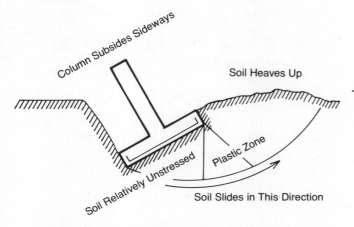

9.2.1. The shear failure of soil generally occurs along a curve. The soil under the footing forms a wedge that produces plastic flow on one side and causes the soil above to heave upwards. As a result the footing is tilted.

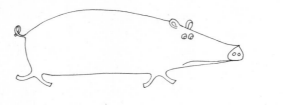

9.2.2. Deficiencies in the foundation are not easily noticed.

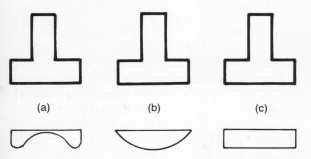

9.2.3. Distribution of soil pressure under a concentrically loaded footing, (a) in an elastic, cohesive soil, (b) in a granular soil, and (c) as assumed in most design calculations.

The pressure in a cohesive soil is highest near the edge of the footing, whereas in a granular soil it is highest under the wall or column. It is generally assumed that the soil pressure is uniform (Fig. 9.2.3). If the load to be transmitted to the foundation is P and the permissible bearing pressure is f, we require a base of area $A = P/f$ (Fig. 9.2.4).

9.3 Footings for Walls and Single Columns

Footings are almost invariably built of brick, natural stone, plain concrete, or reinforced concrete. Steel grillage foundations, consisting of layers of steel beams at right angles to one another, are now rarely used.

The reinforcement in concrete needs at least 65 mm (2½ in.) of cover; more is needed if the soil contains corrosive material. Plain concrete is often sufficient for small buildings founded on a strong foundation material.

The slab of a footing for a wall behaves like a cantilever (Fig. 9.3.1). If the tensile stress due to the bending moment caused by the overhang of the slab (Section 5.8), which is

$$f = \frac{M}{S}$$

exceeds the maximum permissible tensile stress, we must use either a thicker slab or reinforcement (Fig. 9.3.2). The reinforcement is then calculated from Eq. (5.11), in Section 5.8.

A footing for a single column behaves similarly, except that it bends in two directions (Fig. 9.3.3) and main reinforcement must be provided in both directions (Fig. 9.3.4).

In addition, columns may punch through the foundation slab (Fig. 9.3.5); this is the reverse of a shear failure in a flat slab (Fig. 5.5.5). Shear reinforcement (Section 6.1) is uneconomical in single-column footings. It is better to increase the thickness of the slab.

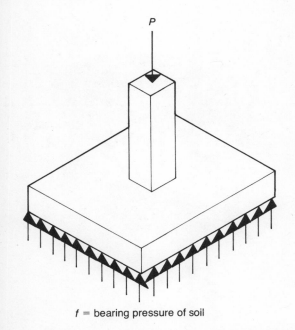

f = bearing pressure of soil

9.2.4. The footing requires an area $A = P/f$ in contact with the foundation material, if the load on the footing is P and the permissible bearing pressure is f.

For small buildings a permissible bearing pressure, based on a general soil classification, is sufficient. This is usually specified in building codes.

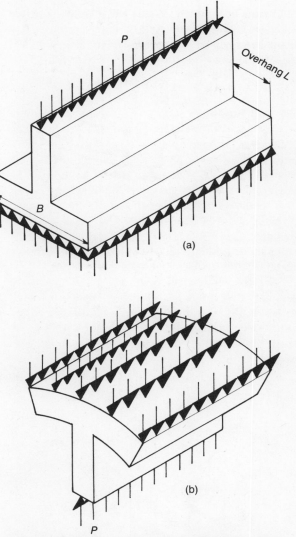

9.3.1. Plain concrete foundation (a) for a loadbearing wall. If the load is P per meter run of wall and the permissible foundation pressure is f, we require a width of footing $B = P/f$. The overhang of the foundation slab sets up a cantilever moment $M = \frac{1}{2} f L^2$ per meter run of the concrete slab, which is more easily visualized if we turn the foundation upside down (b), when it is seen to be a double cantilever slab, as in Fig. 4.5.2; the soil pressure is the "load," and the load on the wall P is the "reaction" of the upturned cantilever.

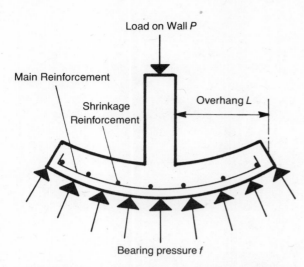

9.3.2. Reinforced concrete foundation for loadbearing wall, with deformation of foundation slab shown grossly exaggerated. The main reinforcement is designed for a cantilever moment $M = \frac{1}{2} f L^2$ per meter run.

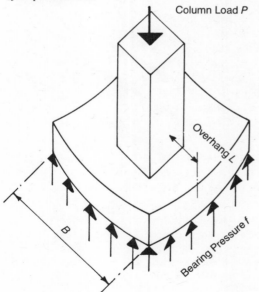

9.3.3. Foundation for column load P, with deformation of foundation slab shown grossly exaggerated. If the permissible bearing pressure is f, we require an area $A = B^2 = P/f$. This gives the overhang L, which determines the cantilever bending moment in the slab.

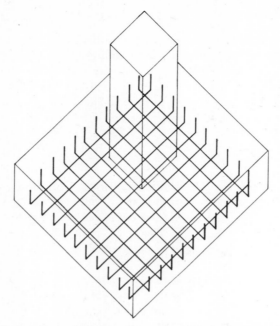

9.3.4. The foundation slab of the column footing bends in both directions, and main reinforcement is required in both directions.

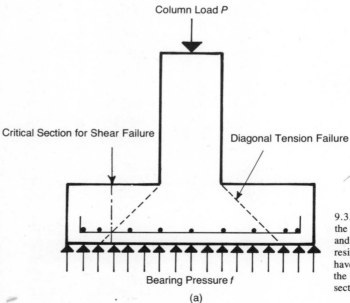

(a)

For a reinforced concrete column footing, the bottommost column-reinforcing bars are cast into the slab; the column-reinforcing cage is then spliced to this reinforcement (Fig. 9.3.6).

A steel column is fitted with a base plate, which is usually a thick plate of steel welded to the bottom of the column. Bolts are cast into the concrete slab, and these pass through holes in the base plate (Fig. 9.3.7).

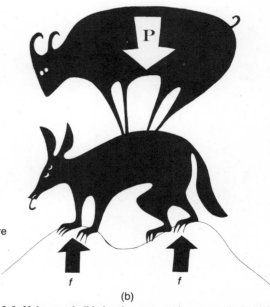

(b)

9.3.5. If the permissible bearing pressure is comparatively high, the cantilever overhang of the foundation slab is relatively small, and therefore the depth of the slab required for flexural resistance is also small. Under those circumstances the slab may have inadequate shear resistance. If the column punches through the slab, it produces a diagonal tension failure; the critical section for shear is therefore not at the column face.

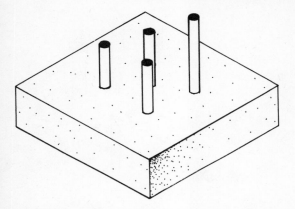

9.3.6. The reinforcing bars for the base of the column are cast into the foundation slab, and the reinforcing cage is then spliced to these projecting bars.

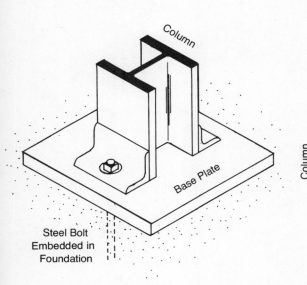

9.3.7. Bolts for fixing the steel column are cast into the foundation slab, and these pass through holes in the column base plate.

9.4 Combined Column Footings, Raft Foundations, and Piled Foundations

If a building extends to the boundary of the site, there may not be space for the cantilever slab of a single-column footing. It may not be advisable to seek permission to extend the column footing beyond the boundary, or it may not be possible to do so because the neighboring building extends to the boundary. It is then necessary to combine the footing of two columns (Fig. 9.4.1).

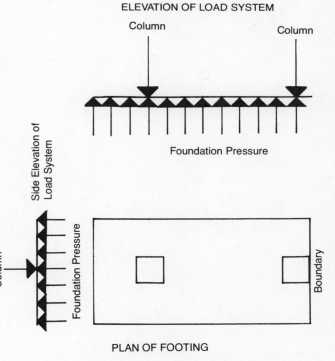

9.4.1. Combined footing for two columns, one close to the boundary. Because there is insufficient room for the cantilever slab of a single-column footing, two columns are placed in a combined footing. The distribution of bending moment in the footing slab is similar to that of a simply supported beam with a cantilever overhang (Fig. 4.6.7) turned upside down.

As the building gets taller, the loads that must be transmitted to the foundation material get bigger and bigger. There is no escape from the fact (Fig. 9.4.2) that the entire weight of the building and the load acting on it must be transmitted to the ground (see Sections 8.2 and 8.3, Eq. 8.2). The complexity of the foundation depends both on the height of the building and on the quality of the foundation material. A ten-story building on rock may need a less elaborate foundation than a five-story building on a compressible clay.

Individual column footings are suitable only for low-rise buildings on high-strength foundation material. As the cantilevers of the footings get longer, they become less efficient, and it is more economical to join them into a continuous beam. This is called a *combined footing,* or *strip footing* (Fig. 9.4.3).

As the strips become wider in order to meet the need for sufficient area to transmit the column loads to the foundation material, it becomes economical to joint them into a single foundation slab. This is called a *raft foundation* (Fig. 9.4.4).

Most tall buildings need raft foundations. In addition, they need basements whose walls also retain the soil (Section 9.5). This closed basement box displaces an appreciable amount of foundation material. This has no significant effect on the bearing capacity of rock, which is a solid material, but it increases the bearing capacity of soil.

Soil behaves in some respects like a sticky liquid (Section 9.5). The displacement of the soil in the basement therefore creates a buoyancy for the building that *reduces* the pressure on the foundation (Fig. 9.4.5); but we would need an uneconomically large basement to float the entire building on its raft foundation.

If the bearing pressure is still too high, it is necessary to use piles. Piles can be used for individual columns (Fig. 9.4.6) or under strip or raft foundations.

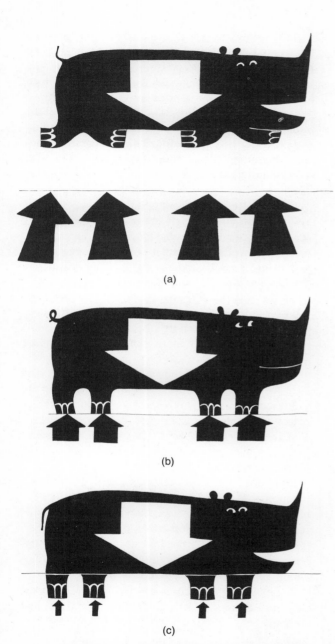

9.4.2. The entire weight of the building and the loads acting on it must be transmitted to the ground.

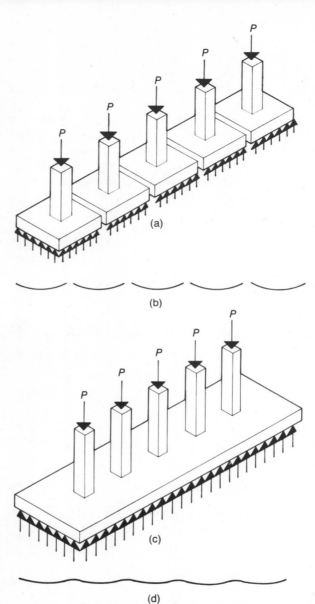

9.4.3. The deflection and the bending moments of a strip footing are lower than those of individual column footings (Fig. 9.3.3) of the same area.
(a) Individual column footings and their deflected shape (b), grossly exaggerated.
(c) Strip footing for the same columns and its deflected shape (d), grossly exaggerated.

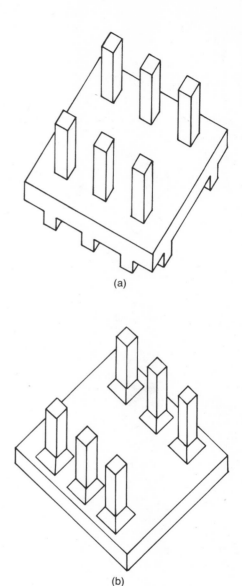

9.4.4. The slab of a raft foundation covers the entire ground surface under the columns. It can be stiffened by beams (a) or by enlarged column heads (b). A raft foundation with beams is a reinforced concrete two-way slab turned upside down (Fig. 5.5.7.a). A raft foundation without stiffening beams is a flat slab turned upside down (Fig. 5.5.3).

Timber piles, which were already employed by the ancient Romans, are still used today; but piles for large buildings are of concrete. They are either precast reinforced or prestressed concrete (Section 5.6), or cast-in-place plain or reinforced concrete.

Precast piles are driven into the ground with a pile driver, which uses a weight dropped from a height or a power-operated hammer. None of the soil is removed; the soil is therefore compressed by the pile-driving operation.

Alternatively, we bore a hole in the ground, remove the soil, and replace it with site-cast plain or reinforced concrete. The hole can be bored with an auger (a big "corkscrew"), or a steel casing can be driven into the ground with a pile driver. The casing can be left in the ground (which produces a stronger pile) or withdrawn (which produces a cheaper pile). In either case the concrete can be rammed into the hole to give it a bulbous foot, which increases the bearing capacity (Fig. 9.4.7).

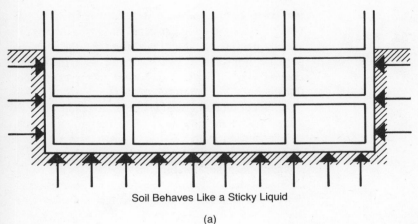

Soil Behaves Like a Sticky Liquid

(a)

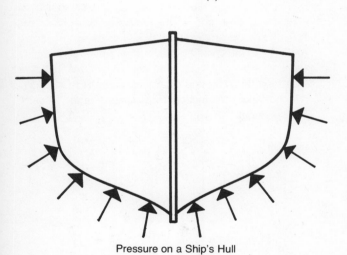

Pressure on a Ship's Hull

(b)

9.4.5. A ship floats because the weight of the volume of water displaced equals its own weight. The line to which a ship settles in the water depends on its weight and the weight of its contents.

A raft foundation (a) differs from a ship (b) in two respects. Water has no bearing pressure, whereas soil has. Consequently it is not necessary to displace a weight of soil sufficient to ensure flotation of the building. Second, the buoyancy in soil is less because the horizontal pressure of soil is less than that of water (Section 9.5). However, the buoyancy of the raft foundation makes a useful and important contribution by reducing the load transmitted to the foundation.

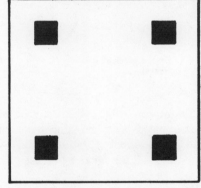

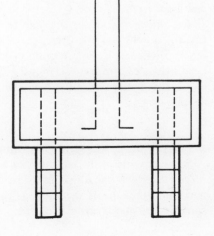

9.4.6. Pile cap for single reinforced concrete column supported on four piles. The pile cap transmits the load from the column to the four piles. The bottommost bars of the column reinforcement are cast into the pile cap, and the column-reinforcing cage is spliced to them.

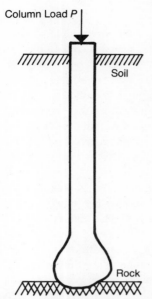

9.4.7. Site-cast pile with enlarged foot, produced by ramming the wet concrete into the bored hole, after the casing has been partly withdrawn.

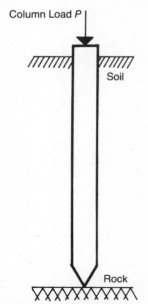

9.4.8. Precast pile driven to solid rock, so that it becomes a loadbearing pile.

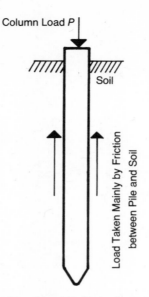

9.4.9. Precast pile supported by friction between the surface of the pile and the soil.

Endbearing piles, which can be either precast or cast-in-place, are taken through the entire layer of soil down to solid rock (Figs. 9.4.7 and 9.4.8). This is the safer method, but it cannot be used if the layer of soil is very deep.

Precast friction piles rely for their loadbearing capacity on the friction between the surface of the pile and the surrounding soil (Fig. 9.4.9).

The penetration of the pile into the soil during the final blow with the pile driver gives an indication of the bearing capacity of a friction pile, but there have been instances of piles refusing to go any deeper and later settling under a smaller load than that exerted by the pile driver.

9.5 Retaining Walls and Basement Walls

Soil has some of the attributes of a liquid. A mass of soil exerts a horizontal pressure. In a liquid the horizontal pressure is the same as the vertical pressure. In soil the horizontal pressure is less than the vertical pressure; the ratio depends on the physical properties of the soil. For sand and gravel the horizontal pressure is about one third of the vertical pressure; for clay it varies greatly with the water content. As in a liquid, the pressure increases proportionately with the depth (Fig. 9.5.1).

A dam or retaining wall is needed to retain water at a level above the general level of the ground. Rock can be cut vertically and does not need a retaining wall unless it is weathered. Soil can be graded to a stable slope, whose inclination depends on the particle size of the soil and its water content; alternatively, if we wish to make an abrupt change in the level of the soil, we can use a retaining wall (Fig. 9.5.2).

The wall of the basement of a building also acts as a retaining wall, and it must be designed to resist the horizontal pressure exerted by soil or by waterbearing rock. The wall must be made strong enough to resist the horizontal pressure, but the foundation benefits from the buoyancy of the basement (Fig. 9.4.5), which reduces the loadbearing capacity of the floor of the basement.

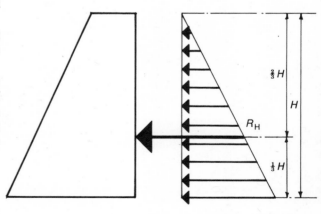

9.5.1. The soil pressure behind a retaining wall increases proportionately with depth. The distribution of pressure is therefore triangular, and the resultant horizontal pressure R_H acts at the center of gravity of the triangle, that is, at a depth $\frac{2}{3}H$.

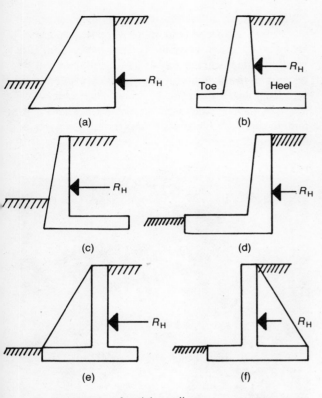

9.5.2. Various types of retaining wall:
(a) Gravity retaining wall built of plain concrete or masonry.
(b) Normal reinforced concrete retaining wall. The vertical cantilever, which retains the soil, is joined to two other cantilevers, called the heel and the toe. The soil on the heel weighs down the structure and thus increases its stability. The toe increases the moment required to overturn the wall (see Fig. 9.6.3.).
(c) Reinforced concrete retaining wall without toe. This is used if the soil on the left of the retaining wall is part of another property.
(d) Reinforced concrete retaining wall without heel. This is the normal type of retaining wall for a basement (Fig. 9.4.5.a).
(e) Reinforced concrete retaining wall with buttresses. As the height of the wall increases, the bending moment in the cantilever also increases, and additional depth is needed. The buttress provides this depth.
(f) Reinforced concrete retaining wall with counterforts. The buttresses are unsightly and obstruct the area in front of the wall. Counterforts do not raise these objections, but as the soil pressure pushes the vertical wall away from the counterforts, tension reinforcement must be provided.

9.6 Stability of Tall Buildings

Buildings and towers can overturn either as a result of uneven settlement (Fig. 9.6.1) or because the resultant force falls outside the base. The latter is so gross an error that it has probably never happened, but it is at least theoretically possible. If the foundation has adequate strength and does not settle, the resultant of the foundation pressure should lie within the middle third of the foundation to avoid uplift (Fig. 9.6.2). However, the building does not actually overturn until the resultant falls outside the foundation (Fig. 9.6.3). The remedy is simple. The foundation can be widened to bring the resultant within the middle third, or it can be deepened to give a moment resistance to the vertical cantilever of the tall building (Fig. 4.8.2).

9.6.1.

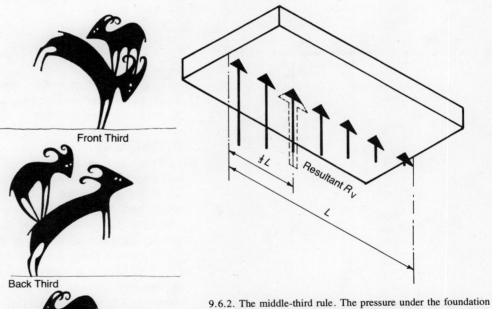

9.6.2. The middle-third rule. The pressure under the foundation varies uniformly. The limit is reached when the pressure at one side becomes zero. The pressure distribution is then triangular, and the resultant of the soil pressure acts at the center of gravity of the triangle, that is, at $\tfrac{1}{3}L$. The resultant can therefore approach no further to the edge of the foundation than $\tfrac{1}{3}L$; that is, it should lie within the middle third of the foundation. If the resultant falls outside the middle third, one edge of the foundation does not bear on the ground.

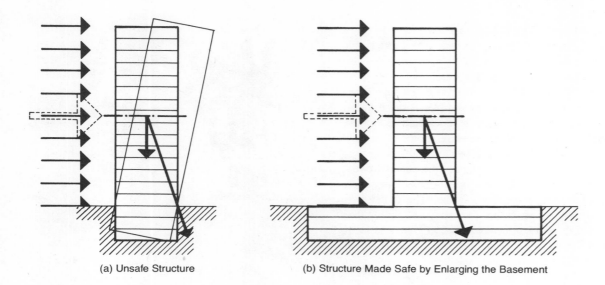

(a) Unsafe Structure

(b) Structure Made Safe by Enlarging the Basement

9.6.3. A tall building overturns under the action of the horizontal wind pressure if the resultant of that pressure and of the vertical load on the building falls outside the base. However, uplift of the foundation already occurs if the resultant falls outside the middle third.

As we noted in Section 9.2, it is much easier for soil to fail on a slope than on a level site. Landslides are quite common, particularly after heavy rain. If a small building is in the way of the landslide, it will be taken with the slide unless very firmly anchored with wooden piles. In developing countries accidents of this type happen frequently. In localities where high rainfall and unstable soil occurs, the only answer is not to build on steep slopes.

A more likely cause of trouble is uneven settlement of the foundation (Section 9.1). This was not uncommon in towers built before the nineteenth century. Three famous Italian examples are the campanile of S. Marco in Venice, built in the sixteenth century, which collapsed in 1902; the La Garisenda Tower in Bologna, built in the twelfth century, which was partly dismantled in the fourteenth century because it had tilted dangerously, but is still standing; and the Leaning Tower of Pisa, built in the thirteenth century, which survives intact (Fig. 9.6.4). Collapse due to uneven settlement still occurs. During this century a number of grain silos and at least two apartment buildings have collapsed for that reason.

The problem arises only if either the properties of the foundation material vary across the building site (which can be ascertained by a site investigation) or the weight of the building is nonuniform (Fig. 2.5.3). Furthermore, it only happens in buildings founded on highly compressible soil. Uniform settlement does not endanger the safety of a building, although the change in level may cause inconvenience (Section 9.1).

9.6.4. The Leaning Tower of Pisa, started in 1174 and completed in 1271, has a height, excluding the fourteenth-century belfry, of 46 m (151 ft 3 in.). It is now 4.2 m (13 ft 10 in.) out of plumb and leans at an angle of 5°15′.

9.7 Foundations for Long-Span Buildings

Most long-span structures are curved (Chapter 10), and consequently they produce horizontal reactions (Figs. 4.7.2 and 4.7.4). Transmitting these horizontal reactions to the foundations and absorbing them there was one of the great problems in the design of the monumental buildings of previous centuries, and it had a pronounced influence on their appearance, particularly in Gothic architecture. The problem has become much simpler since the development of reinforced concrete in the late nineteenth century, but the abutments of a great bridge are still one of the more expensive parts of the structure.

We cannot tie the abutments together in most arch or suspension bridges, because there is generally a stretch of water or a deep gap between them. However, we can absorb the reactions in many curved long-span buildings, and this greatly simplifies construction and reduces the cost (Figs. 9.7.1 and 10.9.1).

9.7.1. A curved structure produces the usual vertical reactions, R_V, which are easily absorbed by the ground. In addition it produces horizontal reactions, R_H, which result from the tendency of the structure to spread. In a bridge (a) these require expensive abutments. In many modern buildings (b), however, the horizontal reactions occur near floor level, and they can be absorbed by a tie within the floor. This obviates the need for abutments. The tie often takes the form of a cable of high-tensile steel in a duct in the floor structure. Because the cable must be tight even in warm weather, it has to be slightly prestressed to allow for temperature expansion.

A prestressed tie in the floor structure has one other important advantage. It can be tightened to lift the roof structure off its formwork, which facilitates one of the most difficult tasks in the construction of a long-span structure (Section 10.9).

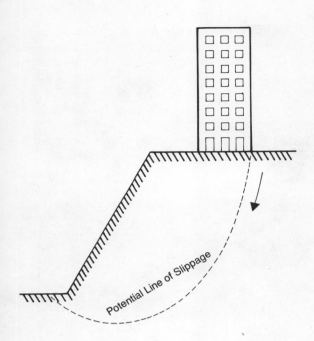

9.6.5. The likelihood of failure of a steep slope due to the weight of a heavy building is increased in regions of high rainfall (which are mostly in the tropics). Good drainage reduces buildup of pore-water pressure in the soil, but piling or reinforcement of the soil with mats is useful only if it cuts through the line of potential slip.

The stability of the slope can be checked by soil mechanics. If there is any risk of a landslide due to the weight of the building, it is advisable to refuse permission to build on the site and use it for a park.

Slopes can be dangerous even to large buildings. In Hong Kong, in 1972, a high-rise luxury apartment building slid down and collided with another building, with great damage to both.

This type of failure can occur not merely when a building is on a slope but also when it is on flat ground adjacent to a slope (Fig. 9.6.5).

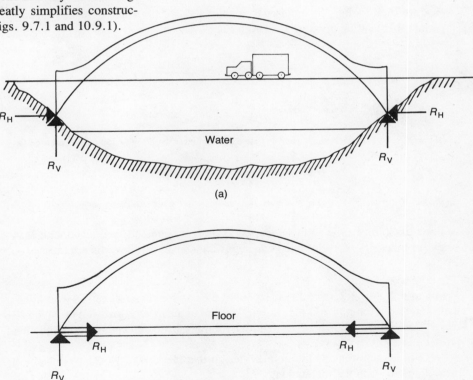

175

9.8 Problems for Chapter 9

If you have not yet read Section 5.9, please do so before attempting the following problems.

We determine the minimum sizes for a wall footing for a two-story house and an individual column base for a three-story building. Next we determine the height at which strip footings cease to be economical and a raft foundation should be used for a building with the same floor plan. We then examine the horizontal pressure on the wall of a basement. Finally, we determine the width required for the basement of a building to ensure its stability under wind pressure.

A method for absorbing the horizontal reactions of an arch in the foundation was considered in Problem 7.8.

Problem 9.1 A loadbearing wall in a two-story building 300 mm *(12 in.)* thick at its base carries a load of 90 kN/m *(6 kips/ft)*. Determine the width of the footing required if the maximum permissible bearing pressure of the soil is 150 kPa *(3000 psf)*.

The width of footing, B, required to support a load P per unit length on a soil with a permissible bearing pressure f is (from Fig. 9.3.1)

$$B = \frac{P}{f}$$

$$B = \frac{90 \text{ kN/m}}{150 \text{ kPa}} = 0.6 \text{ m} = 600 \text{ mm}$$

As the wall is 300 mm thick, this means that there is a cantilever overhang of 150 mm on each side.

$$B = \frac{6 \text{ kips/ft}}{3000 \text{ psf} \times 10^{-3}} = 2 \text{ ft}$$

As the wall is 1 ft thick, this means that there is a cantilever overhang of 6 in. on each side.

For a brick or block wall this widening is most conveniently produced with extra bricks or blocks. For a reinforced concrete wall a small cantilever is appropriate.

Problem 9.2 Fig. 9.8.1 shows the layout of the columns for a three-story building. Each interior column is 500 mm *(20 in.)* square and carries a load of 900 kN *(210 kips)*. Determine the size of its column footings, if the permissible foundation pressure is 240 kPa *(5000 psf)*.

We will assume for this preliminary design that the column does not transmit a bending moment to the foundation.

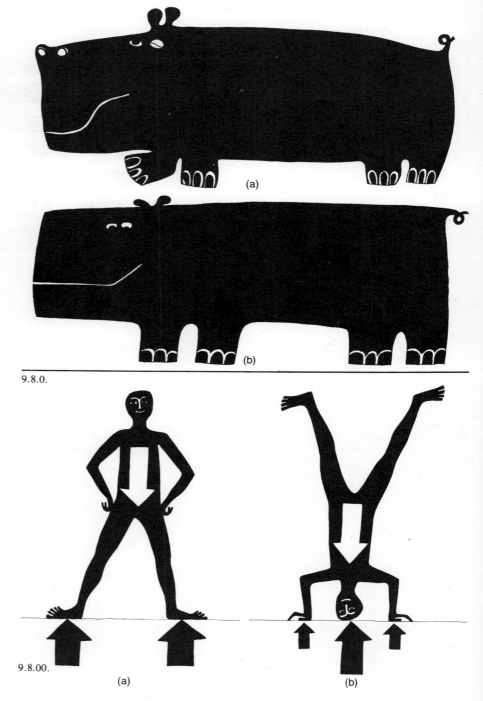

9.8.0.

9.8.00.

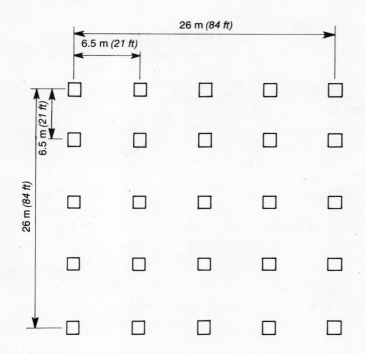

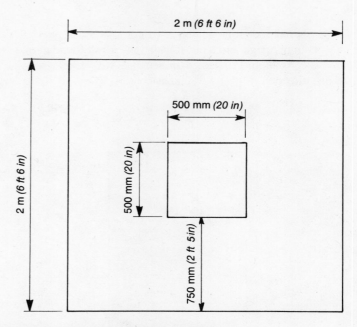

9.8.1. Layout of the columns for Problems 9.2, 9.3, and 9.4. The plan of the superstructure is shown in Fig. 5.11.4.

9.8.2. Dimensions of single-column cantilever footing (Problem 9.2).

From Fig. 9.3.3 the column load is

$$P = fB^2$$

Thus

$$B^2 = \frac{900}{240} = 3.75 \text{ m}^2$$

$$B^2 = \frac{900}{240} = 3.75 \text{ m}^2$$

We will make the footing 2 m square.

$$B^2 = \frac{210}{5000 \times 10^{-3}} = 42 \text{ ft}^2$$

$$B = 6.481 \text{ ft}$$

We will make the footing 6 ft 6 in. square.

Its plan is shown in Fig. 9.8.2. The reinforced concrete slab of the footing has a cantilever span of 750 mm *(2 ft 5 in.)*.

Problem 9.3 The footing of the columns in each of the five rows of the building shown in Fig. 9.8.1 is to be combined into strip footings not more than 3 m *(10 ft)* wide. Determine the highest number of stories for which these strip footings are suitable if each interior column carries a load of 300 kN *(70 kips)* per floor and each exterior column carries half that load.

The maximum permissible foundation pressure is 240 kPa *(5000 psf)*. Each floor contributes a load of $(3 + 2 \times \frac{1}{2}) P$ (Fig. 9.8.3), so that the total load due to n stories transmitted to the foundation by the column is

$$4np = fbB$$

Thus

$$4n \times 300 \text{ kN} = 240 \text{ kPa} \times 3 \text{ m} \times 26 \text{ m}$$

$$n = 15.6$$

$$4n \times 70 \text{ kips} = 5000 \text{ psf} \times 10^{-3} \times 10 \text{ ft} \times 84 \text{ ft}$$

$$n = 15$$

Strip footings should therefore not be used for this floor plan above fifteen stories, as they begin to merge into a raft foundation. A raft foundation, continuous over the entire site, is more efficient than separate strip footings, which act as cantilevers in one direction (Fig. 9.4.3) and as continuous beams in the other.

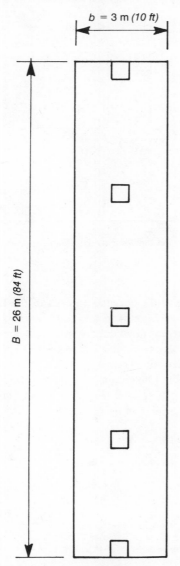

9.8.3. Dimensions of strip footing for five columns (Problem 9.3).

Problem 9.4 Fig. 9.8.4 shows a retaining wall forming part of the basement of a multistory building. Determine the magnitude of the horizontal pressure on the basement wall.

We will assume that the material adjacent to the basement is a granular soil (it need not necessarily be the original foundation material but could be a backfill placed after completion of the basement). In that case the soil pressure increases linearly with depth, as shown in Fig. 9.5.1. The resultant horizontal pressure acts at a depth of

$$\tfrac{2}{3} H = 4 \text{ m}$$

$$\tfrac{2}{3} H = 13 \text{ ft } 4 \text{ in.}$$

The pressure exerted by water is the same in all directions. If the weight of water per unit volume is w_v, the pressure exerted by it at a depth H is

$$w_v H$$

The average pressure over the depth H is half that amount, $\tfrac{1}{2} w_v H$.

The resultant force per unit length of wall is

$$R_H = \tfrac{1}{2} w_v H \cdot H = \tfrac{1}{2} w_v H^2$$

The equivalent fluid weight of a soil can be determined by a test. It is a measure of the horizontal pressure it exerts. For granular soils it is approximately one third of its actual weight.

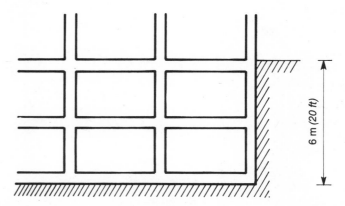

9.8.4. Dimensions of basement retaining wall (Problem 9.4).

We will assume that the equivalent fluid weight of the soil retained by the basement wall is 4.8 kN/m³ *(30 lb/ft³)*.

The resultant force per unit length of wall is

$$R_H = \tfrac{1}{2} w_v H^2$$

$$R_H = \tfrac{1}{2} \times 4.8 \times 6^2 = 86.4 \text{ kN/m}$$

$$R_H = \tfrac{1}{2} \times 30 \times 20^2 = 6000 \text{ lb/ft} = 6 \text{ kips/ft}$$

Problem 9.5 Fig. 9.8.5 shows the enlarged basement of a multistory building. The vertical dead load, including the weight of the building, is 500 kN per meter run of building *(34 kips per foot run of building)*, and the wind pressure is 1 kPa *(20 psf)*. Determine the minimum width B of basement required if the resultant of the dead load and wind load is to lie within the middle third of the bottom of the basement.

The resultant wind force per unit length of building is

$$P = 1 \text{ kPa} \times 60 \text{ m} = 60 \text{ kN/m}$$

$$P = 20 \text{ psf} \times 200 \text{ ft} = 4000 \text{ lb/ft} = 4 \text{ kips/ft}$$

As Fig. 9.8.6 shows, the triangle formed by the forces W and P and their resultant is similar to a triangle formed by the center line and base line of the building and the resultant striking the bottom of the basement. If this is to be at one end of the middle third, then

$$\frac{\tfrac{1}{6} B}{\tfrac{1}{2} H + h} = \frac{P}{W}$$

$$B = \frac{6 P (\tfrac{1}{2} H + h)}{W}$$

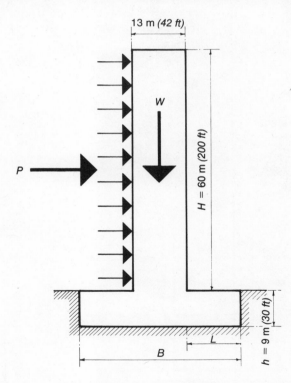

9.8.5. Dimensions of enlarged basement for tower block. The floor plan of the superstructure is shown in Fig. 5.11.4.

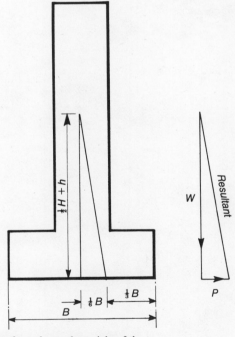

9.8.6. The resultant of the force due to the weight of the building, W, and the force due to the wind, P, must fall within the middle third of the base of the building. The edge of the middle third is one sixth of the length of the base B from the center of the building.

$$B = \frac{6 \times 60\,(30 + 9)}{500} = 28.08 \text{ m}$$

The base must project at least 7.54 m beyond the facade of the tower block.

$$B = \frac{6 \times 4\,(100 + 30)}{34} = 91.77 \text{ ft}$$

The base must project at least L = 24.89 ft beyond the facade of the tower block.

Reference

9.1 PERMAFROST AND BUILDINGS by J. A. Pihlainen. *Better Building Bulletin* No. 5, Division of Building Research, National Research Council of Canada, Ottawa, 1955. 27 pp.

Chapter 10 Curved Structures and Long-Span Buildings

A mere copier of nature can never produce anything great. (Sir Joshua Reynolds, Discourses)

We examine the various types of structure suitable for curved roofs and for long-span roofs.

Because we are working near the limit of what is possible, long-span roofs are often curved, to give the greatest possible lever arm to the resistance moment; but curved structures produce thrusts, which must be transmitted to the ground. Domes solve both problems, and that is why the longest-spanning structures, both past and present, are domes.

Newer structures, notably suspension roofs, space frames, and membranes supported by air pressure, are now challenging the supremacy of domes for long spans.

10.1 The Simple Theory of Vaults and Domes

Vaults and domes rank with lintels (Section 5.4), trusses (Section 7.3), and arches (Section 7.7) as the great classical forms of structure.

The cylindrical vault, continuously supported, is essentially an arch extended in the third dimension. Hence its correct shape for pure compression is an upturned catenary. The semicircular vault (Fig. 10.1.1), which is much easier to construct, develops some bending.

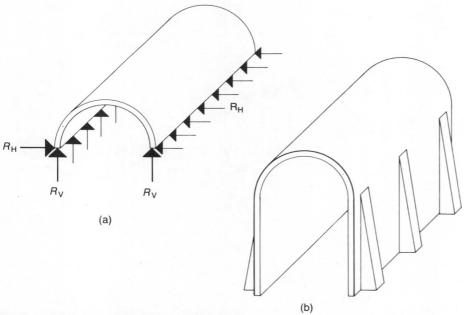

10.1.1. A semicircular vault is easier to construct than a moment-free catenary vault, but it develops some bending moment in addition to the compressive force (Section 7.7.).

Vaults have horizontal reactions, R_H, as well as vertical reactions, R_V (a). The vertical reactions are easily absorbed by loadbearing walls. In traditional masonry construction the horizontal reaction was absorbed by buttresses (b), unless the wall was so thick that it acted like a buttress.

In modern architecture it is simpler to construct a cylindrical shell roof in which the reactions are absorbed by ties between the columns (Section 10.3).

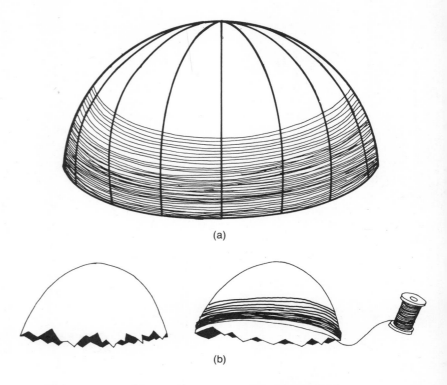

10.1.2. In a hemispherical dome carrying a vertical load the meridional forces are entirely compressive. The hoop forces are compressive in the upper portion of the dome and tensile in the lower portion (shown by the dotted rings). The change from compression to tension occurs at an angle $\theta = 51°50'$ from the crown (Fig. 10.1.3).

The internal forces in a hemispherical dome are divided into two sets at right angles to one another. It is convenient to use the familiar coordinates of the earth's surface (Fig. 10.4.1). The forces along the meridians (called *meridional forces*) are, in fact, like those in a series of arches intersecting at the crown; under vertical loading they are always compressive. The forces along the circles of latitude (called *hoop forces*) are compressive near the crown of the semicircular dome and tensile near the springings of the dome (Fig. 10.1.2).

The simple theory of domes is statically determinate (Section 7.2); that is, it utilizes only the equations of equilibrium (Section 4.4) in three dimensions. It assumes that the dome is thin relative to its diameter and that the edges are free to move; we will examine the limitations of these assumptions in Section 10.3. The mathematical derivation is given in a number of books on shells (for example, Ref. 10.1). Because the shell is curved, it is necessary to use differential geometry, but readers who are well acquainted with differential calculus should have no undue difficulty in following it; the solution is given in Fig. 10.1.3.

10.2 Problems in the Construction of Masonry Vaults and Domes

The Japanese found a way of overcoming the impermanence of timber: They rebuilt their most important structures in precisely the same form at regular intervals. However, most of the great buildings of Europe's past have either vaulted or domed roofs, because this was the only way of using durable materials over any but the shortest spans prior to the eighteenth century.

The great problem with vaulted roofs is the absorption of the horizontal reaction (Fig. 10.1.1). As we noted in Section 7.7, the adoption of the pointed arch greatly reduces the bending moment, and it also gives much greater freedom to vary the ratio of height to span, which in a semicircle must be 1:2 (Fig. 7.7.2.a); but there is still a horizontal reaction.

In Moslem architecture this was frequently absorbed by an iron or, over a small span, a wooden tie rod (Fig. 10.2.1.a). In medieval Italy, whose architecture was influenced by Moslem structures, ties were also used occasionally (Fig. 10.2.2).

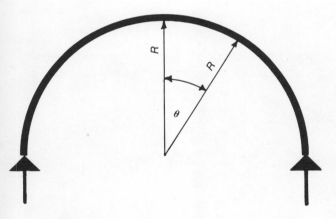

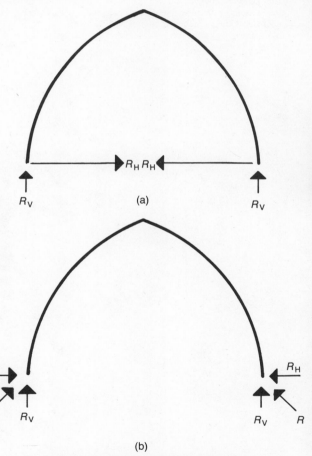

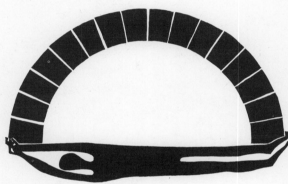

10.2.2. Tie rod for absorbing horizontal reactions.

10.1.3. Forces in a hemispherical dome.

Meridional forces: $N_\theta = -wR \dfrac{1}{1 + \cos\theta}$ (10.1)

Hoop forces: $N_\phi = wR \left(\dfrac{1}{1 + \cos\theta} - \cos\theta \right)$ (10.2)

The shear forces in the dome are zero throughout.

w = vertical load per unit area of dome
R = radius of curvature (or half the diameter or span of the hemispherical dome)
θ = angle subtended by the point of the dome under consideration with the crown

10.2.1. Vaulted construction produces horizontal reactions. These can be absorbed by a tie rod (a) or by buttresses (b).

In Roman and Romanesque architecture the walls were often so thick that they acted as buttresses. When the walls became much thinner during the Gothic era, buttresses that transferred the inclined reaction, R, to the ground, became a prominent feature of the great cathedrals of northern Europe (Figs. 10.2.1.b and 10.2.3).

The hemispherical dome has a great advantage over the vault; its reactions are purely vertical, and therefore there is no need for buttresses or for unsightly interior ties. The horizontal restraint is in fact provided by the hoop tension near the springings of the dome; this makes the structure self-contained (Fig. 10.1.2).

183

10.2.3. If the inclined reaction is brought to the ground within the middle third of the buttress, the masonry is entirely in compression (Fig. 9.6.2). The weight of the buttress itself greatly reduces its depth, because it turns the reaction from the vault closer to the wall.

The amount of masonry required for a buttress can be further reduced by using a flying buttress (c and d). This is particularly important when the inclined thrust, R, is large and high above the ground. The flying buttress transmits the thrust R to an outer buttress; it "flies" over an air space that would otherwise have to be filled with masonry for a single buttress.

In the nineteenth century, during the Gothic revival, the middle-third rule was known, but structural mechanics was still a comparatively new science, and designers were inclined to be careful; tension in masonry was not acceptable. Consequently most Neo-Gothic cathedrals were designed by the middle-third rule (d), and some medieval cathedrals were strengthened to conform to it.

The medieval masterbuilders did not know the middle-third rule, and many medieval buttresses do not satisfy it. A buttress does not in fact collapse as long as the resultant reaction remains inside it.

As medieval masonry joints had negligible tensile strength, the joints opened up as soon as tension developed. Since the construction of the cathedrals took a long time, sometimes centuries, this opening up was a slow process, and the joints were presumably filled by repointing during routine maintenance operations.

Many modern masonry building codes accept opening up of joints over one fifth of the depth (Fig. 8.9.3) as a result of wind load but not as a result of dead load. The horizontal thrust R of the Gothic roof is due partly to wind and partly to the thrust of the vault.

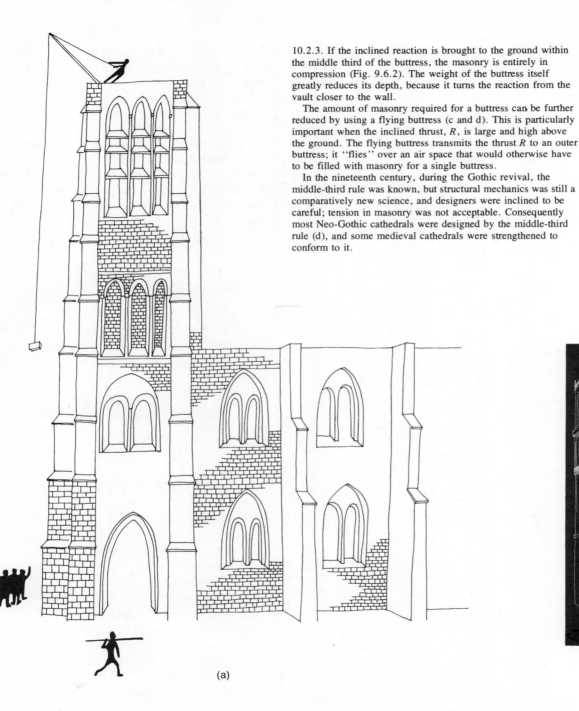

(a)

(b)

The Romans absorbed this hoop tension with a great thickness of material (Fig. 10.2.4). In the Duomo of Florence, Filippo Brunelleschi used chains to absorb it (Fig. 10.2.5), and chains were used in many subsequent masonry domes. In the nineteenth century iron domes, many of which were in fact intersecting arches, began to replace masonry domes (Section 7.7).

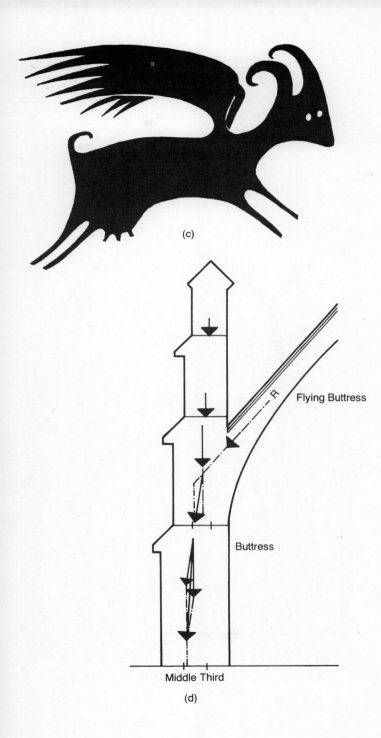

10.2.4. The Pantheon in Rome, completed in 123 A.D., has a hemispherical plain concrete dome 44 m (143 ft) in diameter. The dome is very thick at its springings (7 m or 23 ft), so that the tensile stresses in the concrete are low, even though the tensile *forces* are high. The dome begins to get thicker at approximately the angle where it is now known that the hoop forces change from compression to tension, so that the Romans probably had an empirical knowledge of that fact from previous experience with cracks in domes.

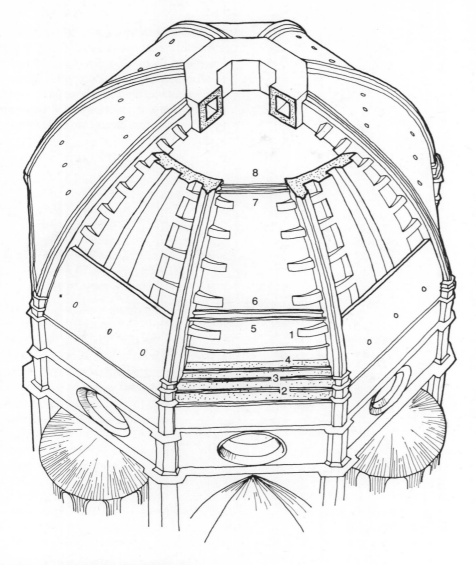

10.2.5. Brunelleschi's structural design for the Duomo of Florence (the Cathedral of S. Maria del Fiore) was particularly important because it was the first and largest of all the masonry domes of the Renaissance. Brunelleschi used a timber chain (1) and seven stone chains (2-8) to resist the hoop tension. The timber chain consists of logs of timber joined with iron bolts through fish plates. The stone chains consist of large blocks of very strong sandstone joined with iron U-clamps.

The designers of the Hagia Sophia in Istanbul, Anthemios and Isidoros, solved the problem of the hoop tension by cutting off the portion of the dome that develops tension. Whether they had observed tension cracks in older domes, or whether it was a coincidence, we do not know. This overcomes one difficulty in masonry construction, but it creates another, because buttresses are now needed (Fig. 10.2.6). The buttresses of the Hagia Sophia predate the buttresses of the first Gothic cathedral by five centuries.

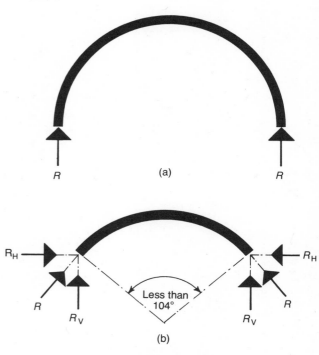

10.2.6. (a) In a hemispherical dome the reactions, R, are entirely vertical, and buttresses are not required. (b) In a dome that is less than hemispherical the reactions are inclined to the vertical, and the horizontal component, R_H, must be absorbed with buttresses. The shallower the dome, the greater the horizontal reactions. If the angle subtended by the shallow spherical dome at the center of curvature is less than 104° (see also Fig. 10.1.2), there is no hoop tension; the stresses in the dome are then entirely compressive.

10.3 Reinforced Concrete Shell Domes and Cylindrical Vaults

In modern steel and reinforced concrete construction, neither the hoop tension nor the absorption of the horizontal reactions present any difficulties. Structural steel and concrete-reinforcing steel solve both problems by their strength. Similar considerations apply to vaults.

Thus concrete shell domes can be made as thin as possible provided the forces act within the shell (Fig. 10.3.1). This minimum thickness varies from 40 mm (1½ in.) to 80 mm (3¼ in.), depending on climate, labor

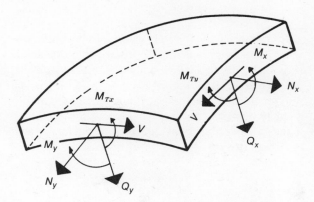

10.3.2. A "thick" concrete shell is needed when bending moments and transverse shear forces occur. In addition to the membrane forces, N_x, N_y, and V, it is possible to have two types of transverse shear force, Q_x, Q_y (which tend to cut through the shell instead of distorting it), and four types of moments (M_x and M_y, which tend to bend the shell, and M_{Tx} and M_{Ty}, which tend to twist the shell). If any one of them is present, the shell must have enough thickness (at least 100 mm or 4 in.) to provide a lever arm for a resistance moment (Fig. 5.5.11).

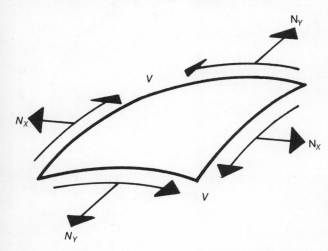

10.3.1. A "thin" concrete shell is possible only if the forces lie within the surface of the shell and there are no bending moments. The forces that can be resisted by a thin shell are two forces (tensile or compressive) at right angles to one another and a shear force *within* the surface of the shell. These forces are called the membrane forces, and their existence in a thin membrane is easily demonstrated with a soap film: Dip a loop of wire in a concentrated solution of detergent, thus producing a membrane with boundaries. With care we can compress it, subject it to tension, and distort (or shear) it without breaking the membrane. As soon as we apply shear or bending across the membrane, it bursts.

In a dome the membrane forces are the hoop force, N_x, and the meridional force, N_y; there is no membrane shear. In a cylindrical shell the membrane forces are horizontal and vertical (beam action and vault action), and there is also a membrane shear force, V.

costs, and quality of workmanship. Thin concrete is more difficult to produce with an accurate cover to the reinforcement, and it is more permeable to heavy rain, so that low rainfall intensity, low labor costs, and high quality of workmanship all encourage thinner concrete shells (see also Section 10.7).

Where bending moments occur (Fig. 10.3.2), that is, mainly near the supports, a greater thickness of concrete is required, sufficient to accommodate a lever arm for the resistance moment (Fig. 5.5.11).

Domes are today as well suited to long-span construction as they have been in the past, although "long spans" are now much longer. The longest span* today (a steel dome; Fig. 10.4.3) is 207 m (680

---

*The longest-spanning reinforced concrete structure has a span only 1 m less. It is also a dome; its span is 206 m (676 ft) (Fig. 10.3.3).

ft). The longest span prior to the nineteenth century was the Pantheon (Fig. 10.2.4), a plain concrete dome spanning 44 m (143 ft), and the longest span of the period between ancient Rome and the industrial revolution was the Duomo of Florence (Fig. 10.2.5), a double-shelled masonry dome spanning 42 m (138 ft). The dome has three essential advantages for long spans:

1. It has curvature, which is essential for long spans (Figs. 4.7.2 and 4.7.4).
2. It has curvature in two directions at right angles to one another, which gives it greater rigidity than a singly curved structure.
3. Its supports are entirely at the lowest level of the structure (which does not apply to some other shell forms), so that it is relatively easy to bring the reactions down to the ground.

The concept of the double-shelled dome, first used by Brunelleschi in the Duomo of Florence, has recently been revived. It is not certain whether Brunelleschi appreciated the extra resistance to bending moment of two shells separated by a lever arm, but he mentioned in his submission to his client, the *Opera del Duomo*, that the space between the two shells would drain off any water that got through the masonry and thus preserve the frescoes. This is still a strong argument in favor of double-shell construction today (Fig. 10.3.3). It is possible to use a shell thickness for each leaf that would be inappropriate for a single shell under the same conditions. The use of a double shell also restrains buckling of the thin sheets of concrete (Section 5.3), because it is possible to place stiffeners between the two shells. However, double shells have not been used often, because there are few structures large enough to justify their use.

Most of the domes built in the twentieth century have been shallow, but this is for functional reasons, not because a shallow dome is easier to construct than a hemispherical one. The great classical domes roofed cathedrals and other ceremonial buildings; these were intended to impress by their height, and they were not designed to be heated. Most modern domes have a specific function, such as roofing a sporting arena; this does not benefit from height, and the extra volume of air enclosed by a higher dome would add to the operating cost of a heated or air conditioned building.

In a hemispherical dome, hoop tension reinforcement has to be provided in the lower part of the dome (Fig 10.1.2). In a shallow dome that makes an angle of less that 52° at the springings (Fig. 10.2.6), tension reinforcement is theoretically not required, but the dome nevertheless needs a small amount of reinforcement in both directions to resist temperature and shrinkage stresses.

A shallow dome has inclined reactions, and these can be absorbed in modern architecture by buttresses (Fig. 10.3.4), as was common in Byzantine and Moslem architecture. It is cheaper to use a tension ring (Fig. 10.3.5). However, this tension ring creates bending stresses in the dome (Fig. 10.3.6).

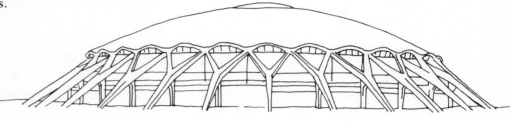

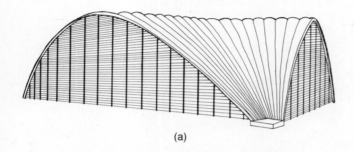

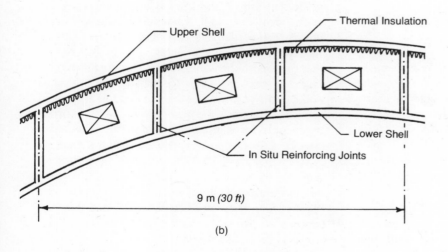

10.3.3. Double shell construction (b) for the CNIT Exhibition Hall (a), the world's longest-spanning concrete dome. Each shell is 60 mm (2.36 in.) thick, and therefore the total thickness of the concrete is only 120 mm. The space between the shells is 3.75 m (12 ft 3 in.), which is an ample space for maintenance and provides a large lever arm for the resistance moment. The stiffening diaphragms are at 9 m (30 ft) centers.

10.3.4. Absorption of the inclined reactions of a shallow spherical dome by buttresses (Palazetto dello Sport, Rome, built by Pier Luigi Nervi in 1957). The 58.5 m (192 ft) diameter dome is of lamella construction, that is, it has diagonally intersecting ribs (Fig. 10.4.10) that converge on the buttresses.

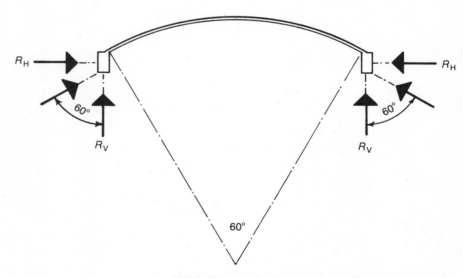

10.3.5. The vertical component, R_V, of the inclined reaction at the edge of a shallow dome is absorbed by columns or loadbearing walls. The horizontal component, R_H, is commonly absorbed by a tension ring.

We have so far considered only the effect of distributed vertical loads, that is, the weight of the dome itself and a snow load. Wind loads are not large because of the shape of the dome. They induce shear stresses (which vertical loads do not do), but these are low; a small amount of additional reinforcement may be needed.

However, temperature changes can produce large stresses in domes near the springings. The dome expands in hot weather and contracts in cold weather (Fig. 10.3.7). Near the springings the expansion is resisted, and this sets up bending moments. The thickness of the dome must be increased in this region, and reinforcement must be provided on both faces for the reversible temperature stresses.

Solutions have been obtained for domes of various nonspherical shapes (Ref.10.1). Domes of oval or elliptical shape were used to great advantage in Baroque architecture, but there have been few such buildings in reinforced concrete, although they present no structural difficulties.

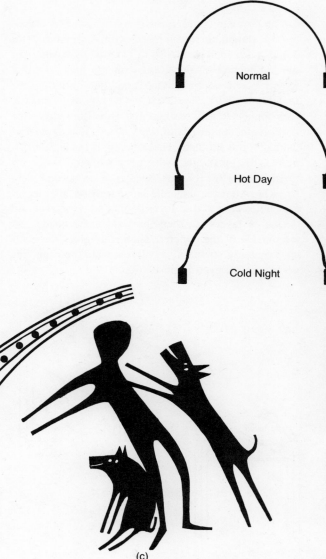

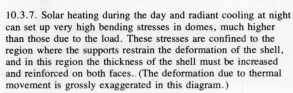

10.3.7. Solar heating during the day and radiant cooling at night can set up very high bending stresses in domes, much higher than those due to the load. These stresses are confined to the region where the supports restrain the deformation of the shell, and in this region the thickness of the shell must be increased and reinforced on both faces.. (The deformation due to thermal movement is grossly exaggerated in this diagram.)

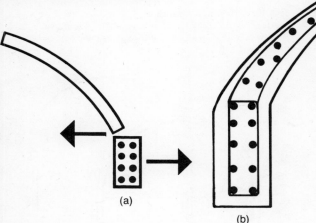

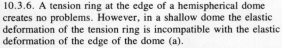

10.3.6. A tension ring at the edge of a hemispherical dome creates no problems. However, in a shallow dome the elastic deformation of the tension ring is incompatible with the elastic deformation of the edge of the dome (a).

The hoop forces in a *shallow* dome are entirely compressive. Therefore the diameter of the dome becomes smaller as the dome is loaded; that is, the edge of the dome moves inwards.

The meridional forces are also compressive (Figs. 10.1.2 and 10.1.3), and the resulting horizontal reactions push the tension ring outwards.

Since the shell and the tension ring are cast in one piece, the shell bends the ring and the ring bends the shell (c). The edge of the shell would be damaged by the bending action unless it is suitably thickened and reinforced (b).

For commercial and industrial buildings square interior spaces are more useful than circular spaces. A dome can be squared (Fig. 10.3.8) provided that the circular tie at the edge of the shell is replaced by straight ties, which prevent the dome from spreading under the influence of the horizontal reactions.

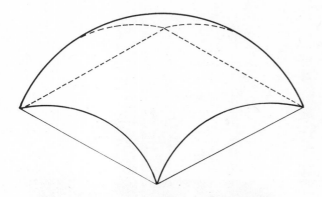

10.3.8. Square dome. Ties are needed along all four edges.

The structural design of small domes presents few complications today, but there are constructional problems. The surface of domes is not ruled (Section 10.5), and there is consequently no simple way of constructing the formwork for a small or medium-sized concrete dome from straight pieces. This was no great problem in the construction of masonry domes built from blocks of stone, which in any case had to be cut to shape. It is still no great problem in the construction of long-span domes, because there is little difference between a continuous curve of 100 m (330 ft) and one formed from straight pieces of timber 3 m (10 ft) long. For domes of small span, however, a great deal of cutting is necessary to produce the timber formwork, and this adds to the cost.

This can be avoided by casting the dome on an air-inflated membrane. Early attempts to do this were unsuccessful, because the membrane was inflated first and the concrete then applied with a concrete gun. This operation had to start somewhere, and the weight of the concrete deformed the membrane and slightly altered the geometry of the dome. Although this makes little difference to strength, it is disturbing to the eye, which easily detects a slight departure from the spherical shape.

The problem is overcome in a technique developed by the Italian architect Dante Bini, who has built domes with spans up to 36 m (118 ft) by casting the entire concrete on the ground on top of a carefully folded neoprene membrane and then inflating the membrane with the wet concrete on top, using electric fans connected to pipes laid under the floor slab. The reinforcement consists of steel springs that expand as the balloon is inflated. The concrete is made 40 to 70 mm (1½ to 2¾ in.) thick, which is not enough either for thermal comfort or for resistance to temperature stresses. After the concrete has hardened, the necessary holes are cut into it for doors and windows, and insulating layers are then sprayed on the inside and the outside of the concrete shell; this gives the shell the necessary stiffness. Finally, a waterproofing finish is sprayed over the insulation.

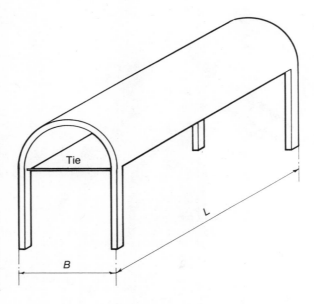

10.3.9. A cylindrical shell needs ties in one direction (B) to prevent the curved slab from flattening out. Edge beams in the other direction (L) are useful to stiffen the shell, but they are not essential.

Cylindrical shells are the modern version of the barrel vault. They must have ties to prevent the curved slab of concrete from spreading (Fig. 10.3.9). Alternatively, buttresses (Fig. 10.1.1) or rigid frames (Fig. 10.3.10) can be used for the same purpose.

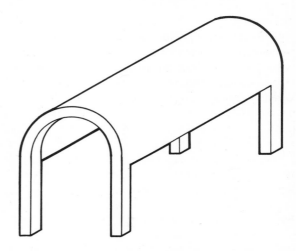

10.3.10. A rigid frame can be used instead of a tie if the visual interference caused by the tie is undesirable. The frame performs the same purpose, but it requires more structural material.

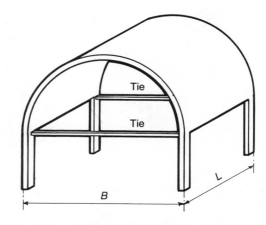

10.3.11. The vault action predominates if the width of the shell, B, is much greater than its span, L.

190

The structural behavior of the cylindrical shell is a blend of vault action and beam action. The vault (Fig. 10.1.1) is an arch extended in the third dimension. If the width, B, of the shell is 2 or 3 times as great as its span, L, the vault action predominates (Fig. 10.3.11). This is accentuated by adding arch ribs to the shell (Fig. 10.3.12).

If the span, L, of the shell is 2 or 3 times as great as its width, B, the beam action predominates; this is accentuated by adding edge stiffeners along the straight edges. The shell then behaves like a beam of curved cross section (Figs. 10.3.13 and 10.3.14).

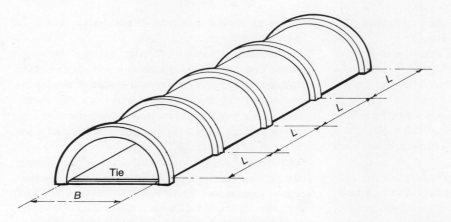

10.3.12. A series of arch ribs accentuate the vault action. This is a composite shell consisting of five spans.

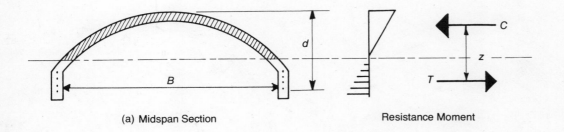

(a) Midspan Section

Resistance Moment

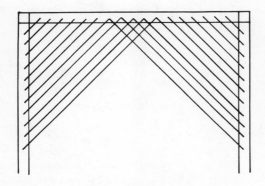

(b) Plan

10.3.13. The beam action predominates if the span of the shell, L, is much greater than its width, B. The shell then behaves like a reinforced concrete beam with a curved compression flange. The beam action is accentuated by a stiffener along the straight edge, and this is also useful for accommodating the large amount of tension (T) reinforcement required to balance the compression (C) in the concrete shell; the tensile force must equal the compressive force for horizontal equilibrium (see also Fig. 10.9.3).

The bars that resist the bending moment at mid-span are bent up into the shell near the supports to serve as shear reinforcement (Fig. 6.1.5.a), as shown in the plan (b).

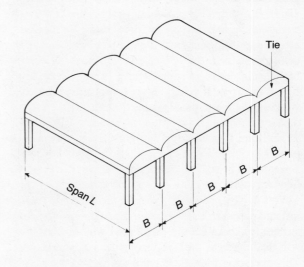

10.3.14. Whereas the composite shell in Fig. 10.3.12 has the appearance of a curved roof, a composite of several beam-action cylindrical shells has the appearance of a flat roof with curved roof segments. It is particularly suitable for an industrial building that requires a fire-resistant roof structure and large open spaces.

10.4 Schwedler Domes, Geodesic Domes, Lattice Vaults, and Lamella Roofs

Domes and cylindrical shells can also be built of metal, either structural steel or aluminum. Aluminum is more expensive, but it is lighter and does not require painting.

Although it is possible to make a structure from sheet metal, which also provides a roof surface, it is more practical to build a lattice structure and then cover it.

We noted in Section 7.3 that plane frames are statically determinate if they are assembled from a number of triangles. This is also true for a curved surface. Fewer members would turn the frame into a mechanism, so that it would not be stable. More members would make it statically indeterminate. Triangulated domes and vaults are more rigid than triangulated plane frames because of their three-dimensional character, and there is rarely any need to use the extra, redundant members needed to make them statically indeterminate.

There are two sets of coordinates on a terrestrial globe. The meridians of longitude are all the same size. They are *great circles:* the circles of largest diameter that can be drawn on a globe (see Glossary). The equator is also a great circle, but all the other circles of latitude have a smaller diameter, and they are called *small circles*.

There are two principal methods for triangulating domes. One uses both great and small circles; the other uses great circles only. The first was developed by the German structural engineer J. W. A., Schwedler more than a century ago, and it is named after him (Fig. 10.4.1). *Schwedler domes* have three sets of members. One set follows circles of latitude, one follows circles of longitude, and the third completes the triangulation.

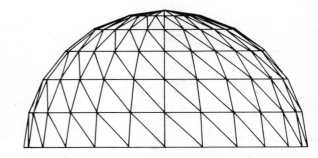

10.4.1. In the Schwedler domes there are three sets of members: One follows circles of longitude (a); one follows circles of latitude (b); the third completes the triangulation (c).

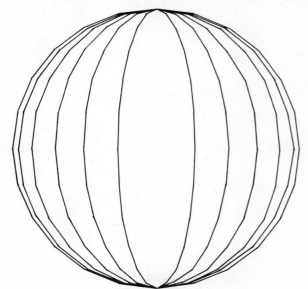

Longitudinal Meridians: All are the same size—they are the circles of largest diameter that can be drawn on a globe.

(a)

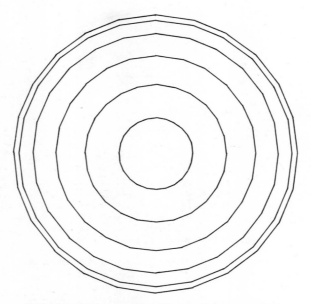

Latitudinal Circles: All except the equator are smaller in size than meridians.

(b)

Schwedler Dome: A triangulation of great and small circles

(c)

Although there is a widespread impression that the geodesic system is superior to the Schwedler system, the world's two longest-spanning steel-framed buildings, the Louisiana Superdome, and the Houston Astrodome, are double-lattice Schwedler domes (Fig. 10.4.3).

The *Gitterkuppel* is a variation on the Schwedler dome in which the circles of longitude are replaced by equilateral triangles (Fig. 10.4.2).

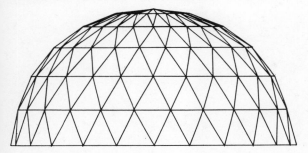

10.4.2. *Gitterkuppel*, a variation of the Schwedler dome in which circles of longitude are replaced by symmetrical triangles.

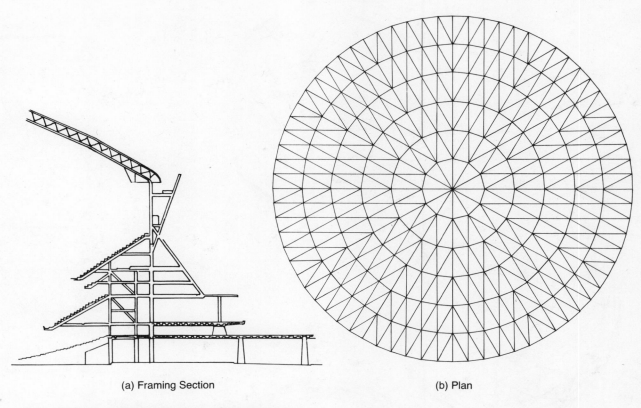

(a) Framing Section

(b) Plan

10.4.3. The Louisiana Superdome, the longest-spanning steel structure at present (207.3 m or 680 ft), is a double-lattice Schwedler dome.

In the *geodesic dome* (Fig. 10.4.4) all members follow great circles. There are three sets, intersecting at angles of 60°. Geodesic domes were built before R. Buckminster Fuller became interested in them, but he has designed more than any other person, including the largest and best known ones. A double lattice may be used for extra stiffness (Fig. 10.4.5). There are many ways in which a spherical surface can be subdivided into regular plane figures, and any of these can be used for lattice domes. If a system uses polygons with more than three sides, it requires rigid joints; welded joints are satisfactory. Some possible combinations are shown in Fig. 10.4.6. The merits, if any, of a nontriangulated system are purely visual.

193

10.4.4. In a geodesic dome all members follow great circles. All members intersect at an angle of 60° and form a series of equilateral spherical triangles.

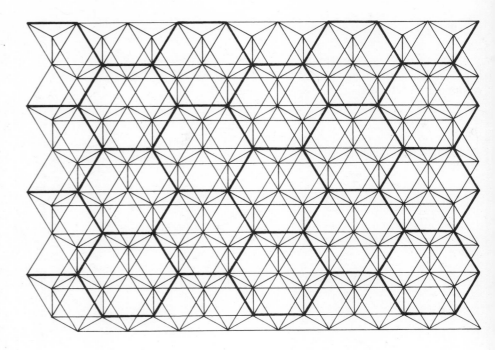

10.4.5. Double-lattice geodesic dome with rigid joints, emphasizing hexagons formed by an assembly of six triangles. This type of structure has been used by R. Buckminster Fuller for several of his domes.

Barrel vaults can also be triangulated by the Schwedler system (Fig. 10.4.7) or the geodesic system (Fig. 10.4.8), or polygons with welded joints can be used (Fig. 10.4.9).

Timber and concrete are less frequently employed in curved space frames. Their most common use is in lamella roofs (Fig. 10.4.10). Several long-span lamella timber roofs have been built in North America.

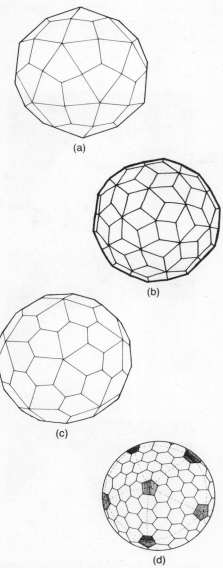

10.4.6. Rigid-jointed lattice domes with nontriangulated systems:
(a) Unequal quadrilaterals.
(b) Parallelograms.
(c) Pentagons.
(d) Pentagons and hexagons.

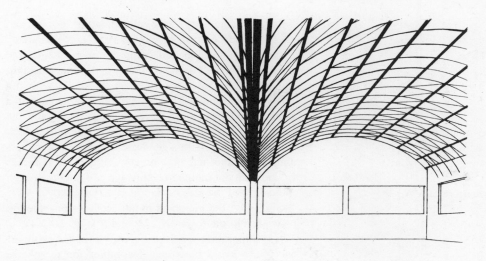

10.4.7. Schwedler-type lattice vault.

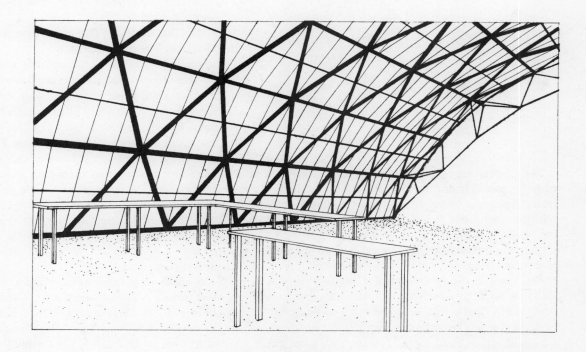

10.4.8. Geodesic lattice vault.

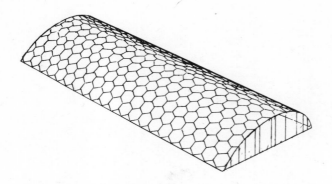

10.4.9. Rigid-jointed hexagonal lattice vault.

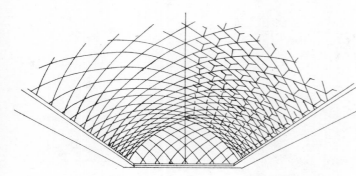

10.4.10. Lamella roof.

In Italy, Pier Luigi Nervi has constructed a number of long-span cylindrical roofs and domes of the lamella type, assembled from precast concrete units with site-cast joints (which are at the intersection of the ribs); the dome in Fig. 10.3.4 is of this type. As these space frames have rigid joints, they do not need to be triangulated.

We noted in Section 5.1 that a plane truss was really an assembly of tension and compression members, and its final design had to be based on the theory of plane frames. For a preliminary design, however, we could regard it as a deep "beam with holes cut into it to make it lighter" (Fig. 4.7.7).

In the same way, triangulated lattice domes and shells are space frames (Section 10.12), which are an assembly of tension and compression members and must be designed as such. For a preliminary design, however, we can think of them as shells with big holes cut into them. Thus the latitudinal members in a Schwedler dome resist the hoop forces, N_x, and the longitudinal members resist the meridional forces, N_y, in Fig. 10.3.1. In a geodesic dome we can make a similar calculation by taking components of the forces in line with the members. Thus we can obtain a rough indication of the size of the members needed.

Similarly, the members in lattice shells provide the resistance to the forces N_x, N_y and V (Fig. 10.3.1).

It is important to make the mesh of the lattice small enough to avoid buckling of compression members. This is dependent on their slenderness ratio (Section 5.3) and therefore on their length between joints.

10.5 The Geometry of Curved Surfaces and the General Theory of Shells

The geometry of the dome and the vault is well known, but other curved surfaces require some explanation. We distinguish between *singly curved* surfaces, such as cylindrical shells, and *doubly curved* surfaces. Doubly curved surfaces can be divided into the categories of domes and saddle shells. If we cut a dome in two directions at right angles to one another, both cuts are convex curves. If we cut a saddle in the same way, one curve is convex and the other is concave.

If we think of a mountain range, a dome corresponds to a mountain peak, a cylindrical shell to a ridge, and a saddle shell to a pass between two mountains (Fig. 10.5.1).

Some shells have ruled surfaces and some do not. A *ruled surface* is one on which a set of straight lines can be drawn. Two sets of straight lines can be drawn on a doubly ruled surface. The distinction has constructional significance, because both steel sections for structural steelwork and timber for concrete formwork come in straight pieces. Thus a ruled surface is easier to construct, and a doubly ruled surface is even better.

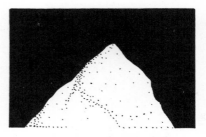

(a) Mountain Top

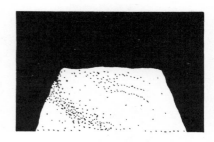

(b) Mountain Ridge

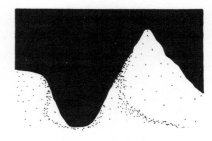

(c) Mountain Pass

10.5.1. Classification of shells:
(a) Dome (mountain top)
(b) Cylindrical shell (mountain ridge)
(c) Saddle shell (mountain pass)

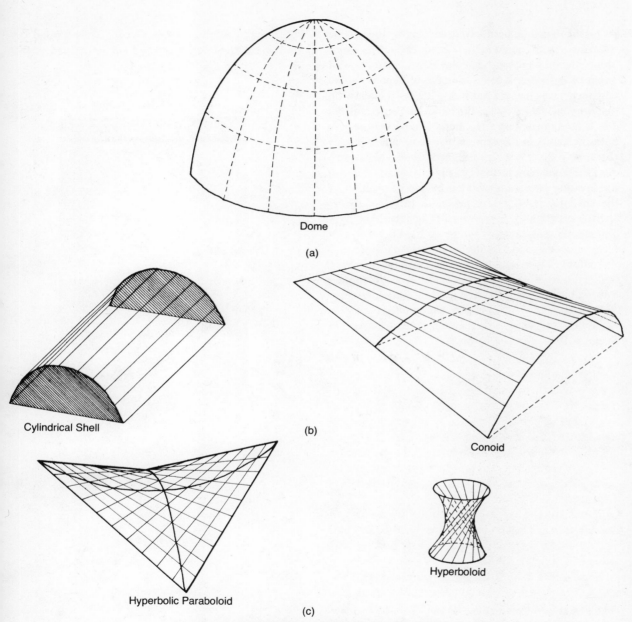

10.5.2. It is impossible to draw a straight line on the surface of a dome (a). It is possible to draw one set of straight lines on a cylindrical shell and on a conoid (b). Two sets of straight lines can be drawn on a hyperbolic paraboloid and a hyperboloid (c).

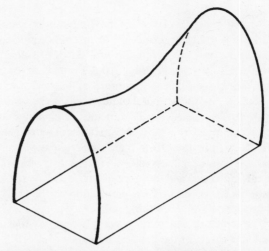

10.5.3. We bend a wire to the shape of the boundaries of a cylindrical shell and dip it into a strong solution of detergent. The soap film formed is a minimal surface. It is a saddle shape, not a cylindrical surface.

Among common shell forms, domes are not ruled surfaces, and this is why they present constructional problems (Section 10.3). Cylindrical shells and conoids have singly ruled surfaces, and hyperboloids and hyperbolic paraboloids have doubly ruled surfaces (Fig. 10.5.2).

It is also worthwhile distinguishing between *minimal surfaces* and those that are not minimal. A minimal surface is the equivalent of a straight line. If we pull a string tight between two points, we get a straight line, which is the shortest path between those two points. If we bend a piece of wire to the shape of the boundaries of a shell, and dip this wire into a concentrated solution of detergent, we obtain a soap skin, which naturally adopts the form of the minimal surface corresponding to those boundaries. A minimal surface is always a saddle surface or a plane surface (Fig. 10.5.3). Since saddle shells are structurally efficient (Fig. 10.6.3), resistant to buckling, and in some cases ruled, this experiment is a useful starting point for devising a suitable shell form of a nonstandard type.

197

Although the theory of domes discussed in Section 10.1 (Fig. 10.1.3) goes back to the nineteenth century, it is now considered part of a series of solutions for membrane shells (Fig. 10.3.1) that cover all standard cases, that is, domes, cylindrical shells, conoids, hyperbolic paraboloids, and hyperboloids (Ref. 10.1). The theory is statically determinate and simple to apply. It is quite satisfactory for small shells that carry little apart from their own weight, provided that the area near the shell boundaries is given additional thickness and reinforcement to allow for the disturbances caused by the connection of the shell to its supports (Fig. 10.3.6.b).

For larger shells, more accurate solutions that include the effect of the bending moments (Fig. 10.3.2) are needed. Until a few years ago these solutions were available for only a limited range of standard geometric shapes. However, since the development of the finite element method in the 1960s, this limitation has largely disappeared. The finite element method is essentially an application of the matrix-displacement method (Section 8.7) to continuous surfaces, which are for this purpose subdivided into a large series of interconnected elements. Most shells can now be analyzed with a computer at a reasonable cost.

One might have expected that this important advance in shell theory would have resulted in the construction of more shell structures. In fact, the reverse has happened (see also Section 10.7).

10.6 Hypars and other Nontraditional Surfaces

The hyperbolic paraboloid, usually referred to with the shortened term *hypar*, has the equation

$$z = k\,x\,y \qquad (10.3)$$

where x, y, and z are the three Cartesian coordinates and k is a constant. No equation in solid geometry could be simpler; the shape of the surface (Fig. 10.6.1) was first obtained by the nineteenth-century professor of mathematics who decided to investigate what Eq. (10.3) meant in geometric terms. The membrane theory of hypars and the first construction of hypars was the work of F. Aimond, a French railroad engineer in the 1920s, but it was Félix Candela who made hypar shells well known to architects. In the 1950s he built a large number of hypars that were both ingenious in conception and economical in cost (Fig. 10.6.2). Candela designed these shells with Aimond's membrane theory (Fig. 10.6.3), although the bending stresses generated by the connection of the shell to its stiffeners penetrate further into the shell than in domes. However, all Candela shells have a moderate span (about 15 m or 50 ft). Larger hypars have been designed recently by others.

The only other doubly ruled surface, the hyperboloid (Fig. 10.6.4), is familiar from its use in cooling towers, but it has rarely been used for buildings.

Conoids (Fig. 10.5.2.b) are mainly used for north- or south-light shells and are therefore discussed in Section 10.8.

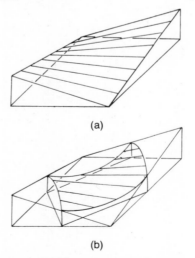

10.6.1. A hypar is formed by one straight line moving over two other straight lines at an angle to one another (a). If we cut this surface at 45°, we obtain a surface of entirely different appearance (b). It resembles the minimal surface in Fig. 10.5.3, although its mathematical equation is different.

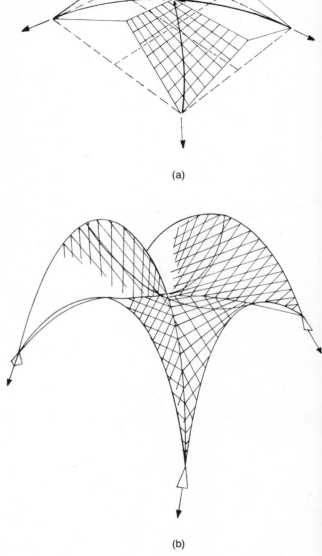

10.6.2. Hypars can be combined into compound surfaces. These have been used by Félix Candela to great advantage for a variety of structural forms (Ref. 10.2).

A large number of different forms can be produced by combining simple shells (Figs. 10.3.3 and 10.6.2). However, it is important to consider the implications of sharp changes of curvature, which may occur when portions of different surfaces are combined.

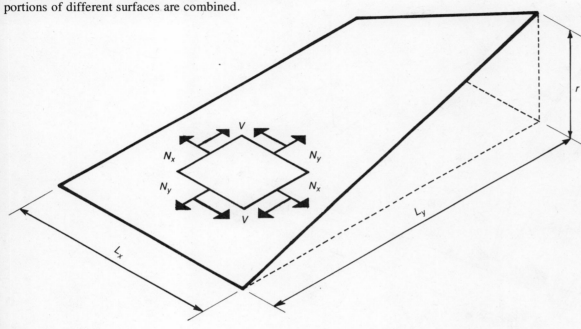

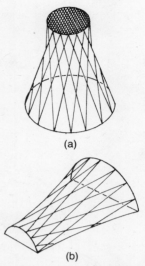

10.6.4. Hyperboloids (a) used vertically, which is the familiar cooling-tower shape, and (b) used horizontally.

10.6.3. The membrane theory for the hypar shell is very simple. For a uniformly distributed vertical load, the two direct membrane forces are zero, and the membrane shear is constant:

$$N_x = N_y = 0 \tag{10.4}$$

$$V = \frac{w' L_x L_y}{2r} \tag{10.5}$$

where w' is the vertical load per unit area, measured on the *plan* of the shell.

The membrane shear can be resolved at 45° into a diagonal tension and a diagonal compression (Fig. 10.5.2.c). Thus the hypar can be considered a close-meshed net of intersecting "arches" and "suspension cables." Any "slackening" of the "cables" increases the compression in the "arches." The hypar is thus an efficient structure and is resistant to buckling because the "cables" restrain the "arches."

Saddle-shaped suspension structures (Fig. 10.10.2) are based on the same structural principle, except that a steel net is substituted for reinforced concrete.

Sharp corners and discontinuities, whether in shells or folded plates (Section 10.7), produce high bending stresses. These are not significant in shells of small span, because they can be absorbed in a normal thickness of concrete; but they create serious problems in long-span structures.

The roof surface of the Sydney Opera House (Fig. 10.6.5) is a combination of sectors of spherical domes (see Glossary). The structure presents many problems: it is tall; its spans are moderately large (up to 40 m or 130 ft); and the roof surface has discontinuities and sharp corners. Joern Utzon conceived it as a thin shell; when he arrived in Sydney after having won the competition, he announced that "the roof of the Opera House will be made of concrete a few inches thick." But an early analysis already showed the existence of large bending moments. The structure was eventually built as an assembly of concrete arch ribs about 2 m (6 ft 6 in.) deep (Figs. 10.6.6 and 10.6.7). It is a superb sculpture, but not an efficient structure.

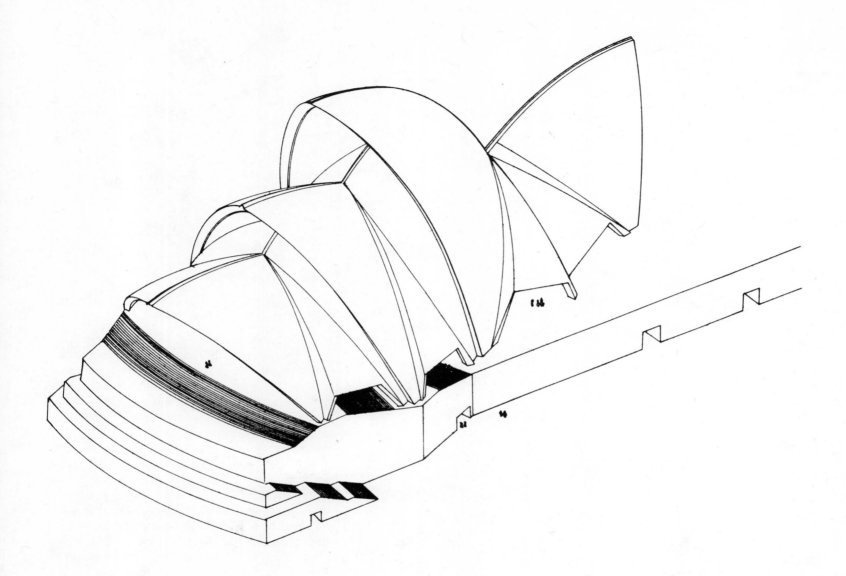

10.6.5. Isometric view of the roof of the Major Hall of the Sydney Opera House. All the surfaces are spherical, with a radius of 75 m (246 ft), that is, the roof is an assembly of dome sectors. Each "shell" consists of a series of concrete arch ribs assembled with prestressing cables (Fig. 10.6.7). The structure was analyzed by computer as a rigid three-dimensional space framework. (*From Ref. 10.5.*)

A: Concrete structure (Fig. 10.6.7)
B: Void between exterior concrete structure and interior steel-plywood structure
C: Steel frame for plywood panels
D: Plywood panels for ceiling and interior lining
E: Foyer
F: Concert hall
G: Drama theater
H: Recording studio

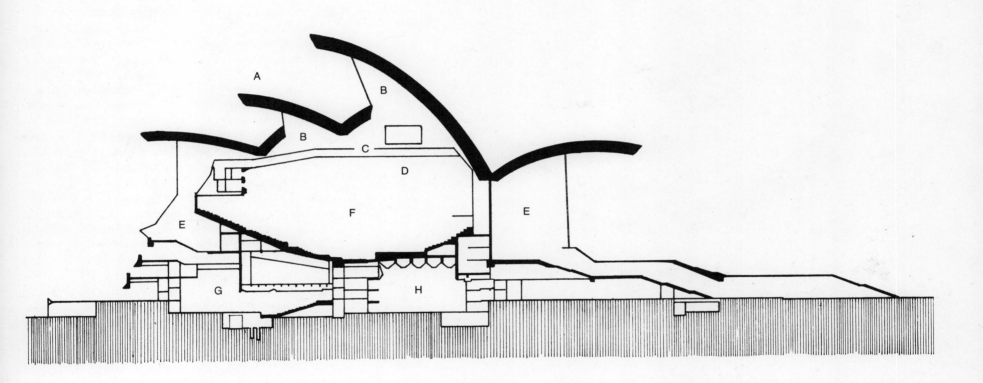

10.6.6. Longitudinal section through the Major Hall of the Sydney Opera House. (*By courtesy of Mr. David Littlemore, one of the three architects who completed the design of the building.*)

201

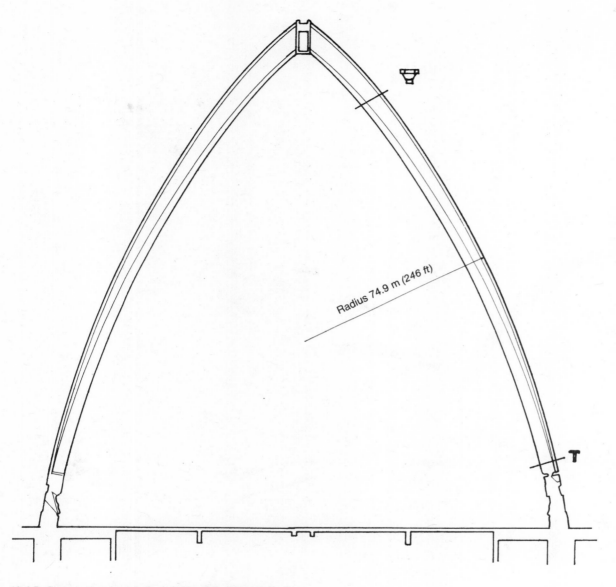

10.6.7. Cross section through the "shell" of the Major Hall of the Sydney Opera House, showing one of the ribs. It varies from a T-shaped cross section to a Y-shaped cross section and has an average depth of about 2 m (6 ft 6 in.). The ceiling of the Major Hall itself, a plywood shell supported by a steel frame, is not shown. (*From Ref. 10.5.*)

10.7 Folded-Plate Roofs

Any sharp corner in a structure must produce bending stresses. Thus folded plates, unlike thin shells, are subject to bending (Fig. 10.3.2.). However, their formwork consists of flat surfaces and can be constructed from hardboard or waterproof plywood.

The great increase in strength obtained by folding a flat surface is easily demonstrated (Fig. 10.7.1).

The greater thickness of concrete required to accommodate a lever arm for the resistance moment (Fig. 5.5.11) is not necessarily a disadvantage for moderate spans. The minimum thickness of a curved or sloping concrete surface that a contractor is able to cast at a reasonable cost with the requisite dimensional accuracy, and the requirements regarding waterproofing, vary in different parts of the world. In countries such as the United States, Great Britain, and Australia labor costs are high, and clients insist on perfect waterproofing. Under those circumstances the minimum thickness of concrete is 80 mm ($3\frac{1}{4}$ in.), and a structure of this depth is capable of resisting some bending. Many types of folded plate structures could be built with this minimum thickness with spans up to about 20 m (65 ft).

Thus in developed countries shells have become mainly structures for longer spans. In developing countries, such as those of Latin America and south Asia, thinner shells are acceptable and the cost of concreting them is less; shorter spans are therefore economical for shells.

We can produce in folded-plate construction the visual equivalent of a cylindrical shell, a circular dome, a square dome, or a rigid frame (Fig. 10.7.2). However, the triangulated shape is a distinctive folded-plate form (Fig. 10.7.3).

Since flat plates have flexural resistance, the horizontal reactions could be absorbed by bending within the shell. However, this would increase the amount of structural material. It is advisable to use ties or rigid frames at the columns wherever possible (Figs. 10.7.1.d, 10.7.2, and 10.7.3).

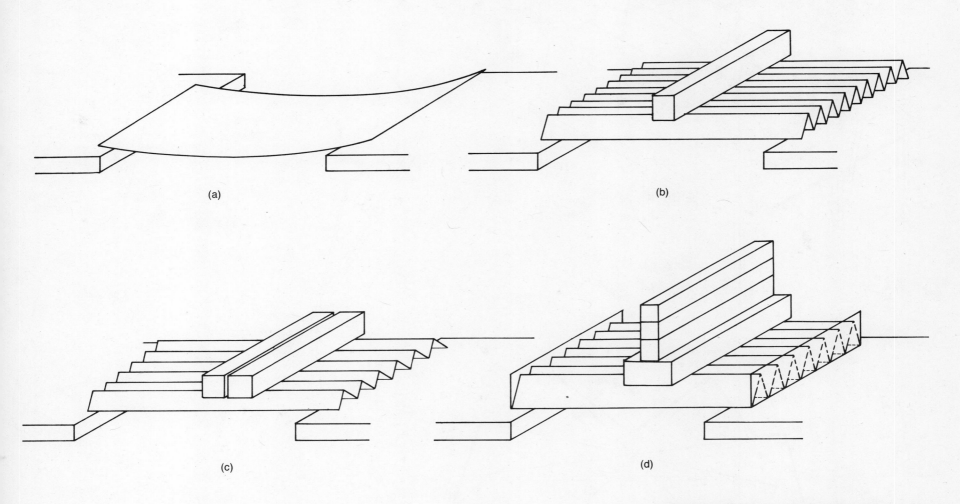

10.7.1. Folding a piece of paper greatly increases its strength; folding a reinforced concrete plate does the same.
(a) A flat sheet of paper will not carry even its own weight.
(b) A folded piece of paper can carry its weight and a superimposed load as well.
(c) As the load is increased, the folds straighten out and the structure collapses.
(d) Therefore the structure can be greatly strengthened by gluing stiffeners to its ends.

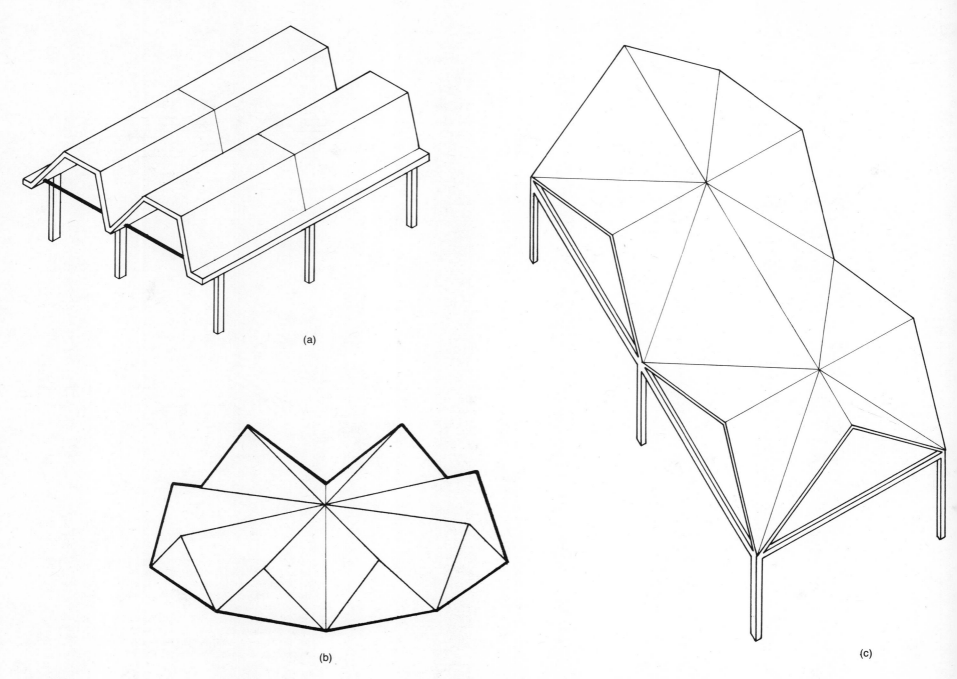

204

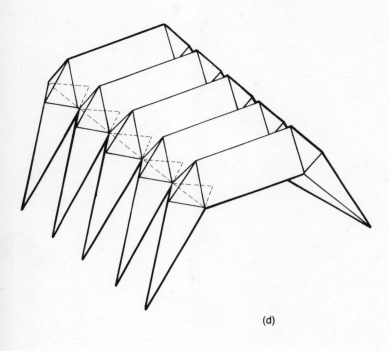

10.7.2. Folded plates can be designed to serve the same purpose as the various shell forms:
(a) Cylindrical folded plate.
(b) Circular folded plate.
(c) Square folded plate.
(d) Folded plate portal frame.
The ties are not essential, since folded plates have flexural strength, but they greatly reduce the bending stresses.

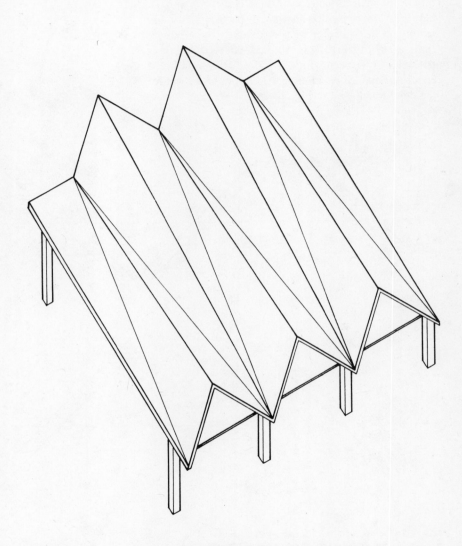

10.7.3. Folded-plate roofs formed from triangles are both attractive in appearance and easy to construct. The formwork consists of a number of identical triangles.

The folded plate is a rigid frame in three dimensions. It can be analyzed by any of the standard methods for rigid frames, such as moment distribution or the matrix-displacement method (Section 8.7). However, many folded plates can be designed more simply. If the folded plate spans mainly in one direction, as in Figs. 10.7.1, 10.7.2a, and 10.7.3, it is possible to treat it (Fig. 10.7.4) as a reinforced concrete beam of peculiar cross section, as we did for the cylindrical shell in Fig. 10.3.13.

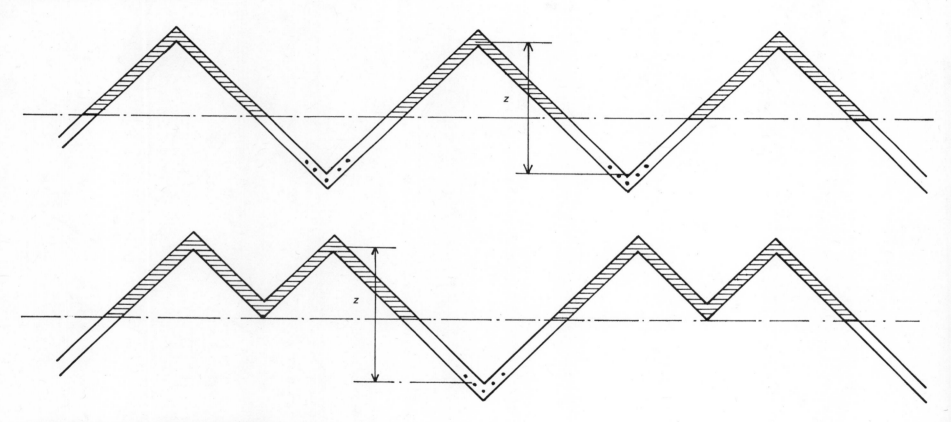

10.7.4. Two cross sections of the triangulated folded plate structure in Fig. 10.7.3. The structure can be designed as a simply supported beam with the cross sections shown. The resistance moment is given by Eq. (5.11) in Section 5.8; z is the lever arm.

10.8 Side-Lit and Top-Lit Roof Structures

We noted in Figs. 7.3.3 and 7.6.3 how roof trusses and rigid frames can be adapted to give indirect lighting (from the north or the south, depending on the hemisphere). The same can be done with shell roofs and folded-plate roofs (Figs. 10.8.1 and 10.8.2).

The conoid (Fig. 10.5.2.b) is particularly suitable for north-light (south-light) roofs (Fig. 10.8.3), because it gives the greatest amount of light halfway between the columns and has the greatest depth where the structure requires it. Once the window frames are in position, it is possible to form the shell with straight pieces of timber between adjacent window frames.

It is possible to cut quite large openings into shell and folded-plate structures without endangering the structure, although it is necessary to check that these are in regions that are not too highly stressed, and extra reinforcement must be placed around the opening. The use of lanterns in domes for top lighting is traditional (Fig. 10.8.4).

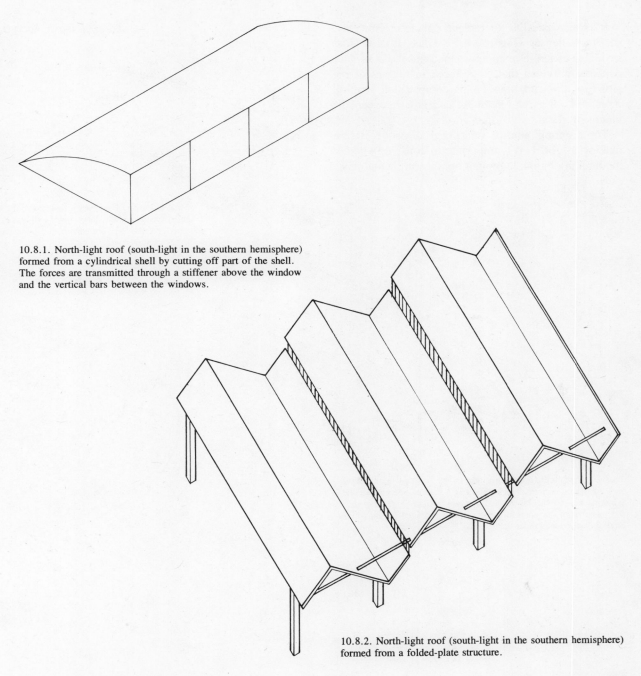

10.8.1. North-light roof (south-light in the southern hemisphere) formed from a cylindrical shell by cutting off part of the shell. The forces are transmitted through a stiffener above the window and the vertical bars between the windows.

10.8.2. North-light roof (south-light in the southern hemisphere) formed from a folded-plate structure.

It is also possible to replace concrete with glass blocks or lenses (Fig. 10.8.5). Glass is almost as strong as concrete in compression, and it can replace concrete provided that it is placed in regions of the structure that are stressed in compression in both directions. The reinforcement can pass between the blocks of glass.

Top lighting admits solar heat directly. Before making use of it, it is important to consider whether the building might be too hot on a warm, sunny day.

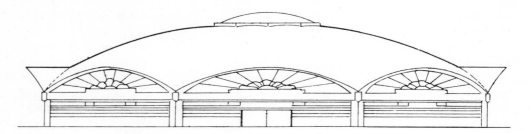

Glass Lantern to Admit Daylight

10.8.4. Lanterns are easily inserted at the crown of a dome. For most types of shell, large holes can be cut to admit daylight, provided that they are surrounded by stiffeners and additional reinforcement.

10.8.3. North-light roof (south-light in the northern hemisphere) formed from a conoid. The conoid has a natural opening for the window, and no special vertical bars are needed to transmit the forces between the window panes. It also provides the maximum daylight where it is most needed.

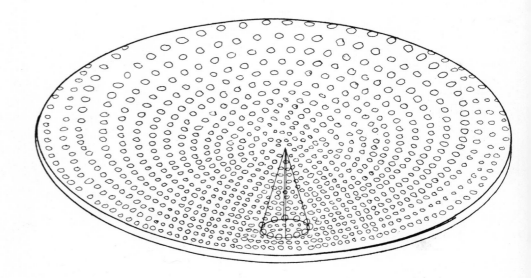

10.8.5. Daylight can also be admitted through glass blocks or lenses placed in the concrete. This is admissible in any part of the shell that is stressed in compression in both directions. Only normal reinforcement is required, and this runs between the glass blocks.

10.9 Prestressed Shells

Prestressing is most advantageous for long spans. There are three ways in which a shell structure can be prestressed:

1. through the foundation slab or through an external tie (Fig. 10.9.1);
2. through the ties at the edge of a shell (Fig. 10.9.2);
3. within the shell (Fig. 10.9.3).

The first method is applicable to all shells. It obviates the need for large abutments in a long-span shell (Fig. 9.7.1); it is, in fact, a substitution of a tie for a buttress. It is necessary to have at least a small prestress to ensure that the tie remains tight and to counter temperature movement. This can be considerable; for the tie in Fig. 10.9.1, which is 206 m (676 ft) long, a temperature change of 40° C (72° F) produces a change in length of 96 mm (4 in.).

If a larger prestress is appplied, it is also possible to lift the structure off its formwork. This can save a great deal of trouble and expense in the removal of the formwork.

The second method is suitable for all shells with straight boundaries, such as squared domes (Fig. 10.3.8), cylindrical shells (Fig. 10.3.9), hypars (Fig. 10.9.2), and conoids (Fig. 10.8.3), but it is worthwhile only for the larger spans. Since prestressing steel is about five times as strong as ordinary reinforcing steel and since the tensile force in shells can be quite large (for example, in Fig. 10.3.13), the space required for the steel, and therefore the size of the edge member, is greatly reduced.

The alternative method is to bend the prestressing cables up into the shell (Fig. 10.9.3) and use them for load balancing (Figs. 5.6.1 and 5.6.2). The shell is then lifted off the formwork during the prestressing operation, and the prestressing can be continued until the weight of the shell is balanced by the prestress. This reduces the structural sizes and also simplifies the removal of the formwork.

A fourth method may be mentioned here. Concrete planks can be laid on top of suspension cables to form the roof surface; the cables are then prestressed with prestressing jacks, or simply by the weight of the concrete membrane. Although these structures have prestressed cables and a concrete shell, they are essentially suspension structures with a concrete roof surface.

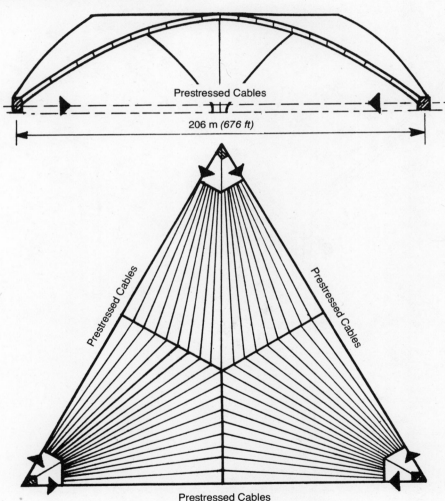

10.9.1. Prestressed cables between abutments in the CNIT Exhibition Hall in Paris (see also Fig. 10.3.3).

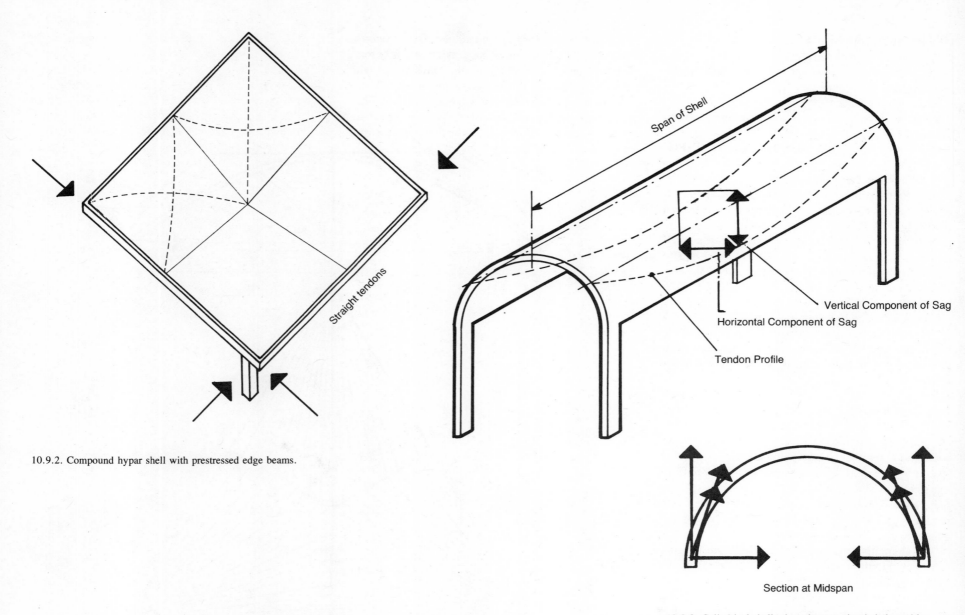

10.9.2. Compound hypar shell with prestressed edge beams.

10.9.3. Cylindrical shell whose beam action is balanced by parabolic prestressed cables. The cables run within the body of the shell. The edge beams normally required in a reinforced concrete shell of comparable span can be eliminated (see also Fig. 10.3.13).

10.10 Suspension Structures: Why We Do Not Use Them More Often

In 1826 the Menai Suspension Bridge in North Wales became the first suspension structure to set a record for span, and the record has been held by suspension bridges ever since, except for a period of 40 years. Today the longest suspension bridge has a span of 1410 m (4626 ft); the longest suspension roof, the Burgo Paper Mill in Mantua, Italy, has a span of 163 m (535 ft) (Fig. 10.10.1). This is much less than the longest-spanning dome (207 m or 680 ft).

The cable is so efficient for long-span bridges because it is flexible. It is therefore purely in tension, and there is no need to provide thickness to resist bending. The lever arm of the resistance moment is the sag of the cable (Figs. 4.7.1 and 4.7.2). This flexibility is not wholly advantageous in a roof structure. Excessive vibrations cannot be tolerated in a building. Waterproofing of the roof is difficult if it moves too much. Most suspension roofs are therefore prestressed to reduce their flexibility, and some also have concrete roofs for the same reason.

The first modern cable roof was an arena at the State Fair in Raleigh, North Carolina, completed in 1953 (Fig. 10.10.2). The loadbearing cables are suspended from two intersecting arches, which are anchored against one another (Fig. 10.10.3). At right angles to the loadbearing cables are secondary cables, prestressed to ensure that even on a hot day they remain taut. Corrugated sheets of galvanized iron are supported on this cable network.

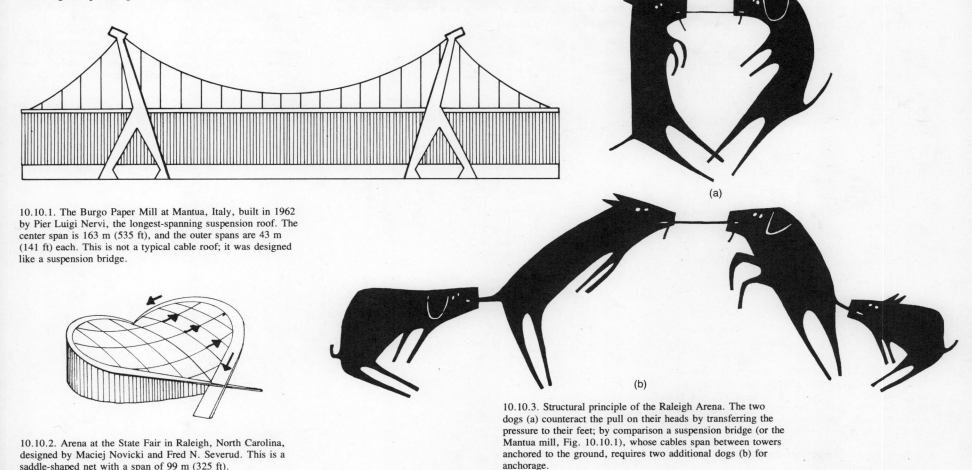

10.10.1. The Burgo Paper Mill at Mantua, Italy, built in 1962 by Pier Luigi Nervi, the longest-spanning suspension roof. The center span is 163 m (535 ft), and the outer spans are 43 m (141 ft) each. This is not a typical cable roof; it was designed like a suspension bridge.

10.10.2. Arena at the State Fair in Raleigh, North Carolina, designed by Maciej Novicki and Fred N. Severud. This is a saddle-shaped net with a span of 99 m (325 ft).

10.10.3. Structural principle of the Raleigh Arena. The two dogs (a) counteract the pull on their heads by transferring the pressure to their feet; by comparison a suspension bridge (or the Mantua mill, Fig. 10.10.1), whose cables span between towers anchored to the ground, requires two additional dogs (b) for anchorage.

The building produced some unforeseen complications that added to the cost and defeated the original objective of an economical structure. In spite of their great weight, the concrete arches were, in fact, long and slender compression members that showed a tendency to buckle. Guy wires with strong ground anchorages had to be added to ensure their stability. Internal wires were fixed to the flatter portion of the roof to prevent buckling, and damping springs were inserted at the cable connections to avoid flutter of the roof structure.

The problems posed by this important pioneering structure can be solved in one of two ways. We can make the roof much heavier by placing concrete slabs on the cables, or we can interconnect two sets of cables so that, as one slackens, the other is automatically tightened. A typical example of the second method is shown in Fig. 10.10.4, and a typical example of the first in Fig. 10.10.5.

A number of structures, mostly with concrete roofs, have been built with parallel cables. They can be anchored to the ground (Fig. 10.10.6); a neater-looking and more efficient method is to anchor the cables to the reinforced concrete structure of the banked seats, which are then tied together with pre-stressed cables (Fig. 10.10.7).

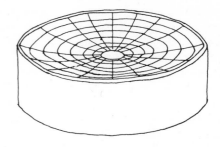

10.10.5. Municipal Stadium in Montevideo, Uruguay, built in 1957. Its span is 94 m (308 ft). The roof is formed by a single layer of cables, covered with precast concrete slabs. These were loaded temporarily with a large weight of building materials to prestress the cables, and the joints between the concrete slabs were then filled with cement mortar to anchor the prestress. The weight on the roof was removed after the mortar had hardened.

This is an economical roof structure, but a central drain pipe is necessary to remove the rainwater, and this spoils the interior. In a later building of a similar type, the Oakland-Alameda County Auditorium, built in 1967 with a span of 128 m (420 ft), the problem was solved by pumping the rainwater off the roof. There is a safety device to protect the structure against collapse, whereby water is dumped on the arena floor if an excess collects because of exceptionally heavy rain or a failure of the pumps.

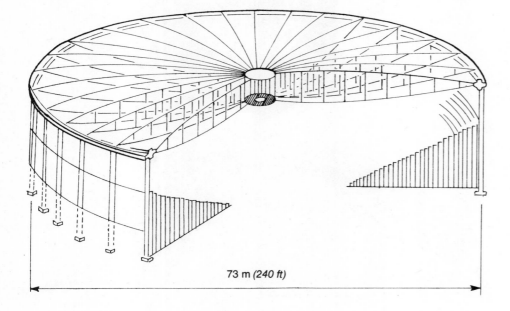

10.10.4. Memorial Auditorium in Utica, New York, designed by Lev Zetlin and completed in 1959. Its span is 73 m (240 ft). There are two sets of cables, separated by struts that cause them to act in conjunction. The upper and the lower cables were given a different amount of prestress, so that they have different natural frequencies. Vibrations in one set of cables are therefore out of phase with the other, and the opposing forces damp the vibration of the structure.

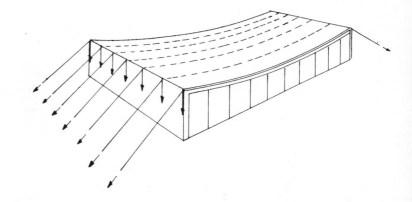

10.10.6. Suspension roof with parallel cables anchored to the ground.

Structures using suspended cables have a functional advantage for arenas, because the shape is better suited to an array of banked seats than that of a dome (Fig. 10.10.8).

Frei Otto has designed more suspension structures than any other person, although not all have been built (Ref. 10.3). Among his most successful structures are a number of large, tentlike roofs supported by prestressed cables (Fig. 10.10.9).

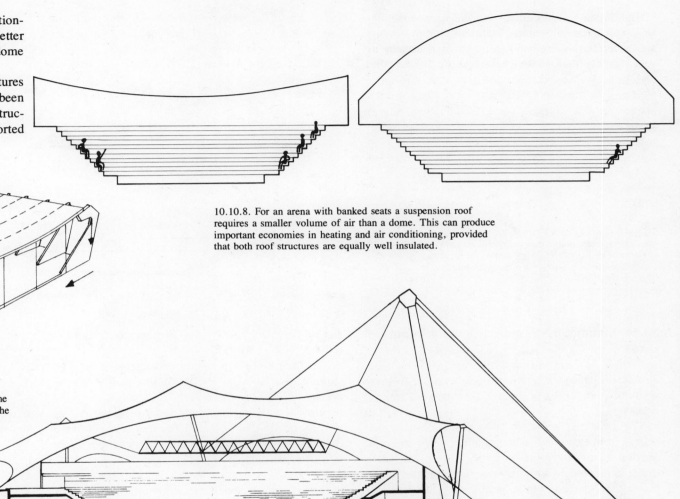

10.10.8. For an arena with banked seats a suspension roof requires a smaller volume of air than a dome. This can produce important economies in heating and air conditioning, provided that both roof structures are equally well insulated.

10.10.7. Suspension roof with parallel cables anchored to the reinforced concrete structure supporting the banked seats. The horizontal reaction is absorbed by cables buried in the floor structure.

10.10.9. Roof over sports arena for the 1972 Olympic Games in Munich, designed by Frei Otto. It is not possible to assign an exact span to this structure, but it is approximately 130 m (430 ft).

The tentlike simplicity of this prestressed cable structure is deceptive. The roof over the entire sports arena cost about $48 million (U.S. dollars). The design required a great deal of theory as well as a model analysis, and the consultants included two eminent engineers, Fritz Leonhardt and John H. Argyris.

Cables can be used to increase the span of cantilevers, and this is particularly useful for aircraft hangars and other buildings that require large entrances as well as unobstructed interior spaces (Fig. 10.10.10).

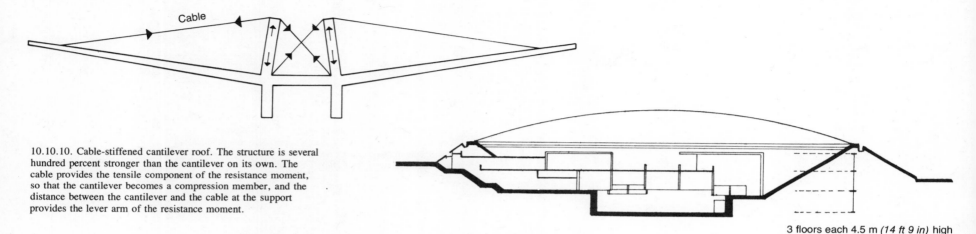

10.10.10. Cable-stiffened cantilever roof. The structure is several hundred percent stronger than the cantilever on its own. The cable provides the tensile component of the resistance moment, so that the cantilever becomes a compression member, and the distance between the cantilever and the cable at the support provides the lever arm of the resistance moment.

10.11.1. The United States Pavilion at the Osaka International Exposition of 1970, since demolished. The membrane covered an oval 142 m by 83 m (466 ft by 273 ft), with a rise of only 6.1 m (20 ft). The membrane consisted of a glass fiber fabric coated with PVC (polyvinyl chloride) and reinforced with a net of 32 cables. The weight of the fabric was 15 tonnes (17 U.S. tons) and that of the cables 45 tonnes (50 U.S. tons).

This is the second-largest pneumatic dome built at the time of writing (1978). The largest, completed in 1976, is the Pontiac Silverdome near Detroit, which seats 80 500 spectators.

10.11 Pneumatic Structures: Span Without Limit

Cables also play an important part in pneumatic structures. They are almost invariably used to give additional strength to large, air-supported membranes.

The structural principle of pneumatic structures differs from that of all other structures. We noted in Section 4.7 that vertical loads on a horizontal span create a bending moment, and the structure must resist that moment. In a pneumatic structure the vertical load is resisted by an opposite air pressure. Consequently there is, in theory, no limit to the span of a pneumatic structure.

In practice, excess pressure is required to allow for the effect of wind and (in cold climates) snow. When there is no snow and the wind produces an uplift, the excess internal pressure produces tension in the roof membrane, which eventually limits the span. Dynamic effects on large spans have not yet been fully explored. They may limit the span further, but at present it seems much greater than we could usefully require.

The weight of the membrane roof, even when it is stiffened by cables, is very small, and a quite low air pressure is sufficient to balance it. The building may be entered through revolving doors so as to lose as little air pressure as possible. The loss of air pressure through the doors and through the membrane is made up with an air fan.

The idea was patented by F. W. Lanchester, one of the pioneers of the British automobile industry, in 1917 as a method for erecting field hospitals, but World War I ended before one could be built. It was resurrected in 1946 by Walter W. Bird in the United States.

The excess air pressure needed is, in fact, quite small. The second-largest pneumatic structure built so far, the American Pavilion at the Osaka International Exposition (Fig. 10.11.1), has an excess air pressure of 0.23% above the external air pressure (240 Pa, 5 psf or 25 mm of water).

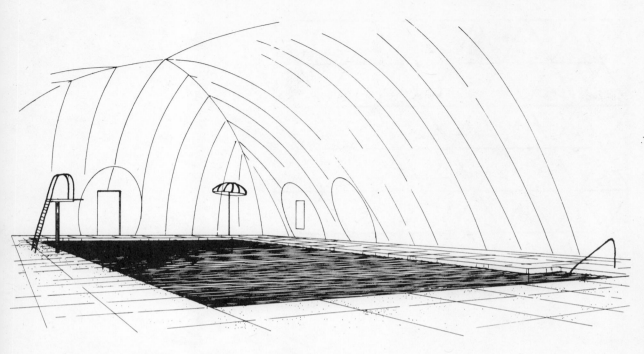

10.11.2. Roof over swimming pool, one of many built by Birdair Structures.

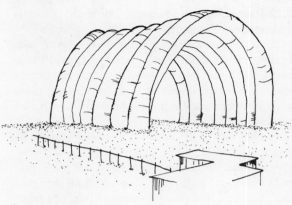

10.11.3. The Fuji Pavilion at the Osaka International Exposition of 1970, under construction. The structure, since demolished, consisted of 16 arched tubes inflated with a pressure of 10 kPa (205 psf, 1000 mm of water or 10% above atmospheric pressure). The tubes were connected to a turbo-compressor that automatically increased the internal air pressure in a high wind, up to a maximum of 25 kPa in storm conditions. The membrane consisted of PVA (polyvinyl acetate) coated internally with PVC and externally with Mypalon. The lower ends were anchored in steel tubes.

This excess air pressure can be maintained with a small air compressor. A standby compressor using a different source of power ensures that the air supply continues if the compressor fails or its power is cut off.

There are two types of pneumatic roof. In one the people are inside the "balloon," entering and leaving it through an airlock. The structure is made from an unstiffened fabric if the span is small (Fig. 10.11.2) or from a cable-stiffened fabric if it is large (Fig. 10.11.1).

Alternatively, the roof is a closed membrane that people do not enter. This can form a complete building (Fig. 10.11.3) or merely a canopy (Fig. 10.11.4).

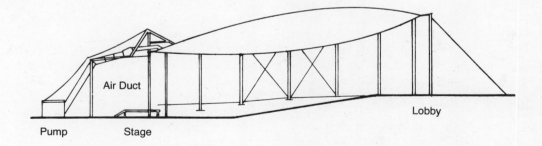

10.11.4. Roof of the Boston Arts Center, Boston. Lack of funds suggested a tent, but this did not offer a large enough area unobstructed by supports. The 44 m (144 ft) diameter and 6 m (20 ft) deep roof was designed by Paul Weidlinger and built by Birdair Structures in 1959. The air pressure is 480 Pa (10 psf).

10.12 Planar Space Frames: Three-Dimensional Trusses

The term "space frame," without a prefix, is usually reserved for trusses with parallel or approximately parallel chords and diagonal members running in three dimensions (Fig. 10.12.1). Although plane frames interconnected in the third dimension (Chapter 7) and lattice domes (Section 10.4) are also frames in space, they are not usually considered under this heading.

We can think of a space frame as a flat plate (Fig. 5.5.1) with many holes in it. The top chord then provides the compressive component of the resistance moment, the bottom chord provides the tensile component, and the diagonal members provide the lever arm of the resistance moment and the shear resistance (Fig. 10.12.2).

Most space frames are supported on columns with approximately equal spacing in the two directions at right angles to one another. If the spacing is precisely the same in both directions, the bending moment due to the vertical load acting on the roof is divided equally in both directions, as in a flat plate. As in the case of a flat plate, the structure performs the function of both a floor slab and the supporting beams, and this approximately doubles the bending moment (see Section 5.5 and Problem 10.7).

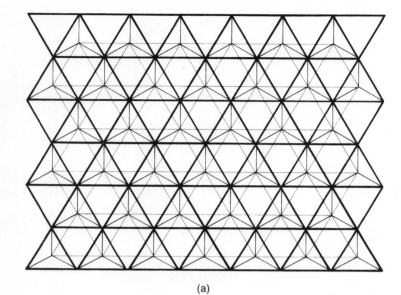

(a)

A space frame can be considered a flat plate with many holes drilled in it.

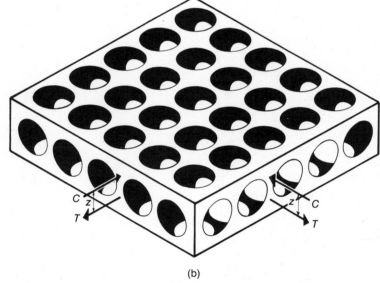

(b)

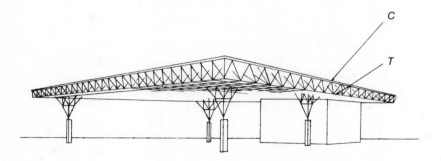

10.12.1. Planar space frame: a very deep flat plate with many holes in it to reduce the weight.

10.12.2. If the columns are equidistant in both directions, the members of the space frame (a) in each direction resist equal bending moments. The top flanges provide the compressive resistance forces, C, the bottom flanges provide the tensile resistance forces, T, and the diagonals keep them apart to provide a lever arm (b).

Plane frames, if made too deep, are liable to sideways buckling (Fig. 10.12.3). Space frames avoid this problem because of their three-dimensional character. They can therefore be made very deep, and because the resistance moment has a long lever arm, we need less material in the top and bottom chords. Space frames are therefore suitable for large spans, for example, for aircraft hangars (Fig. 10.12.4).

The final structural analysis of space frames was difficult until recently, because the only available methods produced a large number of simultaneous equations, which are laborious to solve manually. But, they can easily be written in matrix form and evaluated by computer.

However, the main problem remains to be solved: There is no wholly satisfactory method for connecting the large number of structural members, which meet at various spatial angles; far more members meet at a space-frame joint than at a plane-frame joint. The joint requires a great deal of fabrication, either on the building site (Fig. 10.12.5) or in a factory (Fig. 10.12.6).

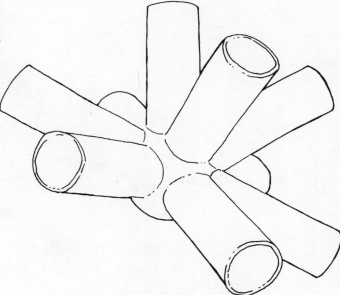

10.12.5. Welded joint for nine members meeting at a point.

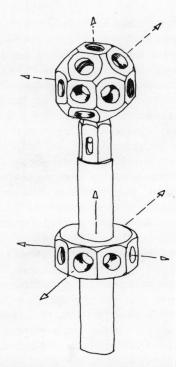

10.12.6. MERO patent joint for up to 18 members, developed by Max Mengeringhausen in Germany in 1932 and still in use.

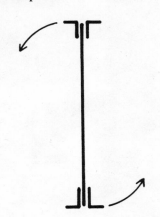

10.12.3. Plane (two-dimensional) frames can buckle sideways if they are long and deep and not suitably restrained. Planar space frames cannot buckle sideways, because diagonal members restrain them in all directions.

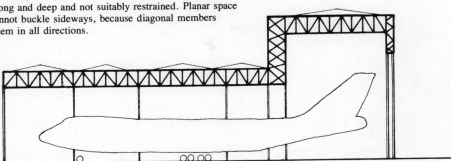

10.12.4. Space frames are particularly suitable for large spans.

10.13 A Postscript

We have not included plane trusses in this chapter, because they have already been considered in Sections 7.3, 7.4, and 7.5. However, trusses can be used for medium, if not for long spans, provided that there is sufficient room overhead to make them deep enough or that they can run the full height of a floor (Fig. 7.4.2). Deep trusses must be suitably restrained to prevent sideways buckling (Fig. 10.12.3). For some interiors (for example, school assembly halls) trusses can be left exposed in the interior.

Parallel plane trusses are simple, cheap, and versatile. They can also be used to produce curved roof structures. The architects of ancient Rome and of the Renaissance used trusses for curved roofs; for example, Wren used a truss to give the dome of St. Paul's Cathedral in London its shape with great economy in material and labor (Fig. 7.7.6). There are occasions when a structural fake is an economical and sensible answer to a problem.

For long spans, however, the structure dominates the form, because we are working near the limit of what is possible. Domes have held the record for interior span almost continuously for two thousand years. We explained the reasons for their success in Section 10.3.

In spite of the tendency for structures to grow bigger and for records to be broken, it is possible that spans will not increase much further. Sporting arenas and exhibition halls have continually set new records for the last one hundred years, but it is difficult to think of a useful purpose that a bigger span would serve. Even in existing arenas people at the back are so far from the action that they cannot see it properly. They would perhaps enjoy the event more in front of a color television set at home, and they would save a lot of gasoline. The present limit on the size of single-cell buildings, as on tall buildings (Section 8.10), is a function of transportation and usefulness rather than of structure.

Just as tall buildings cost more than low buildings, so long spans cost more than short spans. Consider when you make your preliminary design whether a few columns or loadbearing walls to shorten the span would do any harm!

10.14.0.

10.14 Problems for Chapter 10

If you have not yet read Section 5.9, please do so before attempting the following problems.

We examine the design of two small reinforced concrete domes, one hemispherical and one shallow. We next consider the amount of reinforcement required for a cylindrical concrete shell reinforced with normal reinforcing steel, or prestressed with high-tensile steel. We then determine the reinforcement required for a reinforced concrete folded-plate cantilever roof.

There are two problems on cables, one for a suspension bridge connecting two buildings and one for a circular suspension roof. Finally, we estimate the depth required for an aluminum space frame.

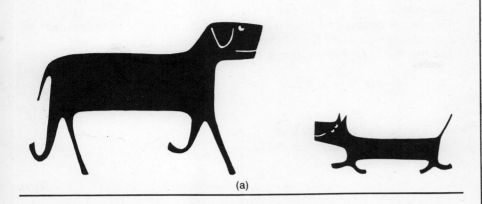
(a)

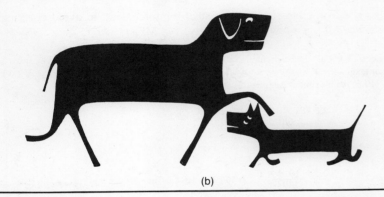
(b)

(c)

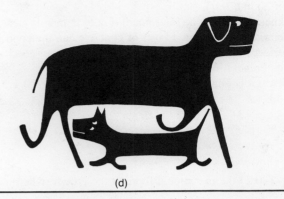
(d)

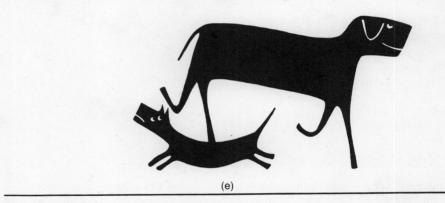
(e)

(f)

Problem 10.1 Determine the forces in a hemispherical reinforced concrete dome for a span of 20 m *(70 ft)*.

The thickness of the shell (for a small dome) is mainly determined by construction considerations. How thin a layer of concrete can the contractor put on a hemispherical surface? How heavy is the rainfall, and how good are available waterproofing membranes? We will use a 75 m *(3 in.)* shell.

The forces in a hemispherical dome are given in Fig. 10.1.3. The radius of curvature is

$R = 10$ m

$R = 35$ ft

From Eq. (10.1) the meridional force at the crown, where $\theta = 0°$ and therefore $\cos \theta = 1$, is

$$N_\theta = -wR \frac{1}{1 + \cos \theta} = -\tfrac{1}{2} wR$$

The negative sign denotes a compressive force.

From Eq. (10.2) the hoop force at the crown is

$$N_\phi = wR \left(\frac{1}{1 + \cos \theta} - \cos \theta \right) = -\tfrac{1}{2} wR$$

At the springings, where $\theta = 90°$ and $\cos \theta = 0$, the forces are

$$N_\theta = -wR$$

and

$$N_\phi = +wR$$

The positive sign denotes a tensile force.

Taking some intermediate values for 30°, 45°, and 60°, we have

	N_θ	N_ϕ
30°	$-0.536\,wR$	$-0.330\,wR$
45°	$-0.586\,wR$	$-0.121\,wR$
60°	$-0.667\,wR$	$+0.167\,wR$

Evidently the meridional forces increase from the crown to the springings, where they reach a maximum. They behave exactly like the forces in a three-pinned arch (Problems 7.6 and 7.7).

The hoop forces change from compressive to tensile between 45° and 60°. In fact, this happens at $\theta = 51°50'$ (Section 10.1).

The greatest compressive force is the meridional force at the springings, and the greatest tensile force is the hoop force at the springings.

Thus the maximum compressive stress in the concrete is

$$f_c = \frac{-wR}{D} \tag{10.6}$$

where D is the thickness of the shell.

The area of reinforcement required is

$$A_s = \frac{+wR}{f_s} \tag{10.7}$$

where f_s is the permissible stress in the steel reinforcement; to avoid cracking of the concrete we will use a low stress, namely, 140 MPa *(20 ksi)*.

Concrete weighs 24 kN/m³, and therefore a 75 mm shell weighs 1.8 kPa. The live load caused by a possible repair team and the wind load amount to another 0.5 kPa, making a total load of $w = 2.3$ kPa.

From Eq. (10.6) the maximum compressive stress in the concrete is

$$f_c = \frac{-wR}{D} = \frac{2.3 \text{ kPa} \times 10^{-3} \times 10 \text{ m}}{0.075 \text{ m}} = 0.3067 \text{ MPa}$$

From Eq. (10.7) the area of reinforcement required is

$$A_s = \frac{+wR}{f_s} = \frac{2.3 \times 10^{-3} \times 10}{140} = 164.3 \times 10^{-6} \text{ m}^2/\text{m} = 164.3 \text{ mm}^2/\text{m}$$

Concrete weighs 150 lb/ft³, and therefore a 3 in. shell weighs 37.5 psf. The live load caused by a possible repair team and the wind load amount to another 10 psf, making a total of $w = 47.5$ psf.
From Eq. (10.6) the maximum compressive stress in the concrete is

$$f_c = \frac{-wR}{D} = \frac{47.5 \text{ psf} \times 35 \text{ ft} \times 12}{3 \text{ in.}} = 6650 \text{ psf} = 46.18 \text{ psi}$$

From Eq. (10.7) the area of reinforcement required is

$$A_s = \frac{+wR}{f_s} = \frac{47.5 \text{ psf} \times 35 \text{ ft}}{20\,000 \text{ psi}} = 0.083 \text{ in.}^2/\text{ft}$$

The stresses in the concrete are only a fraction of the maximum permissible stresses, and the amount of reinforcement calculated is far less than the minimum required to resist shrinkage and temperature stresses (Fig. 10.3.7). Therefore the design of the dome is governed not by the membrane forces in it but by shrinkage and temperature stresses. However, bending stresses are set up unless support conditions allow the shell to contract and expand in accordance with the hoop force calculated for the springings. These are examined in Problem 10.2.

Problem 10.2 Determine the principal structural sizes for a shallow reinforced dome for a span of 20 m *(70 ft)*. The springings of the dome subtend an angle of 60° at the center of curvature, as shown in Fig. 10.14.1.

The span and the two radii of curvature at the springings (Fig. 10.14.1) form an equilateral triangle, and therefore

R = span = 20 m

R = *span = 70 ft*

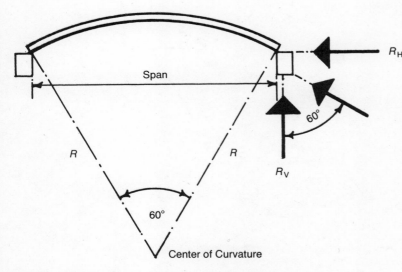

10.14.1. Span, radius of curvature, and support reactions of the shallow dome considered in Problem 10.2.

From the calculations made in Problem 10.1, it is evident that the membrane forces are not critical for the thickness or reinforcement of the shell, except at the boundaries. The meridional forces press outwards at the edge of the dome and their horizontal component has to be absorbed. This can be done with buttresses (Fig. 10.3.4) or with a tension ring (Fig. 10.3.5). In a hemispherical dome the boundary problem is obviated because the hoop tension within the dome provides the horizontal restraint.

In problem 10.1 we calculated the magnitude of the meridional force at the angle $\theta = 30°$ as

$$N_\theta = -0.536\, w\, R$$

Its horizontal component (Fig. 10.14.1) is

$$R_H = 0.536\, w\, R \cos 30°$$

If we absorb this with a tension ring, we have the force R_H (Fig. 10.14.1) pressing outwards as in a water pipe under pressure. We can use the pressure pipe as an analogy and say that a uniform hydraulic pressure R_H is resisted by two forces T when the pipe is (hypothetically) cut through its diameter (Fig. 10.14.2). Consequently

$$2\,T = R_H \cdot 20 \text{ m } (70\, ft)$$

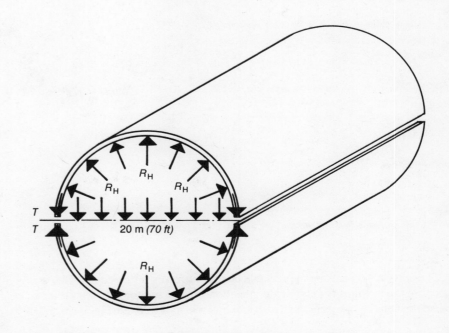

10.14.2. The horizontal reaction R_H presses on the tension ring, as a fluid does in a pressure pipe (Problem 10.2).

From Problem 10.1, $w = 2.3$ kPa, so that

$$R_H = 0.536 \times 2.3 \times 20 \times \cos 30° = 21.35 \text{ kN/m}$$

and

$$T = \tfrac{1}{2} \times 21.35 \times 20 = 213.5 \text{ kN}$$

Taking the maximum permissible stress in the steel reinforcement as 140 MPa, as in Problem 10.1, we require an area of reinforcement of

$$A_S = \frac{213.5 \times 10^{-3}}{140} = 1.525 \times 10^{-3} \text{ m}^2 = 1525 \text{ mm}^2$$

If we use the reinforcing bars, the diameter is

$$\sqrt{\frac{1525}{4} \times \frac{4}{\pi}} = 22.0 \text{ mm}$$

Canadian No. 25 bars (25.2 mm in diameter)* are suitable.

From Problem 10.1, $w = 47.5$ psf, so that

$$R_H = 0.536 \times 47.5 \times 70 \times \cos 30° = 1543 \text{ lb/ft} = 1.543 \text{ kips/ft}$$

and

$$T = \tfrac{1}{2} \times 1.543 \times 70 = 54.02 \text{ kips}$$

Taking the maximum permissible stress in the steel reinforcement as 20 ksi, as in Problem 10.1, we require an area of reinforcement of

$$A_S = \frac{54.02}{20} = 2.701 \text{ in.}^2$$

If we use four reinforcing bars, the diameter is

$$\sqrt{\frac{2.701}{4} \times \frac{4}{\pi}} = 0.927 \text{ in.}$$

No. 8 (1 in. diameter) bars are suitable.

The tension ring sets up appreciable bending stresses in the dome. The shell adjacent to the ring needs to be made much thicker and reinforced on both faces, because the bending moment changes from positive to negative (Fig. 10.3.6.b).

---

*The corresponding British size of bar is 25 mm in diameter, and the corresponding Australian size is 24 mm in diameter.

Problem 10.3 Estimate the amount of reinforcement needed for the cylindrical shell roof shown in Fig. 10.14.3.

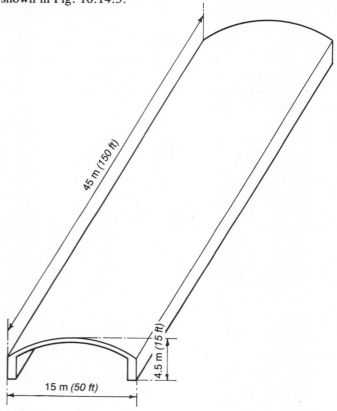

10.14.3. Dimensions of cylindrical concrete shell for Problem 10.3.

We noted in Section 10.3 and Fig. 10.3.13 that a cylindrical shell whose straight lines lie in the longer direction can be designed as a beam of curved cross section. There is ample concrete in the structure, but tension reinforcement is needed to provide the tensile component of the resistance moment.

If the roof consists of a single shell, this may be regarded as simply supported, and from Table 4.1, line 9, we have

$$M = \tfrac{1}{8} W L$$

If there are several shells cast monolithically, end to end, the structure may be regarded as a continuous beam, and the bending moment coefficients in Table 4.2 apply.

The tensile force is

$$T = \frac{M}{z}$$

where z is the lever arm of the resistance moment. The area of steel reinforcement required is

$$A_S = \frac{T}{f_s}$$

where f_s is the maximum permissible steel stress; in order to avoid cracks in the concrete, we will use a low steel stress, namely 140 MPa (*20 ksi*). We will assume that the lever arm $z = \frac{2}{3}$ (overall depth of shell). We will assume that the concrete shell is 75 mm (*3 in.*) thick, but the edge beams will need to be thicker to accommodate the reinforcement.

Concrete weighs 24 kN/m³, and a 75 mm shell weighs 1.8 kPa. We will add 0.4 kPa for the curvature and the edge beams. The live load and wind load amount to about 0.5 kPa, making a unit load of w = 2.7 kPa.

The total load is

$$W = 2.7 \times 15 \times 45 = 1823 \text{ kN}$$

and the maximum bending moment is

$$M = \tfrac{1}{8} \times 1823 \times 45 = 10\,250 \text{ kN m} = 10.25 \text{ MN m}$$

The lever arm is

$$z = \tfrac{2}{3} \times 4.5 = 3 \text{ m}$$

and the tensile force is

$$T = \frac{M}{z} = \frac{10.25}{3} = 3.417 \text{ MN}$$

This is divided between two edge beams; the force in each is 1.709 MN.

The area of reinforcement required is

$$A_S = \frac{1.709 \text{ MN}}{140 \text{ MPa}} = 0.0122 \text{ m}^2 = 12\,200 \text{ mm}^2$$

The cross-sectional area of a Canadian No. 35 (35.7 mm diameter) bar is 1000 mm², so that thirteen No. 35 bars are required.

Concrete weighs 150 lb/ft³, and a 3 in. shell weighs 37.5 psf. We will add 7.5 psf for the curvature and the edge beams. The live and wind load amount to about 10 psf, making a unit load of w = 55 psf.

The total load is

$$W = 55 \times 50 \times 150 = 412\,500 \text{ lb} = 412.5 \text{ kips}$$

and the maximum bending moment is

$$M = \tfrac{1}{8} \times 412.5 \times 150 = 7734 \text{ kips ft}$$

The lever arm is

$$z = \tfrac{2}{3} \times 15 = 10 \text{ ft}$$

and the tensile force is

$$T = \frac{M}{z} = \frac{7734}{10} = 773.4 \text{ kips}$$

This is divided between two edge beams; the force in each is 386.7 kips. The area of reinforcement required is

$$A_S = \frac{386.7 \text{ kips}}{20 \text{ ksi}} = 19.34 \text{ in.}^2$$

The cross-sectional area of a No. 12 (1½ in. diameter) bar is 1.767 in.², so that eleven No. 12 bars are required.

This is a great deal of reinforcement, which requires a large edge beam.

It is often more advantageous to prestress the edge beams. Assuming (Problem 5.11) a maximum permissible steel stress of 700 MPa (*100 ksi*) and allowing 15% for loss of prestress due to shrinkage and creep, we obtain a cross-sectional area of steel:

$$A_S = \frac{1.709 \text{ MN}}{0.85 \times 700} = 2872 \times 10^{-6} \text{ m}^2 = 2872 \text{ mm}^2$$

$$A_S = \frac{386.7}{0.85 \times 100} = 4.550 \text{ in.}^2$$

This is less than a quarter of the steel needed for the reinforced concrete shell.

Problem 10.4 A reinforced concrete folded-plate roof is cantilevered 3 m (*10 ft*) as shown in Fig. 10.14.4. Determine the amount of reinforcement needed.

We will make the concrete plate 100 mm (*4 in.*) thick. This is the normal minimum for satisfactory construction, and it is sufficient to resist the bending moments, provided that we round the internal corners and increase the thickness of the concrete nearby.

From Fig. 10.7.4

$$M = T z$$

where T is the tensile force in the steel reinforcement and z is the lever arm of the resistance moment. As a rough approximation we assume

$$z = \tfrac{1}{2} D \text{ (overall depth of folded plate)}$$

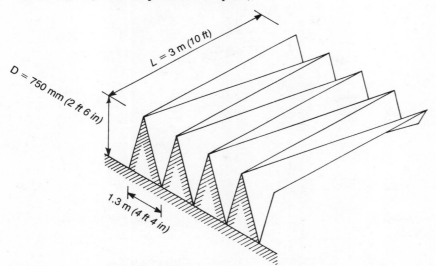

10.14.4. Dimensions of folded-plate concrete cantilever roof for Problem 10.4.

Note that the cantilever bending moment (M) has its maximum at the support, and it is negative. The full reinforcement is required at the support near the top of the folded plate. From Table 4.1, line 3,

$$M = -\tfrac{1}{2} W L$$

where W is the total load.

To avoid cracking of the concrete we will use a low stress for the steel reinforcement, namely 140 MPa (*20 ksi*).

We will consider one unit of the folded plate, 1.3 m wide. Concrete weighs 24 kN/m³, and as the plate is steeply sloping, we will assume that its weight is 1.5 times that of a flat 100 mm plate, that is, 3.6 kPa. The live load caused by a possible repair team and the snow load (assuming moderate snow fall only) amounts to 1.5 kPa, making a total of 5.1 kPa. The total load per unit is

$$W = 5.1 \times 3 \times 1.3 = 19.89 \text{ kN}$$

The maximum bending moment is

$$M = -\tfrac{1}{2} W L = \tfrac{1}{2} \times 19.89 \times 3 = 29.84 \text{ kN m (negative moment, reinforcement on top)}$$

The tensile force required for the resistance moment is

$$T = \frac{M}{z} = \frac{29.84}{\tfrac{1}{2} \times 0.75} = 79.57 \text{ kN}$$

The area of reinforcement required is

$$A_s = \frac{79.57 \times 10^{-3} \text{ MN}}{140 \text{ MPa}} = 568.4 \times 10^{-6} \text{ m}^2 = 568.4 \text{ mm}^2$$

The cross-sectional area of Canadian No. 10 (11.3 mm diameter) bars is 100 mm², so that 6 bars are required.

We will consider one unit of the folded plate, 4 ft 4 in. wide. Concrete weighs 150 lb/ft³, and as the plate is steeply sloping, we will assume that its weight is 1.5 times that of a flat 4 in. plate, that is, 75 psf. The live load caused by a possible repair team and the snow load (assuming moderate snow fall only) amount to 30 psf, making a total of 105 psf. The total load per unit is

$$W = 105 \times 10 \times 4.33 = 4547 \text{ lb} = 4.547 \text{ kips}$$

The maximum bending moment is

$$M = -\tfrac{1}{2} W L = \tfrac{1}{2} \times 4.547 \times 10 = 22.74 \text{ kips ft (negative moment, reinforcement on top)}$$

The tensile force required for the resistance moment is

$$T = \frac{M}{z} = \frac{22.74}{\tfrac{1}{2} \times 2.5} = 18.19 \text{ kips}$$

The area of reinforcement required is

$$A_s = \frac{18.19}{20} = 0.910 \text{ in.}^2$$

The cross-sectional area of a No. 4 (½ in. diameter) bar is 0.196 in.², so that 5 bars are required.

Problem 10.5 Two buildings, 15 m (*50 ft*) apart, are to be joined by a flexible suspension bridge, to carry a load of 32 kN/m (*2 kips/ft*). Determine the size of the cables.

Suspension cables are made from high-tensile steel, and the maximum permissible stress of one type is 410 MPa (*60 ksi*).

We will assume a ratio of span to sag of 7.5. This gives a sag $s = 2$ m (*6 ft 8 in.*) for a span L = 15 m (*50 ft*).

The theory of suspension cables was discussed in Section 4.7. The cable tension at mid-span (T) and the horizontal reactions, R_H, at the supports, are given by Eq. (4.6):

$$R_H = T = \frac{WL}{8s}$$

The vertical reaction at the supports is

$$R_V = \tfrac{1}{2}W$$

The resultant of the vertical and the horizontal components of the support reaction must equal the cable tension at the supports

$$R = T_o = \sqrt{R_V^2 + R_H^2}$$

This is the maximum cable tension. The result is analogous to the solution of the arch (Problem 7.7) in which the maximum thrust also occurred at the supports.

There are two cables, and each carries half the load, so that the total load on the cable is

$$W = \tfrac{1}{2} \times 32 \times 15 = 240 \text{ kN}$$

The support reactions are

$$R_V = \tfrac{1}{2} \times 240 = 120 \text{ kN}$$

and

$$R_H = \frac{240 \times 15}{8 \times 2} = 225 \text{ kN}$$

Their resultant is

$$R = \sqrt{120^2 + 225^2} = 255 \text{ kN}$$

This is also the maximum cable tension, and the area of steel required is

$$A_S = \frac{255 \times 10^{-3}}{410} = 0.6220 \times 10^{-3} \text{ m}^2 = 622 \text{ mm}^2$$

There are two cables, and each carries half the total load, so that the total load per cable is

$$W = \tfrac{1}{2} \times 2 \times 50 = 50 \text{ kips}$$

The support reactions are

$$R_V = \tfrac{1}{2} \times 50 = 25 \text{ kips}$$

and

$$R_H = \frac{50 \times 50}{8 \times 6.667} = 46.88 \text{ kips}$$

Their resultant is

$$R = \sqrt{25^2 + 46.88^2} = 53.13 \text{ kips}$$

This is also the maximum cable tension, and the area of steel required is

$$A_S = \frac{53.13}{60} = 0.8855 \text{ in.}^2$$

Suspension cables are twisted from smaller wires, and the cross-sectional area of the steel is approximately $\tfrac{2}{3} \times \tfrac{1}{4} \pi\, d^2$, where d is the diameter of the cable. We thus need a cable 35 mm ($1\tfrac{3}{8}$ *in.*) in diameter.

*Problem 10.6 Determine the size of the cables for a suspension roof of the type illustrated in Fig. 10.10.5. The cables are supported on a vertical cylinder of load-bearing walls with a diameter, L, of 60 m (*200 ft*).

The roof needs a fairly large sag, s, to be effective, and we will use

$$s = \frac{L}{5}$$

The cables are supported by an outer compression ring, which is made of reinforced concrete, and joined to a small inner tension ring, which is made of steel. Each cable carries a triangular slice of the load. This case is not listed in Table 4.1, but it is easily analyzed. The half load $\tfrac{1}{2}W$ acts at the centroid of the load triangle (Fig. 10.14.5), and therefore at $L/6$ from the compression ring. Taking moments about the compression ring, we have

$$T \cdot s = \tfrac{1}{2}W \cdot \tfrac{1}{6}L$$

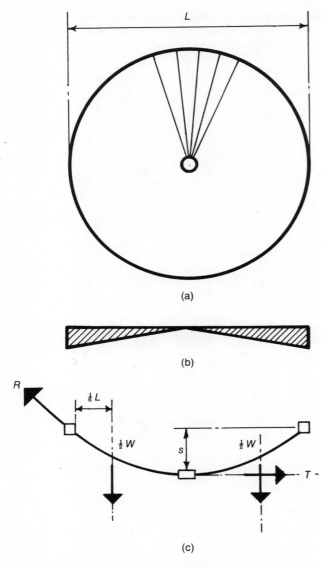

10.14.5. Load and force distribution for the circular suspension roof of Problem 10.6.
(a) Plan of roof.
(b) Distribution of load carried by each pair of cables.
(c) Forces acting on one pair of cables. The tension ring is at the lowest point at the center, and the compression ring is at the outside, at the higher level.

The horizontal support reaction is

$$R_H = T = \frac{WL}{12\,s} \tag{10.8}$$

The vertical support reaction is

$$R_V = \tfrac{1}{2} W \tag{10.9}$$

Their resultant is equal to the maximum cable tension.

The compression ring is precisely the opposite of the tension ring, which we considered in Problem 10.2. The forces are pressing inwards, as in a pipe under suction. We can therefore treat them as if they were fluid pressure, and from Fig. 10.14.6

$$P \cdot L = 2\,C$$

The roof carries only its own weight and a small live load to allow for repairs. Wind suction reduces the weight of the roof. This type of structure could not be economically built in regions subject to heavy snow falls, since the snow would accumulate in the dished roof surface and impose high loads. Water is drained off (Section 10.10). We will assume 5 kPa (*100 psf*) for the load, which allows for a deck sufficiently heavy to provide thermal insulation and dampen vibrations. The maximum permissible stresses are 410 MPa (*60 ksi*) for the cables and 12 MPa (*1.5 ksi*) for the concrete compression ring.

The circumference of the compression ring is $\pi L = 188.5$ m. If we use 62 cables (31 pairs), we obtain a spacing of almost exactly 3 m around the circumference.

The sag is

$$s = \frac{L}{5} = 12 \text{ m}$$

The total load carried by a pair of cables is

$$W = \frac{5 \text{ kPa} \times \tfrac{1}{4}\pi \times 60^2 \text{ m}^2}{31} = 456.0 \text{ kN}$$

From Eqs. (10.8) and (10.9) the reactions are

$$R_H = \frac{456 \text{ kN} \times 60 \text{ m}}{12 \times 12 \text{ m}} = 190.0 \text{ kN}$$

and

$$R_V = \tfrac{1}{2} \times 456 \text{ kN} = 228 \text{ kN}$$

The resultant is

$$R = \sqrt{190^2 + 228^2} = 296.8 \text{ kN} = 0.2968 \text{ MN}$$

The area of steel required is

$$A_s = \frac{0.2968 \text{ MN}}{410 \text{ MPa}} = 723.9 \times 10^{-6} \text{ m}^2 = 723.9 \text{ mm}^2$$

Using the formula given at the end of Problem 10.5, we find that the cable diameter is approximately

$$\sqrt{\frac{6 \times 723.9}{\pi}} = 37.19 \text{ mm (say, 40 mm)}$$

The horizontal reaction is 190 kN for each 3 m of the compression ring, or 63.33 kN/m (Fig. 10.14.6). From Eq. (10.10)

$$C = \tfrac{1}{2} \times 63.33 \times 60 = 1900 \text{ kN} = 1.9 \text{ MN}$$

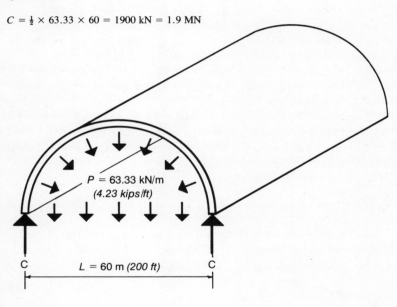

10.14.6. The horizontal reactions R_H pulls on the compression ring, as a vacuum does on a pipe (Problem 10.6).

Assuming a maximum permissible concrete stress of 12 MPa and neglecting the contribution of the reinforcement, we require a compression ring of

$$\frac{1.9 \text{ MN}}{12 \text{ MPa}} = 0.1583 \text{ m}^2$$

A section 300 mm wide by 600 mm deep has a cross-sectional area of 0.18 m².

The compression ring requires reinforcement. We will use $1\tfrac{1}{2}\%$, which adds to its compressive strength but is not allowed for in the above calculation.

$$A_s = 0.015 \times 0.18 \times 10^6 = 2700 \text{ mm}^2$$

A Canadian No. 20 (19.5 mm diameter) bar has a cross-sectional area of 300 mm², and an even number of bars is required (Fig. 10.14.7), so that we will use ten bars.

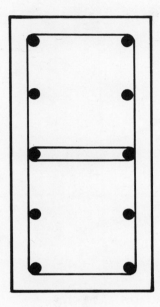

10.14.7. Layout of the reinforcement for the compression ring for the suspension roof of Problem 10.6. As the bars are in compression, ties are required at 300 mm (*12 in.*) centers to stop them from buckling. We need 2 sets of ties for the 10 bars because the American (ACI) concrete code requires that each alternate longitudinal bar be given lateral support by the corner of a tie.

The circumference of the compression ring is $\pi L = 628$ ft. If we use 62 cables (31 pairs), we obtain a spacing of almost exactly 10 ft around the circumference.

The sag is

$$s = \frac{L}{5} = 40 \text{ ft}$$

The total load carried by a pair of cables is

$$W = \frac{100 \text{ psf} \times \frac{1}{4}\pi \times 200^2 \text{ ft}^2}{31} = 101\,400 \text{ lb} = 101.4 \text{ kips}$$

From Eqs. (10.8) and (10.9) the reactions are

$$R_H = \frac{101.4 \text{ kips} \times 200 \text{ ft}}{12 \times 40 \text{ ft}} = 42.25 \text{ kips}$$

and

$$R_V = \tfrac{1}{2} \times 101.4 \text{ kips} = 50.7 \text{ kips}$$

The resultant is

$$R = \sqrt{42.25^2 + 50.7^2} = 66.0 \text{ kips}$$

The area of steel required is

$$A_s = \frac{66.0 \text{ kips}}{60 \text{ ksi}} = 1.10 \text{ in.}^2$$

Using the formula given at the end of Problem 10.5, we find that the cable diameter is approximately

$$\sqrt{\frac{6 \times 1.10}{\pi}} = 1.45 \text{ in. (say } 1\tfrac{1}{2} \text{ in.).}$$

The horizontal reaction is 42.25 kips for each 10 ft of the compression ring, or 4.225 kips/ft (Fig. 10.14.6). From Eq. (10.10)

$$C = \tfrac{1}{2} \times 4.23 \times 200 = 423 \text{ kips}$$

Assuming a maximum permissible concrete stress of 1.5 ksi, and neglecting the contribution of the reinforcement, we require a compression ring of

$$\frac{423 \text{ kips}}{1.5 \text{ ksi}} = 282 \text{ in.}^2$$

A section 12 in. wide by 24 in. deep has a cross-sectional area of 288 in.²

The compression ring requires reinforcement. We will use $1\tfrac{1}{2}$%, which adds to its compressive strength, but is not allowed for in the above calculation.

$$A_S = 0.015 \times 288 = 4.32 \text{ in.}^2$$

A No. 6 ($\tfrac{3}{4}$ in. diameter) bar has a cross-sectional area of 0.441 in.², so that ten bars are required (Fig. 10.14.7).

Problem 10.7 A space frame of the type shown in Fig. 10.12.1 is to be built from aluminum tubes. Determine a suitable arrangement of the tubes, if the column spacing is 40 m (*130 ft*) in both directions.

The frame is very light, and if the roof is not accessible to the public, only to a repair team, the live load is also low apart from the snow load, if any, that can accumulate on the flat roof.

We will assume a roof load of 2.5 kPa (*50 psf*), but this would not be sufficient in regions with heavy snowfall.

We will arrange the tubes in horizontal equilateral triangles at the flat top and flat bottom faces of the space frame, with diagonals connecting the top and the bottom to form further equilateral triangles. The entire space frame is thus formed by a series of regular tetrahedra (triangular pyramids) (Fig. 10.14.8).

We will use 150 mm (*6 in.*) diameter tubes and make the sides of the triangles in the top and bottom faces 3 m (*10 ft*). If the top and the bottom of the space frame contain the same tubes, the compressive forces are critical for the design; because of the low modulus of elasticity and the consequent tendency to buckle, the compressive strength of aluminum is lower than its tensile strength.

The radius of gyration (Section 5.3) of a very thin tube is one quarter of its diameter. For a thick tube it is slightly more than one quarter of the diameter. The slenderness ratio is therefore approximately 70.

We will assume that for the slenderness ratio proposed the aluminum alloy chosen has a permissible compressive stress of $f = 85$ MPa (*12.5 ksi*).

We will assume that the roof is simply supported on the columns, so that the maximum bending moment, from Table 4.1, line 9, is

$$M = \frac{WL}{8}$$

The space frame spans in both directions, and the column spacing is the same in both directions. As in the case of a flat plate, the full bending moment must be provided in each direction, that is,

$$M = \frac{WL}{8}$$

in each direction.

The resistance moment (Fig. 10.12.2) is

$$M = CD$$

where C is the compressive force in the top flange and D is the depth of the space frame.

The depth of the space frame made up from regular tetrahedra measuring t along their sides is given in advanced geometry books:

$$D = \sqrt{\tfrac{2}{3}}\,t = 0.8165\,t$$

This is the distance of one vertex from the centroid of the opposite face, measured along the center lines of the tubes for $t = 3$ m (10 ft).

$D = 2.45$ m

$D = 8.165$ ft

This is used for the purpose of calculating the strength of the frame. To obtain the overall depth we must add the radius of the upper and the lower tubes; this gives 2.6 m (*or 8 ft 8 in.*).

If we neglect the angle of the tubes in the horizontal plane, there are two tubes per 3 m width of frame (Fig. 10.14.8). The load carried by a 3 m width of the frame is

$$W = 2.5 \text{ kPa} \times 3 \text{ m} \times 40 \text{ m} = 300 \text{ kN m}$$

and the maximum bending moment is

$$M = \frac{WL}{8} = \frac{300 \times 40}{8} = 1500 \text{ kN m}$$

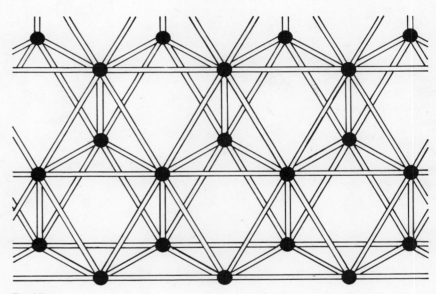

Top View

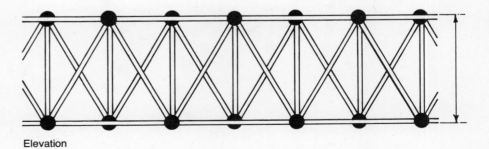

Elevation

(a)   (b)

10.14.8. Layout of space frame in Problem 10.7.

The compressive force due to 2 tubes is

$$C = \frac{M}{D} = \frac{1500 \text{ kN m}}{2.45 \text{ m}} = 612 \text{ kN}$$

and the cross-sectional area required for each tube is

$$A = \frac{C}{2f} = \frac{612 \text{ kN}}{2 \times 85 \text{ MPa}} = 3600 \text{ mm}^2$$

A 150 mm tube 8.5 mm thick has a cross-sectional area of

$$A = \tfrac{1}{4}\pi (150^2 - 133^2) = 3778 \text{ mm}^2.$$

If we neglect the angle of the tubes in the horizontal plane, there are two tubes per 10 ft width of frame (Fig. 10.14.8). The load carried by a 10 ft width of frame is

$$W = 50 \text{ psf} \times 10 \text{ ft} \times 130 \text{ ft} = 65 \text{ kips}$$

and the maximum bending moment is

$$M = \frac{WL}{8} = \frac{65 \times 130}{8} = 1056 \text{ kip ft}$$

The compressive force due to 2 tubes is

$$C = \frac{M}{D} = \frac{1056 \text{ kip ft}}{8.165 \text{ ft}} = 129.4 \text{ kips}$$

and the cross-sectional area required for each tube is

$$A = \frac{C}{2f} = \frac{129.4 \text{ kips}}{2 \times 12.5} = 5.18 \text{ in.}^2$$

A 6 in. tube $\tfrac{5}{16}$ in. thick has a cross-sectional area of

$$A = \tfrac{1}{4}\pi (6^2 - 5.375^2) = 5.583 \text{ in.}^2$$

References

10.1 ELEMENTARY STATICS OF SHELLS, Second Edition, by Alf Pflüger. F.W. Dodge, New York, 1961. 122 pp.

10.2 CANDELA: THE SHELL BUILDER by Colin Faber. Reinhold, New York, 1963. 240 pp.

10.3 TENSILE STRUCTURES, Volume 1: Pneumatic Structures, Volume 2: Cables, Nets and Membranes, by Frei Otto. M.I.T. Press, Cambridge, 1967 and 1969. 320 + 171 pp.

10.4 PNEUMATIC STRUCTURES by Thomas Herzog. Crosby Lockwood Staples, London, 1977. 192 pp.

10.5 "SYDNEY OPERA HOUSE," *Structural Engineer*, Vol. 47, No. 3 (March 1969), by O.N. Arup and G.J. Zunz. pp. 99–132.

Chapter 11 The Structure and the Environment

My object all sublime
I shall achieve in time.
(W. S. Gilbert, The Mikado)

This chapter, by way of a postscript, examines the interaction between the structural and environmental systems. It is intended not as an abridged treatise on environmental problems but merely as a statement of the potential for utilizing the structure.

11.1 The Cost of the Structural System and Its Interaction with the Environmental System Prior to the Nineteenth Century

In the great buildings of the Middle Ages and the Renaissance the walls were generally loadbearing, and therefore part of the structure. Buildings had only primitive heating and no artificial cooling or ventilation. In summer solid masonry buildings were often pleasantly cool on a hot afternoon because of the thermal inertia (see Glossary) of the thick masonry walls and roofs: the structure performed a secondary function by acting as a thermal store (Fig. 11.1.1). In most parts of the world it is only necessary to reduce the temperature a few degrees below the outside shade temperature to achieve pleasant conditions, in summer, so that the thermal inertia was quite effective.

Winter heating presented greater problems, because a much larger temperature change might be required for thermal comfort. The thermal inertia of the walls was equally helpful in reverse, but it was not sufficient in cool climates. The great height of many monumental buildings of the past created additional problems. Even today many medieval cathedrals are unheated and are considered unheatable at a reasonable cost. People dressed warmly, and on cold days they took along a hand or foot warmer. In a thirteenth-century manuscript, preserved in the Bibliothèque Nationale in Paris, the Gothic master builder Villard de Honnecourt described a hand warmer in the form of a brass apple made from two fitting halves (Ref. 11.1, p. 66): "Inside this brass apple there must be six brass rings, each with two pivots. The two pivots must be alternated in such a way that the brazier remains upright, for each ring bears the pivots of the others. If you follow the instruction and the drawings, the coals will never drop out, no matter which way the brazier is turned. A bishop may freely use this device at High Mass; his hands will not get cold as long as the fire lasts."

11.1.1. In traditional architecture, thick masonry walls performed a secondary function as a thermal store.

Residential buildings were heated, but the cost of the heat source was not high; it might be an open fire place or a stove. Only in ancient Rome for about five centuries and in northeastern Asia from late medieval times was hot air conveyed in ducts to heat the building (Refs. 1.4, pp. 91–92, and 2.1, p. 239).

The structure also played its part in sunshading and in natural ventilation. Colonnades that provided summer shading sometimes contained buttresses to absorb horizontal reactions from the roof. Openings in the loadbearing walls were often arranged with great skill in the traditional architecture of Middle Eastern and Mediterranean countries to give the best natural ventilation possible, as surviving buildings testify. Ventilation ducts were not used before the nineteenth century.

Windows for natural lighting were generally openings in loadbearing walls. Artificial lighting prior to the nineteenth century was mainly for the wealthy, because tallow, vegetable oil, and fish oil were all edible and too precious to burn in large quantities (Fig. 11.1.2).

11.1.2. Prior to the nineteenth century most substances available for artificial lighting were also edible.

Passenger elevators were used only in mines (Section 8.1). Therefore the cost of the building was almost identical with the cost of the structure prior to the nineteenth century.

11.2 The Cost of the Structure is No Longer the Largest Part of the Cost of a Building

During the nineteenth and twentieth centuries there was a rapid increase in the standard of living of the majority of the population in most countries of North America, Europe, and Australasia; this resulted in much higher standards of heating and artificial lighting and in the introduction of artificial ventilation, air conditioning, and passenger elevators. Loadbearing walls were for most long-span and multistory buildings replaced by nonloadbearing walls. Structural systems became more efficient and used less structural material.

As a result the cost of the structure no longer dominates the cost of the building as it did prior to the nineteenth century, except for long-span buildings. For multistory buildings, which are the bread-and-butter of most architectural practices, the structure accounts for 15% to 40% of the total cost. The building services frequently cost twice as much as the structure.

The coordination of the structural and environmental systems is therefore important. It is generally not worthwhile to optimize the cost of the structural system if this increases the cost of the environmental system.

The environmental system requires an initial capital outlay of, say, E_c dollars and an annual charge for maintenance and energy (electricity, oil, gas, solar energy) of e_m dollars. This annual charge can be capitalized into an amount of E_m dollars by determining the number of dollars we would need to borrow from a bank today to pay the annual bills. Thus the equivalent capital cost of the environmental system is

$$E = E_c + E_m$$

Similarly we can determine the equivalent cost of the structural system:

$$S = S_c + S_m$$

except that S_m is always much smaller than S_c, whereas E_m could be larger then E_c. To optimize the cost of the building we must consider the sum $S + E$.

Certain structural decisions increase the cost of the environmental system; in particular, the use of a light structural system emphasized by a glass curtain wall adds to the cost of the air conditioning.

The leading architects of the Modern Movement (notably Ludwig Mies van der Rohe, but also Walter Gropius and Le Corbusier) were all attracted to the use of glass, which showed how light a structure could be in the twentieth century (Fig. 11.2.1). Ludwig Hilberseimer, an associate of Mies van der Rohe, first at the Bauhaus and later at the Illinois Institute of Technology, claimed that in Mies's buildings, "the disunity between architecture and engineering has been overcome; the engineer, once the servant of the architect, is now his equal" (Ref. 11.3, p. 21).

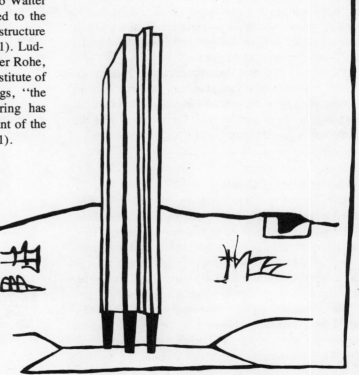

11.2.1. Glass showed how light a structure could be in the twentieth century. *(Drawing after Le Corbusier.)*

This applies only to the structure. The environmental design is subordinated to the visual effect of the metal-and-glass curtain wall, the essential ingredient of the Miesian style. The perfection of air conditioning arrived just in time. Indeed, it is doubtful whether glass-walled buildings could have been successful without it.

The decision to use glass walls should be made for the right reason; it is an aesthetic, not a technical decision.

Sir Henry Wotton in the seventeenth century introduced his *Elements of Architecture* with the statement, "Well building has three conditions. Commodity, Firmness, and Delight" (Ref. 11.2 p. 1). It is perfectly proper for the designer of a building to modify the commodity and the firmness of a building to increase the delight, but one should not do so under the guise of emphasizing the firmness.

Prior to 1973 it was only necessary to balance the capital cost and the annual charge of the additional air conditioning required against the increased delight, as long as the client agreed with the architect that the glass wall was a thing of beauty.

11.3 The Energy Crisis

In 1973 political events in the Middle East produced a shortage of oil, followed by a steep increase in the price of oil and other sources of energy. This had a profound effect on governments and on architects (Ref. 11.12). It was generally felt that architects had a moral obligation to conserve energy, and some authorities enacted rules to make it compulsory.

The increased energy consumption of buildings would have caused problems long before the end of this century even if the 1973 crisis had not happened (Fig. 11.3.1). In developed countries buildings are by far the largest consumers of energy. In 1972 buildings accounted for about 50% of the total energy consumption in the United States (Ref. 11.4, pp. ix and 117) and Great Britain (Ref. 11.5). In Australia, because of its milder climate, the figure was only about 20% (Ref. 11.6), but this is still a very large proportion of the total energy consumption. Furthermore, the energy consumption in North America, Western Europe, and Australasia was increasing in 1972 at the rate of about 5% per year; at that rate consumption *doubles* every fourteen years.

Since energy resources are limited, this rate of increase could not continue indefinitely, and the greatest scope for energy conservation is probably in buildings. Solar energy has been much discussed during the last few years, and the correct choice of structure and of the fabric of the building can make a significant contribution (Ref. 11.13).

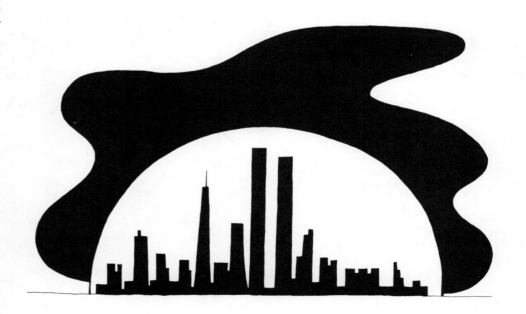

11.3.1. The increased energy consumption of buildings would in any case have caused problems before the end of this century.

11.4 Solar Energy

The use of solar energy can be divided into four main categories (Fig. 11.4.1):

1. Solar energy is the basis of most other forms of energy used today. For example, the sun evaporates water, which falls on high ground and then drives turbines to produce hydroelectric power; solar energy produces wind (Fig. 11.4.2); solar energy makes trees grow, which can then be burned to produce heat. These traditional applications of solar energy are already being exploited, and they can be extended only to a limited extent.

2. Solar energy can be used to produce electricity more directly. For example, the sun's rays could be focused by a large mirror on a container of water to produce high-pressure steam, which then drives a turbine; or a fast-growing crop, such as cassava, could be converted into alcohol, which then fuels an internal combustion engine. In either case the heat engine would then drive an electric generator. These methods will remain more expensive than conventional forms of electricity generation, until there is a further increase in the cost of fossil fuel.

3. Solar collectors can be placed on each building to heat water. The hot water can be used directly as hot water, or it can heat the building, or it can provide the heat source for an air conditioner. The supply of hot water is the most efficient of these operations. It is at present economical in areas where the cost of energy is high, for example, in isolated settlements or in small towns with poor access to transport, provided that there is enough sunshine.

11.4.1 Solar energy.

11.4.2. Solar energy produces wind.

4. Solar energy can be collected through ordinary windows. The heat can also be stored in a mass of material, such as concrete, a rock store, or a water tank, during the day, and the stored heat used during the night. Furthermore, the temperature of the thermal store can be lowered during the night, and the coolness can then be used during the heat of the day.

Only items 3 and 4 are of immediate concern to architects. In the fast-growing literature on solar energy, item 3 (that is, the use of solar collectors) is known as active solar energy; item 4 is known as passive solar energy.

Active solar energy is at the time of writing (1979) more expensive in most parts of the world than conventional forms of energy. It is neither free nor inexhaustible, because the collectors use materials that are not available in unlimited quantities; the materials must be bought, wrought, and installed.

Some passive solar energy must also be paid for by the use of labor and materials, but a certain amount of it can be obtained virtually free by considering the energy system when making structural decisions.

11.5 Thermal Inertia

Thermal inertia is desirable in most buildings because it smoothes out the extremities of the temperature cycle (Fig. 11.5.1). However, there are exceptions; for example, if an apartment is heated electrically for only part of the day because the occupants are out all day, a low thermal inertia permits more rapid heating when the electricity is switched on (Fig. 11.5.2).

The thermal inertia of the building thus provides a heat store (or cold store) that can be drawn on when required. The quantity of heat stored is the product of a material's mass, specific heat, and temperature change. Most structural materials have a high thermal inertia; this is particularly true of concrete and brick because of their massiveness and density. Very simple utilization of this thermal store is possible if the building can be oriented and shaded correctly (Fig. 11.5.3).

Where a large mass of concrete exists in direct contact with the ground (for example, in a basement), this can be utilized through an air or water circulation system (Fig. 11.5.4).

The thermal inertia of the structure of the building is today rarely utilized, because the structure is the province of the structural consultant. However, if passive solar design is contemplated, the use of the structure as a thermal store can lead to appreciable savings. If ducting is needed within the concrete, then due allowance must be made in the structural design for the resulting loss of strength.

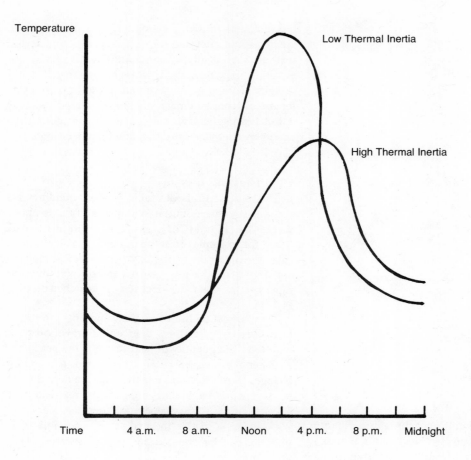

11.5.1. Variation of daily indoor temperature in a building that is unheated or continuously heated. A structure and building fabric with a high thermal inertia smooth out the extremities of the temperature cycle more than those with low thermal inertia.

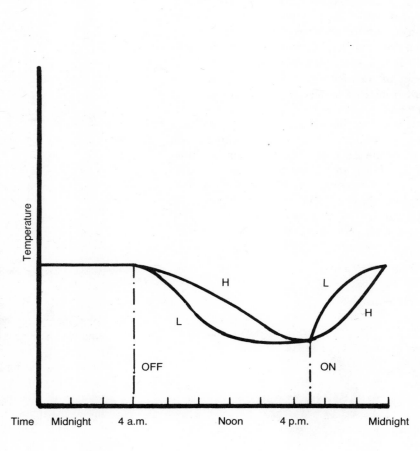

11.5.2. Variation of indoor temperature in an apartment that is heated only from 5 p.m. to 4 a.m. because the occupants are out during the day. If the thermal inertia of the structure and the building fabric are high (H), heating in the evening proceeds more slowly for the same fuel consumption than if the thermal inertia is low (L). Cooling also proceeds more slowly, but this occurs mostly while the occupants are out of the apartment.

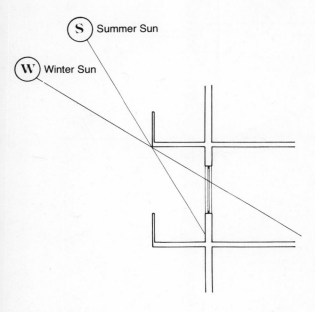

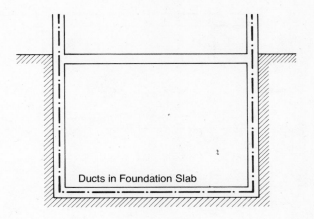

11.5.3. Buildings in or near the subtropics whose principal facades can be oriented north and south are suitable for a simple and direct utilization of solar energy. The floor slab is designed to project beyond the southern (northern in the southern hemisphere) facade by a distance that depends on the latitude (Ref. 11.8). This allows the sun to shine through the window in winter (W) but excludes it in summer (S).

Sunlight can be wholly excluded for about two months during the hottest time of summer, and fully admitted for about two months during the coldest time of winter. Sunlight penetration can be designed to commence at noon one equinox and to terminate at noon at the other. An ordinary large window thus acts as an effective and cheap solar collector.

The solar heating can be enhanced by the thermal inertia of the reinforced concrete floor, provided that a hard floor surface such as concrete, tiles, or terrazzo is acceptable. The solar heat is then stored in the concrete slab and emitted gradually after sunset. The thermal inertia is reduced if wall-to-wall carpeting, which insulates the concrete slab, is used.

As with all solar heating systems, it is necessary to have an alternative heat source for nighttime and for days when the sun is not shining.

11.5.4. Heat or coolness can be stored in a reinforced concrete basement by pumping air or water at a different temperature through ducts buried in the concrete. The additional cost consists of the ducts, a pump, and the extra thickness of concrete necessary because of the loss of cross-sectional area due to the ducts. However, the foundation slab can take the place of a thermal store, and the soil or rock surrounding the foundation further augments its capacity.

11.6 Thermal Insulation

Thermal insulation is probably the most effective way at present to conserve energy in buildings (Ref. 11.14). It was considered a matter of national interest even before 1973 in some countries, which passed legislation requiring minimum standards of insulation for all new buildings. Since then the requirements have been made more stringent in Great Britain (Ref. 11.8) and other European countries, and rules have been introduced in the United States, Canada, and New Zealand (Refs. 11.9, 11.10, and 11.11).

Materials with high thermal inertia do not necessarily provide good thermal insulation and vice versa, as conductivity is related to density. Still air is an excellent thermal insulator, and consequently materials that enclose air in small separate cells are also good insulators. Some useful and inexpensive materials in this category are wood wool, foamed concrete, and mineral wool. These materials have negligible strength, and thermal insulation is therefore generally designed separately from the structure.

However, some structural members supply thermal insulation that can be utilized in the environmental design of the building. Since metals are excellent conductors of heat, they are not insulators. Concrete walls or roofs provide some insulation, as do brick and block walls, although it generally needs supplementation. In Great Britain and Australasia walls are often built with two leaves separated by a cavity or air space. The insulation of the cavity wall can be improved by filling the cavity with an insulating foam, but this complicates moistureproofing.

11.7 Sunshading and Natural Lighting

Sunshading devices should ideally allow the sun to penetrate the window glass in winter (except in the tropics) and shade the glass in the summer (Fig. 11.5.3). Normal window glass transmits heat radiated at a high temperature but not at a low temperature. Solar heat is therefore transmitted, but reradiated interior heat is not. This is why glass is so effective for the utilization of solar energy, but it is also why rooms overheat readily in summer if the windows are not shaded.

The ideal shading device does its job without interfering with natural ventilation and without blocking natural light or the view that is one of the attractions of many high-rise buildings. On southern (or northern) facades sunshading devices should therefore be above the window, and on eastern or western facades they need also to be adjacent to the window. In the first case a projecting floor slab may provide the requisite shade (Fig. 11.5.3); in the second a projecting column may give at least partial shading. Because of the angle of inclination of the sun in the morning and afternoon it is almost impossible to provide complete sunshading on eastern or western facades.

A projecting floor slab serving as a sunshade also provides a platform for cleaning windows, and it improves the distribution of bending moment in the floor slab (Fig. 11.7.1).

Correct design of sunshading is important both for the environmental quality of the building and for energy conservation. In many multistory buildings designed in the 1950s and 1960s it has been necessary to keep the blinds almost permanently drawn because too much solar heat was otherwise admitted. As the rooms were often too dark with the blinds drawn, the electric lights were in use even on a bright, sunny day. The artificial lighting produced a great deal of heat, and this had to be removed by the air conditioning system. As this eventuality was generally anticipated, the air conditioning plant was designed accordingly.

Evidently, better sunshading would reduce the size of the air conditioning system and save both capital and running costs; it would reduce the electricity consumption for artificial lighting; and it would restore daylight and a view of the outside world, which most people find agreeable. The structure can often, if not always, be used for this purpose.

11.8 The Structure and Sound Insulation

There are basically two kinds of sound: airborne sound and impact sound (Fig. 11.8.1). Impact sound can be stopped only by floating the floor or by a thick carpet; however, airborne sound can be attenuated (see Glossary) by a thick concrete floor slab.

Insulation against airborne sound depends mainly on the mass of material between the sound source and the ear (Ref. 11.16). Thus the requirements for thermal insulation and for sound insulation are quite different (Fig. 11.8.2), but the requirements for sound insulation and thermal inertia are similar (Section 11.5). A 100 mm (4 in.) reinforced concrete slab does not provide enough sound attenuation for a noisy television set on the floor above, but a 200 mm (8 in.) slab usually does.

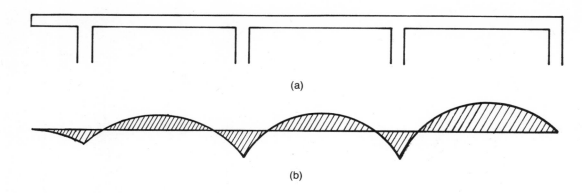

11.7.1. A projecting floor slab used as a sunshade (a) also improves the distribution of bending moment (b) in the slab. The roof slab has three spans and a sunshading roof overhang on the southern facade. The maximum bending moments without the overhang (on the right-hand side) are larger than those on the left-hand side.

11.8.1. Airborne sound (a) and impact sound (b).

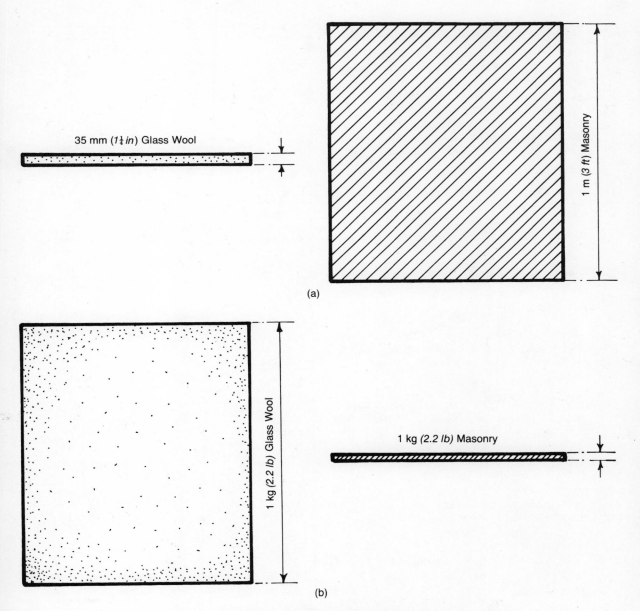

11.8.2. (a) 35 mm (1¼ in.) of glass wool and 1 m (3 ft) of masonry provide the same thermal insulation.
(b) 1 kg (2.2 lb) of glass wool and 1 kg of masonry provide the same insulation against airborne sound.

The sound insulation provided by a thick concrete floor slab is nullified if the windows are open in both apartments. Since open windows are frequently necessary for thermal comfort in the summer when there is no air conditioning, there can be a conflict between aural comfort and thermal comfort.

11.9 The Structure and the Building Services

All high-rise buildings require a service core. To an increasing extent service cores are also used in low-rise and even in single-story buildings, because all pipes and ducts are then in a single unit, which can be prefabricated in a factory.

In tall buildings the elevators dominate the design of the service core. Fire regulations almost invariably require that the elevator shafts be of reinforced concrete. Thus the service core becomes a major component of the structure of the building (Figs. 8.5.5 through 8.5.10). It may even support the entire structure (Fig. 5.2.2).

11.10 Coordination of Structural and Environmental Design

The structural and the environmental systems therefore interact at several points. It is desirable to note these interactions at the preliminary design stage and either propose a solution or bring them to the attention of the structural and the mechanical/electrical consultants so that they may develop a joint approach. This has become a matter of particular importance since the energy crisis of 1973. It is no longer sufficient to optimize "commodity, firmness, and delight"; the energy consumption of the building must also be kept within acceptable levels. Energy conservation will remain an issue in architectural design until a new source of plentiful and reasonably cheap energy is developed.

References

11.1 THE SKETCHBOOK OF VILLARD DE HONNECOURT, edited by T. Bowie. Indiana University Press, Bloomington, 1968. 144 pp.

11.2 THE ELEMENTS OF ARCHITECTURE by Sir Henry Wotton, a Facsimile Reprint of the First Edition (London, 1624) with Introduction and Notes by Frederick Hard. The University Press of Virginia, Charlottesville, 1968. 123 + 139 pp.

11.3 MIES VAN DER ROHE by L. Hilbereimer. Paul Theobald, Chicago, 1956. 200 pp.

11.4 ENERGY CONSERVATION AND ENERGY MANAGEMENT IN BUILDINGS, edited by A. F. C. Sherratt. Applied Science, London, 1976. 330 pp.

11.5 A.F.C. Sherratt in Ref. 11.4, p. ix, and F.S. Dubin in Ref. 11.4, p. 117.

11.6 ENERGY CONSERVATION: A STUDY OF ENERGY CONSUMPTION IN BUILDINGS AND POSSIBLE MEANS OF SAVING ENERGY IN HOUSING—A BRE WORKING PARTY REPORT. Building Research Establishment Current Paper CP 56/75, Garston (England), June 1975. 64 pp.

11.7 B. Rawlings in: REPORT OF THE TASK FORCE ON ENERGY OF THE INSTITUTION OF ENGINEERS, AUSTRALIA. Conference on Energy 1977: Towards an Energy Policy for Australia, Submissions of Working Parties. Institution of Engineers, Australia, Canberra, 1977, p. 217.

11.8 BUILDINGS IN THE TROPICS by Georg Lippsmeier. Callwey, Munich, 1969. pp. 57–71.

11.9 THE BUILDING (SECOND AMENDMENT) REGULATIONS 1976. Great Britain, Statutory Instrument No. 1676. H. M. Stationery Office, London, 1976. 307 pp.

11.10 ENERGY CONSERVATION IN NEW BUILDING DESIGN. ASHRAE Standard 90–75, New York, 1975. 53 pp. *This is a National Voluntary Consensus Standard for new buildings, made mandatory for many government buildings.*

11.11 CANADIAN CODE FOR ENERGY CONSERVATION IN NEW BUILDINGS. Associate Committee for the National Building Code, National Research Council. Draft for Public Comment, Ottawa, June, 1977. 45 pp.

11.12 MODEL BUILDING BYLAW NZS 1900, Chapter 4, 1977. Standards Association of New Zealand, Wellington, 1977.

11.13 ARCHITECTURE AND ENERGY—CONSERVING ENERGY THROUGH RATIONAL DESIGN by Richard G. Stein. Doubleday, New York, 1977. 322 pp.

11.14 SOLAR ENERGY: FUNDAMENTALS IN BUILDING DESIGN, by Bruce Anderson. McGraw-Hill, New York, 1977. 374 pp.

11.15 THERMAL PERFORMANCE OF BUILDINGS by J.F. van Straaten. Elsevier, London, 1967. 311 pp.

11.16 MECHANICAL AND ELECTRICAL EQUIPMENT FOR BUILDINGS by W.J. McGuinness and B. Stein. Wiley, New York, 1970. 1011 pp.

11.17 ACOUSTICS, NOISE AND BUILDINGS by P.H. Parkin, H.R. Humphrey, and J.R. Crowell. Faber, London, 1979. 297 pp.

Suggestions for Futher Reading

Only titles are given for books previously quoted as references. Readers requiring full details should turn to the end of the appropriate chapter.

12.1 General Descriptive Books on Architectural Structures

CONCEPTS OF STRUCTURE by W. Zuk. Reinhold, New York, 1963. 80 pp.

BUILDING STRUCTURES PRIMER by J.E. Ambrose. Wiley, New York, 1967. 123 pp.

STRUCTURE by H.W. Rosenthal. Macmillan, London, 1972. 154 pp.

STRUCTURE: THE ESSENCE OF ARCHITECTURE by F. Wilson. Studio Vista, London, and Van Nostrand Reinhold, New York, 1971. 96 pp.

WHAT IT FEELS LIKE TO BE A BUILDING by F. Wilson. Doubleday, New York, 1969. 62 pp.

THE ELEMENTS OF STRUCTURE by W. Morgan, Second Edition, edited by I. Buckle. Pitman, London, 1977. 252 pp.

STRUCTURE IN ARCHITECTURE, Second Edition, by M. Salvadori and R. Heller. Prentice-Hall, Englewood Cliffs, New Jersey, 1975. 414 pp.

THE MASTERBUILDERS (Ref. 1.4).

SCIENCE AND BUILDING (Ref. 2.1).

AN HISTORICAL OUTLINE OF ARCHITECTURAL SCIENCE, Second Edition, by H.J. Cowan. Applied Science, London, and Elsevier, New York, 1977. 202 pp.

STRUCTURE AND FORM IN MODERN ARCHITECTURE by C. Siegel. Crosby Lockwood, London, 1961. 308 pp.

PHILOSOPHY OF STRUCTURES by E. Torroja. University of California Press, Berkeley, 1962. 366 pp.

BUILDING FAILURES by T. H. McKaig. McGraw-Hill, New York, 1962. 261 pp.

CONSTRUCTION FAILURE by J. Feld. Wiley, New York, 1968. 399 pp.

STRUCTURE—AN ARCHITECT'S APPROACH by H.S. Howard. McGraw-Hill, New York, 1966. 297 pp.

12.2 Descriptive Books on Specific Aspects of Architectural Structures

SURFACE STRUCTURES IN BUILDING by F. Angerer. Alec Tiranti, London, 1961. 142 pp.

THE TURNING POINT OF BUILDING by K. Wachsmann. Reinhold, New York, 1961. 239 pp.

STEEL SPACE STRUCTURES by Z.S. Makowski. Michael Joseph, London, 1965. 214 pp.

STRUCTURES by P.L. Nervi. McGraw-Hill, New York, 1956. 118 pp.

THE STRUCTURES OF EDUARDO TORROJA. Dodge, New York, 1958. 198 pp.

CANDELA: THE SHELL BUILDER (Ref. 10.2).

SCHALENBAU (Shell Structures, in German, but mostly illustrations) by J. Joedicke. Krämer, Stuttgart, 1962. 304 pp.

THE DYMAXION WORLD OF BUCKMINSTER FULLER by R.W. Marks. Reinhold, New York, 1960. 232 pp.

TENSILE STRUCTURES (Ref. 10.3).

FREI OTTO—STRUCTURES by C. Roland. Longman, London, 1970. 172 pp.

PRINCIPLES OF PNEUMATIC STRUCTURES by R.N. Dent. Architectural Press, London, 1971. 236 pp.

PNEUMATIC STRUCTURES (Ref. 10.4).

See also Section 12.6

12.3 Elementary Mathematical Books on Architectural Structures

ARCHITECTURAL STRUCTURES (Ref. 1.1).

SIMPLIFIED ENGINEERING FOR ARCHITECTS AND BUILDERS (Ref. 1.2).

STRUCTURAL MECHANICS (Ref. 1.3).

INTRODUCTION TO STRUCTURAL MECHANICS by T.J. Reynolds, L.E. Kent and D.W. Lazenby. English Universities Press, London, 1973. 454 pp.

STRUCTURAL DESIGN IN ARCHITECTURE by M. Salvadori and M. Levy. Prentice-Hall, Englewood Cliffs, New Jersey, 1967. 457 pp.

STATICS, STRUCTURES AND STRESS by W. Fisher Cassie. Longman, London, 1973. 557 pp.

STATICS by J.L. Meriam. Wiley, New York, 1966. 324 pp.

PROGRAMMED PRINCIPLES OF STATICS by G.E. Nevill. Wiley, New York, 1960. 184 pp.

12.4 More Advanced Books on the Theory of Structures

STRUCTURAL ENGINEERING by R.N. White, P. Gergely and R.G. Sexsmith. Wiley, New York, 1972–. Four volumes (three volumes published).

BASIC STRUCTURAL ANALYSIS by K.H. Gerstle. Prentice-Hall, Englewood Cliffs, New Jersey, 1974. 498 pp.

THE ANALYSIS OF ENGINEERING STRUCTURES by A.J.S. Pippard and J. Baker. Arnold, London, 1968. 578 pp.

STRUCTURAL ANALYSIS by J.C. McCormac. International Textbook Company, Scranton, Pennsylvania, 1967. 494 pp.

DESIGN OF BUILDING FRAMES (Ref. 8.6).

12.5 Books on the Design of Timber, Steel and Concrete Structures

TIMBER CONSTRUCTION MANUAL (Ref. 5.1).

STEEL BUILDINGS: ANALYSIS AND DESIGN (Ref. 5.2) *(American Code)*

STRUCTURAL STEEL DESIGN (Ref. 8.5) *(American Code)*

STEEL DESIGNERS' MANUAL. Crosby Lockwood Staples, London, 1972. 1089 pp. *(British Code)*

STEEL DESIGNER'S HANDBOOK by R.E. Gorenc and R. Tinyou. New South Wales University Press, Sydney, 1973. 306 pp. *(Australian Code)*

REINFORCED CONCRETE FUNDAMENTALS (Ref. 5.7) *(American Code)*

DESIGN OF PRESTRESSED CONCRETE STRUCTURES by T.Y. Lin. Wiley, New York, 1963. 614 pp.

REINFORCED AND PRESTRESSED CONCRETE by F.K. Kong and R.H. Evans. Nelson, London, 1975. 233 pp. *(British Code)*

THE DESIGN OF REINFORCED CONCRETE IN ACCORDANCE WITH THE METRIC SAA STRUCTURES CODE by H.J. Cowan. Sydney University Press, Sydney, 1976. 220 pp. *(Australian Code)*

12.6 Technical or Mathematical Books on Specific Aspects of Structural Design

PLANNING AND DESIGN OF TALL BUILDINGS published by the American Society of Civil Engineers, New York, 1978–1981. Five Volumes, Approximately 500 pp.

This treatise contains chapters on dead loads, live loads, temperature effects, earthquake loading, wind loading, fire and blast, safety factors, and on all aspects of the design of multistory buildings, including those with load-bearing walls.

WIND LOADING ON BUILDINGS by A.J. MacDonald. Applied Science, London, 1975. 219 pp.

FIRE AND BUILDINGS (Ref. 2.2).

MODELS IN ARCHITECTURE (Ref. 6.2) on model analysis of structures.

COMPUTER METHODS IN STRUCTURAL ANALYSIS (Ref. 8.7).

FUNDAMENTAL FOUNDATIONS by W.F. Cassie. Elsevier, London 1968. 226 pp.

THE DESIGN OF FOUNDATIONS FOR BUILDINGS by S. Johnson and T. Kavanagh. McGraw-Hill, New York, 1968. 416 pp.

FOUNDATION FAILURES by C. Szechy. Concrete Publications, London, 1961. 141 pp.

ELEMENTARY STATICS OF SHELLS (Ref. 10.1).

THIN SHELL CONCRETE STRUCTURES by D.P. Billington. McGraw-Hill, New York, 1965. 332 pp.

DESIGN OF THIN CONCRETE SHELLS by A.M. Haas. Wiley, New York, 1962 and 1967. 129 + 242 pp.

DESIGN AND CONSTRUCTION OF CONCRETE SHELL ROOFS by G.S. Ramaswamy. McGraw-Hill, New York, 1968. 641 pp.

THE ANALYSIS OF BRACED DOMES by B.S. Benjamin. Asia Publishing House, London, 1963. 110 pp.

See also Section 12.2

Glossary

Words in italics are further defined in alphabetical order.

ARCH   A curved structure designed to carry loads across a gap mainly by compression. See also *cable*.

ATTENUATION OF SOUND   Reducing the loudness of sound.

BEAM   A structural member designed to carry loads across a gap by bending.

BEAM-AND-SLAB FLOOR   A floor system, particularly in reinforced concrete, whose floor slab is supported by beams; see also *flat slab*.

BENDING MOMENT *(M)*   The *moment* at any section in a structural member due to all the forces to one side of that section. It produces bending in beams.

BENDING MOMENT DIAGRAM   Diagram showing the variation along the span of a member of the moment that is tending to bend that member. It indicates whether the bending moment is *positive* or *negative* and where the maximum bending moment occurs.

BRACING   The use of *diagonal tension* members to prevent the lateral movement of a structure due to wind or other horizontal forces.

BRITTLENESS   A tendency to crack or fracture without appreciable deformation; the opposite of *ductility*. Concrete is a brittle material.

BUCKLING   A *column* or other compression member is said to buckle if it loses its stability and bends sideways. The structural material is not necessarily damaged by buckling. It is a type of failure associated with high *slenderness ratios*.

BUILT-IN SUPPORT   A support that prevents the ends from rotating.

CABLE   A structural member that is flexible and can therefore resist only tension, but not compression or flexure. See also *arch*.

CANTILEVER A structural member *built in* at one end and unsupported at the other end. See also *simply supported beam*.

CAPILLARY TUBE A tube with a very fine bore. If the tube is dipped in a bucket of water, the water level in it rises above that in the bucket as a result of surface tension.

CATENARY The curve assumed by a freely hanging cable under its own weight.

CENTER OF GRAVITY The point in a cross section through which the weight of a body acts. It can be determined easily by experiment or by calculation.

CENTROID *Center of gravity*.

CLOCKWISE The direction in which the hands of a clock move. The movement of a screwdriver driving a right-handed screw into its hole is clockwise.

COLUMN A vertical member of a frame carrying a compressive load.

COLUMN HEAD Enlargement of the top of a column in *flat slab* construction. It is designed to increase the shear resistance of the flat slab.

COMPOSITE CONSTRUCTION Structural steel sections encased in concrete and designed to allow for the strength of the concrete (which is neglected in most structural-steel design calculations).

CONCRETE Artificial stone made from aggregate (gravel, broken stone, or other suitable material about 20 mm or ¾ in. in size), sand, and *portland cement*.

CONSOLIDATION Gradual compression of the clay in a foundation.

CONTINUOUS BEAM A beam continuous over its supports, as opposed to a *simply supported beam*. Bending moments are transmitted across the supports.

COUNTERCLOCKWISE The opposite of *clockwise*.

COVER In *reinforced* and *prestressed* concrete, the thickness of concrete overlying the steel bars or *tendons* nearest to the surface. An adequate cover is needed to protect the steel from rusting and from overheating in a fire.

CREEP Time-dependent deformation due to load. In concrete a sustained load squeezes water from the pores of the material. This produces deformation that may be two or three times as great as the elastic deformation. See also *elasticity*.

CREEP DEFLECTION Deflection of a beam due to *creep*. Elastic deflection occurs instantly, whereas creep deflection requires time to develop; it may be several months before it becomes noticeable.

CROWN The highest point of an arch or dome. See also *springings*.

DEAD LOAD A load that is acting at all times (for example, the weight of the structure), as opposed to the *live load*.

DIAGONAL COMPRESSION (TENSION) Compression or tension caused by a *shear force*, usually at an angle of 45° to the vertical. *Bracing* is often used to resist the diagonal tension.

DIAGONAL MEMBER Member used for *bracing*.

DOME A vault of double curvature, both curves being convex upwards. Most domes are *spherical*.

DUCTILITY The property of a material that enables it to be appreciably deformed before it breaks; the opposite of *brittleness*. Structural steel is a ductile material.

EFFECTIVE DEPTH In reinforced concrete, the distance of the center of the reinforcement from the compression face of the concrete.

ELASTIC METHOD Structural design based on the assumption that structural materials behave elastically. It requires that the stresses due to the *service loads* should be as close as possible to, but not greater than, the *maximum permissible stresses*. Except for the design of reinforced concrete sections (which is based on the *ultimate load*) most structural design is based on the elastic method.

ELASTICITY The ability of a material to deform instantly under load and recover its original shape when the load is removed. The elastic deformation is directly proportional to the load that causes it (thus twice as much load causes twice as much elastic deformation).

EXPANSION JOINT A joint made to allow the expansion of the structure due to *temperature* or *moisture movement*.

FACTOR OF SAFETY A factor that provides a margin of safety against collapse or serious structural damage.

FIRE ENDURANCE The time taken to cause the failure of a structure in a fire.

FIRE LOAD The combustible material in a building per unit of floor area.

FLAT PLATE A concrete slab reinforced in two directions at right angles to one another and supported directly on the columns. There are no enlarged column heads or supporting beams.

FLAT SLAB A concrete slab reinforced in two directions at right angles to one another and supported directly on enlarged *column heads*, without supporting beams. See also *beam-and-slab floor*.

FLEXURE Bending.

FOLDED PLATE A structure formed from flat plates, usually of reinforced concrete, joined at various angles.

GIRDER A large or primary *beam*.

GREAT CIRCLE The greatest circle that can be drawn on the surface of a sphere. The meridians of longitude and the equator are great circles on the terrestrial sphere. Any circle on the surface of a sphere smaller than a great circle is called a small circle. The parallels of latitude are small circles on the terrestrial sphere, except for the equator.

HIGH-CARBON STEEL A steel that is harder, stronger, and less ductile than structural steel, because it contains more carbon.

HINGE A joint that allows rotation, that is, transmits the *shear force* but not the *bending moment*.

HOOP FORCES Forces in a *dome* that follow horizontal circles, like the circles of latitude on a terrestrial globe.

HYPAR (HYPERBOLIC PARABOLOID) A *saddle surface* generated by moving one straight line over two other straight lines. It is doubly *ruled*.

HYPERBOLOID   A doubly *ruled* surface generated by a straight line at an angle to a vertical axis. It has the familiar "water tower" shape.

ISOSTATIC LINE   *Stress trajectory.*

JOIST   A small or secondary *beam*.

LATTICE DOME or SHELL   A dome or shell composed of straight members, usually arranged in triangles.

LEVER ARM   The distance between the resultant tensile force and the resultant compressive force at a section. These two forces and the lever arm form the *resistance moment*.

LINTEL   A short *beam* placed horizontally across a door or window to support the walls above the opening.

LIVE LOAD   A load that is not permanently acting on the structure, as opposed to a *dead load*.

LOAD BALANCING   A method used in *prestressed concrete* to counteract the effect of the *dead load*. The *tendons* are given the same curved shape as the *bending moment diagram* due to the dead load, and this produces a bending moment equal and opposite to that of the dead load.

MATRIX-DISPLACEMENT METHOD   The computer-based method most commonly used for the design of building frames.

MAXIMUM PERMISSIBLE STRESS   The greatest stress permissible in a structural member under the action of the *service loads*. It is obtained by dividing the limiting strength of the material, for example the *yield stress* of steel, by the *factor of safety*.

MECHANISM   A frame that is capable of movement. In structural mechanics it is defined as a *statically determinate* structure from which one or more members have been removed.

MEMBRANE   A thin structural member that cannot resist bending or transverse shear (at right angles to its surface).

MERIDIONAL FORCES   Forces in a *dome* that follow vertical *great circles*, like the meridians of longitude on a terrestrial globe.

MINIMAL SURFACE   The smallest surface between a given set of boundaries; it is produced experimentally by a soap film.

MODULUS OF ELASTICITY (E)   The ratio of elastic *stress* to elastic *strain*. The higher the modulus of elasticity, the less a material deforms.

MOISTURE MOVEMENT   Deformation of a structural member due to *shrinkage* or *creep* of the material.

MOMENT (M)   The moment of a force about a given point is its turning effect, measured by the product of the force and its perpendicular distance from this point.

MONITOR   A series of windows on both sides of a single-story factory roof that admit daylight and sometimes also provide ventilation. They are intended to exclude direct radiation from the sun.

MONOCHROMATIC   Of one color only.

MONOLITHIC   ("Of one piece of stone.") Used to describe concrete cast in one piece or cast with proper construction joints to ensure that the concrete behaves like one piece of stone.

NEGATIVE BENDING MOMENT   A *bending moment* that produces tensile stresses on top of a beam or slab. In a reinforced concrete building frame it normally occurs over the supports of slabs and beams, and these sections must have reinforcement at the upper face. See also *positive bending moment*.

NEUTRAL AXIS   The line in a cross section of a beam or slab where the flexural stress changes from tension to compression.

NORTH-LIGHT TRUSS   A truss used in the northern hemisphere to limit the admission of direct radiation from the sun. The glazing faces north.

PIN JOINT   A *hinge*.

PLANE FRAME   A frame whose members all lie in the same plane, as opposed to a space frame.

PLASTIC DEFORMATION   Continuous permanent deformation in metals, which occurs above a critical stress, called the *yield stress*.

PORTLAND CEMENT   The most common type of cement used for concrete. It is produced by burning together clay (or shale) and chalk (or limestone). This produces a complex mixture of calcium silicates and calcium aluminates, which set hard when mixed with water. It is then insoluble in water.

POSITIVE BENDING MOMENT   A *bending moment* that produces tensile stresses at the bottom of a beam or slab. In a reinforced concrete frame it normally occurs near mid-span, and these sections must have reinforcement at the lower face. See also *negative bending moment*.

PRATT TRUSS   A truss with regularly spaced vertical and diagonal members, as opposed to a Warren truss, which has diagonal members only.

PRECAST CONCRETE   Concrete cast in a factory or on the building site before being hoisted into position, as opposed to *site-cast concrete*.

PRESTRESSED CONCRETE   Concrete which is precompressed by *tendons* in the zone where tension occurs under load.

PRINCIPAL STRESSES   The greatest and smallest stresses at any point in a structural section. See also *stress trajectory*.

RADIUS OF GYRATION   A geometric property of a section, listed in structural section tables. It is used to determine the *slenderness ratio* of a compression member.

RAFTER   A sloping member at the top of a roof structure.

REACTION   The opposite of an action; for example, the downward pressure of a frame produces an equal reaction from its support.

REDUNDANCY   A structural member or support in excess of those required for a *statically determinate structure*.

REINFORCED CONCRETE   Concrete that contains steel reinforcement to improve its flexural resistance. If the steel is prestressed, the material is called *prestressed concrete*.

RESISTANCE ARM   *Lever arm.*

RESISTANCE MOMENT   The internal moment that resists the bending moment due to the loads. It is exactly equal to the *bending moment* but of opposite sign. See also *reaction.*

RIGID FRAME   A frame with rigid joints.

RULED SURFACE   A surface on which a set of straight lines can be drawn. A doubly ruled surface is one on which two sets of straight lines can be drawn.

SADDLE SURFACE   A surface of double curvature. If two cuts are made at right angles to one another, one produces a convex curve and one a concave curve.

SECTION MODULUS (S)   A geometric property of a section, listed in structural section tables. It is a measure of the flexural strength of the section.

SECTOR   A portion of a sphere cut off by two radii of the sphere and the arc between them. It is a part of an "orange slice."

SERVICE CORE   A vertical unit in a building that contains the vertical runs of most of the services, particularly the elevators. It is usually built of reinforced concrete, and it can then be used as a loadbearing element.

SERVICE LOAD   The actual *dead* and *live load* that the structure is designed to carry, formerly called the working load. See also *ultimate load* and *elastic method.*

SETTLEMENT   Vertical movement of the foundation of a building. It may be produced by *consolidation* of the foundation material, which can be due to a lowering of the water table or to heave of an excavation.

SHEAR FORCE   The resultant of all the vertical forces on one side of any section of a horizontal beam. It is a measure of the tendency of the load system to shear or cut through the beam.

SHEAR WALL or PANEL   A wall or panel that in its own plane resists the horizontal forces due to wind or earthquakes tending to shear or distort the wall.

SHRINKAGE   Contraction of concrete or timber due to the drying out of the materials.

SIMPLY SUPPORTED BEAM   A beam freely resting on two supports and not joined to any adjacent part of the structure, as opposed to a *built-in* or *continuous beam* or a *cantilever.*

SITE-CAST CONCRETE   Concrete cast *monolithically* in its final position in the structure, as opposed to *precast concrete.*

SKELETON FRAME   A frame that supports the walls, as opposed to a frame with loadbearing walls.

SLENDERNESS RATIO   A ratio used to assess the tendency of a compression member to *buckle*. It is the ratio of the effective length of the member to its *radius of gyration*, or sometimes to its depth.

SMALL CIRCLE   See *great circle.*

SOUTH-LIGHT TRUSS   The southern-hemisphere equivalent of a *north-light truss.*

SPACE FRAME   See *plane frame.*

SPANDREL   The part of the wall between the head of the window and the sill of the window above it. The term is frequently used as a synonym for a spandrel beam placed within the spandrel or a structural beam at the edge of a building.

SPHERICAL DOME   A *dome* whose surface forms part of a sphere. A hemispherical dome is a dome whose surface is half a sphere.

SPRINGINGS   The level at which an arch or dome springs from its supports. See also *crown.*

STATICALLY DETERMINATE STRUCTURE   A structure whose internal forces and moments can be determined by the use of statics alone.

STATICALLY INDETERMINATE STRUCTURE   A structure that has more members or restraints than can be determined by statics alone. The extra members or restraints are called *redundancies.*

STATICS   The branch of the science of mechanics that deals with forces in equilibrium.

STRAIN   Deformation per unit length.

STRESS   Internal force per unit area.

STRESS TRAJECTORY   A line indicating the direction of the *principal stresses.*

STRUT   A member in compression as opposed to a *tie.*

TEMPERATURE MOVEMENT   Deformation of structure due to changes in temperature.

TENDON   A bar, wire, strand or cable of high-tensile steel, used for *prestressing* concrete.

THERMAL INERTIA   The delay in the change of the temperature inside a building caused by the heat (or coolness) stored by the structure and the fabric of the building.

THERMAL INSULATION MATERIAL   A material that greatly reduces the passage of heat through a wall, roof, or floor.

THREE-PINNED ARCH (or PORTAL)   An arch (or portal) with two pin joints at the supports and a third pin at the crown (or the center of the beam). The third pin renders the structure *statically determinate.*

THRUST   Compressive force, particularly in a portal frame, arch, dome, or cylindrical vault.

TIE   A member in tension, as opposed to a *strut.*

TRIANGLE OF FORCES   A graphical representation of three forces in equilibrium. If the forces are not in equilibrium, they do not form a triangle.

TRIANGULATED STRUCTURE   A structure composed of members arranged in triangles. This renders it *statically determinate.*

ULTIMATE BENDING MOMENT   The bending moment at the *ultimate load.*

ULTIMATE LOAD   The greatest load a structure can actually support before it fails. To obtain the *service load*, we must divide it by a factor of safety.

VERMICULITE   Collective name for a group of hydrous silicates that exfoliate when slowly heated. Exfoliated vermiculite provides an excellent thermal insulation and is widely used to increase the fire endurance of structural steel.

VIERENDEEL TRUSS An open-frame girder without diagonals. It consists of two horizontal chords rigidly jointed to vertical members.

WARREN TRUSS See *Pratt truss*.

WIND BRACING See *bracing*.

WORKING LOAD *Service load:*

YIELD STRESS (f_y) The stress at which there is a great increase in strain without an increase in stress. In structural design it is generally assumed that steel is *elastic* up to the yield stress and then becomes *plastic*.

Notation

Some of the terms are defined in the Glossary.

A	Area of cross section
A_s	Cross-sectional area of reinforcing steel in concrete
b, B	Width
c	Camber; constant
d, D	Depth; diameter
e	Strain
E	Modulus of elasticity
f	Stress
f'	Stress at failure
f_y	Yield stress of steel
H	Height
k	Constant
L	Span
M	Moment
M'	Moment at failure
n	Number of stories
N	Membrane force in shell (also newton; see "Units of Measurement")
P	Column load; thrust
P'	Column load at failure
Q	Transverse shear force
r	Radius of gyration; rise of arch
R	Reaction; radius of curvature
R_H	Horizontal reaction
R_V	Vertical reaction

s	Sag of cable
S	Section modulus
T	Tensile force; torsion
V	Shear force in beam; membrane shear force
w	Load per unit length, or per unit area
W	Total load
x	Horizontal coordinate
y	Vertical coordinate
z	Lever arm of the resistance moment; coordinate in the third dimension in shells
δ	Deflection
ϵ	Eccentricity
θ	Angle
ϕ	Angle

Units of Measurement

In the conventional American (formerly British) units, mass and weight are numerically the same. Mass is most conveniently determined with a balance. If a person steps on a balance and registers 150 lb (pounds), then that is the person's mass.

The force exerted by the earth on the person due to the earth's gravitational attraction is expressed in pounds-force, and 1 lbf (pound-force) is the force exerted by a mass of 1 lb due to an acceleration g. The magnitude of the acceleration g (with which a body would fall freely due to the earth's gravitational attraction) varies slightly in different locations; it has an average value of 32.2 ft/sec^2. Thus the person's weight is 150 lbf. Most books on architecture use the term "pound" and the abbreviation "lb" both for pound and for pound-force.

In the old metric units, used at present in most European countries, a mass of 1 kg has a weight of 1 kgf (kilogram-force).

In the SI (Système International d'Unités) metric units, which are now in use in Australia, Canada, Great Britain, New Zealand, and South Africa and are being introduced in the United States, mass and weight are not numerically equal. The newton (N) is the unit of force, and this is defined as the force exerted by a mass of 1 kg due to an acceleration of 1 m/sec^2. The average value of g in metric units is 9.807 m/sec^2, and therefore the weight of 1 kg is 9.807 N. A person who registers a mass of 70 kg on the balance has a weight of 686.5 N.

Both American and SI units are used throughout this book. The national standards organizations were unable to agree on a single system of SI units for stress: MPa and N/mm^2 mean precisely the same thing, but sound different. Great Britain is using the second, while Australia, Canada, New Zealand, and South Africa have adopted the first. Most countries of Western Europe are converting to N/mm^2, while the United States has chosen MPa.

In the circumstances we had to make a choice. We use MPa throughout this book, and we regret any inconvenience this causes to British readers.

Abbreviations

ft	foot
in.	inch
in.^{2}	square inch
kg	kilogram
kg/m	kilogram per meter
kip	kilopound = 1000 pounds
kip ft	kilopound foot
kN	kilonewton = 1000 newtons
kN m	kilonewton meter
kN/m	kilonewton per meter
kPa	kilopascal = 1000 pascals
ksi	kilopound per square inch
lb	pound
m	meter
mm	millimeter
mm^2	square millimeter
MN	meganewton = 1 000 000 newtons
MPa	meganewton = 1 000 000 pascals (1 MPa = 1 MN/m^2 = 1 N/mm^2)
N	newton
Pa	pascal (1 Pa = 1 N/m^2)
psf	pound per square foot
psi	pound per square inch
sec	second
μm	micrometer = 0.000 001 meters = 0.001 millimeters
°C	degrees Celsius (Centigrade)
°F	degree Fahrenheit

Conversion from American to Metric Units

Length
1 ft = 0.3048 m
1 in. = 25.40 mm

Area
1 in.^{2} = 645.2 mm^2

Section modulus
1 in.^{3} = 16 390 mm^3

Mass
1 lb = 0.4536 kg

Force, Weight
1 lb = 4.448 N
1 kip = 4.448 kN

Weight per unit length
1 lb/ft = 0.014 6 kN/m

Weight per unit area
1 psf = 0.047 88 kPa = 0.047 88 kN/m^2

Moment
1 kip ft = 1.356 kN m

Stress
1 psi = 6.895 kPa = 6.895 kN/m^2
1 ksi = 6.895 MPa = 6.895 N/mm^2

Temperature
°F = 1.8 (°C) + 32

Conversion from Metric to American Units

Length
1 m = 3.280 84 ft
1 mm = 0.039 37 in.

Area
1 mm^2 = 0.001 55 in.^{2}

Section modulus
1 mm^3 = 61.0237 × 10^{-6} in.^{3}

Mass
1 kg = 2.204 62 lb

Force, Weight
1 N = 0.224 809 lb
1 kN = 0.224 809 kip

Weight per unit length
1 kN/m = 68.036 lb/ft

Weight per unit area
1 kPa = 1 kN/m^2 = 20.89 psf

Moment
1 kN m = 0.737 562 kip ft

Stress
1 MPa = 1 N/mm^2 = 0.145 038 ksi = 145.038 psi
1 kPa = 0.145 038 psi

Temperature
°C = $\frac{5}{9}$(°F − 32)

Tables

The sections in Tables 13.1 and 13.2 are identical; only the units of measurement differ.

Table 13.1. Properties of a Selection of Canadian Metric Wide-Flange Steel Sections for Use in the Problems

Nominal Depth (mm)	Mass (kg/m)	Actual Depth (mm)	Actual Width (mm)	Cross-Sectional Area, A (mm²)	Section Modulus, S (mm³)
460	89	463	192	11 400	1 770 000
	82	460	191	10 500	1 610 000
	74	457	190	9 480	1 460 000
	68	459	154	8 710	1 290 000
	60	455	153	7 610	1 120 000
	52	450	152	6 650	943 000
310	253	356	319	32 300	3 830 000
	226	348	317	28 900	3 420 000
	202	341	315	25 700	3 040 000
	179	333	313	22 800	2 680 000
	158	327	310	20 700	2 360 000
	143	323	309	18 200	2 150 000
	129	318	308	16 500	1 940 000
	117	314	307	15 000	1 750 000
	107	311	306	13 600	1 590 000
	97	308	305	12 300	1 440 000
	86	310	254	11 000	1 280 000
	79	306	254	10 100	1 160 000
	74	310	205	9 480	1 060 000
	67	306	204	8 520	949 000
	60	303	203	7 610	849 000
	52	317	167	6 650	747 000
	45	313	166	5 670	634 000
	39	310	165	4 940	549 000
	33	313	102	4 180	415 000
	28	309	102	3 610	351 000
	24	305	101	3 040	280 000
200	42	205	166	5 310	399 000
	36	201	165	4 580	342 000
	31	210	134	4 000	299 000
	27	207	133	3 390	249 000
	22	206	102	2 860	194 000
	19	203	102	2 480	163 000
150	24	160	102	3 060	168 000
	18	153	102	2 290	120 000
	14	150	100	1 730	91 500

Table 13.2. Properties of a Selection of American New Wide-Flange Steel Sections for Use in the Problems

Nominal Depth (in.)	Mass (lb/ft)	Actual Depth (in.)	Actual Width (in.)	Cross-Sectional Area, A (in.²)	Section Modulus, S (in.³)
18	60	18.24	7.555	17.6	108
	55	18.11	7.530	16.2	98.3
	50	17.99	7.495	14.7	88.9
	46	18.06	6.060	13.5	78.8
	40	17.90	6.015	11.8	68.4
	35	17.70	6.000	10.3	57.6
12	170	14.03	12.570	50.0	235
	152	13.71	12.480	44.7	209
	136	13.41	12.400	39.9	186
	120	13.12	12.320	35.3	163
	106	12.89	12.220	31.2	145
	96	12.71	12.160	28.2	131
	87	12.53	12.125	25.6	118
	79	12.38	12.080	23.2	107
	72	12.25	12.040	21.1	97.4
	65	12.12	12.000	19.1	87.9
	58	12.19	10.010	17.0	78.0
	53	12.06	9.995	15.6	70.6
	50	12.19	8.080	14.7	64.7
	45	12.06	8.045	13.2	58.1
	40	11.94	8.005	11.8	51.9
	35	12.50	6.560	10.3	45.6
	30	12.34	6.520	8.79	38.6
	26	12.22	6.490	7.65	33.4
	22	12.31	4.030	6.48	25.4
	19	12.16	4.005	5.57	21.3
	16	11.99	3.990	4.71	17.1
8	28	8.06	6.535	8.25	24.3
	24	7.93	6.495	7.08	20.9
	21	8.28	5.270	6.18	18.2
	18	8.14	5.250	5.26	15.2
	15	8.11	4.015	4.44	11.8
	13	7.99	4.000	3.84	9.91
6	16	6.28	4.030	4.74	10.2
	12	6.03	4.000	3.55	7.31
	9	5.90	3.940	2.68	5.56

Index

Aimond, F., 198
Air conditioning, 233, 234, 238
Aluminum
   beams, 78–79
   buckling failure, 76
   jointing of, 87
American ACI Code, 139
Anthemios, 186
Arches, 123–127
   expansion, 27–28
   problems on, 131–136
Archimedes, 53
Assemblies. *See* Structural assemblies

Basement walls, 172–173
Bauhaus, 233
Beam-and-slab floors, 81–82
Beams
   bending moments, 58, 59, 63–65
   maximum permissible load, 88
   problems on, 90–97
   shear, 102–103
   span, 58–59
   stress, 105, 106–107

theory, 76–79
   torsion, 104–105
Bending moment, 54
   beams, 58, 59, 63–65
   floor construction, 80–81
   portal frames, 120, 122
   prestressed concrete, 84
   spans, 57, 69
Bini, Dante, 190
Bird, Walter W., 214
Bolts and bolting, 87
Bracing members, 73
Brick
   failure of, 30
   See also masonry
Brittleness, 33–36
   climate and, 106
   concrete, 83
   cracking and, 46
   cure for, 35–36
Brunelleschi, Filippo, 185, 187
Buckling, 75–76
Building codes, 15–16
Built-in beams, 64

Butler, Samuel, 31

Cables, 65, 68
Candela, Félix, 198
Cantilevers, 25, 54, 57, 58–59, 63
Carroll, Lewis, 20
Catenary arch, 126
Chicago fire, 51, 138
Climate
   concrete casting, 79
   domestic construction, 41
   foundations, 165–166
   loads, 23
   stress concentrations, 106
CNIT Exhibition Hall (Paris, France), 14
Codes, 15–16
Colossus of Rhodes, 49
Columns
   combined footings, 169
   footings for, 167–168
   maximum permissible load, 88
   problems on, 98–100
Commercial buildings, 42

Communications, 152
Composite construction, 147
Compression members, 75–76
Compression structures, 68
Computer, 112–114, 138, 149
Concrete, 13
   See also Masonry; Prestressed concrete; Reinforced concrete
Concrete beams
   shear, 103
   torsion, 105
Concrete shell structures. *See* Domes
Conoids, 198, 207
Cooling, 232
   See also Air Conditioning
Corbusier, Le, 233
Corrosion, 79
Cor-Ten steel, 36, 38
Cracking, 46
Creep, 46–47, 79–80, 85
Curtain walls, 234
Curved structures, 65–69
Cylindrical vaults, 182

Da Vinci, Leonardo, 76
Dead loads, 21
Demolition, 85
Domes, 13, 15
   expansion, 28
   iron, 127
   masonry construction of, 183–186
   reinforced concrete, 187–191
   theory of, 182–183
   types of, 192–196
Domestic construction, 40–42
Ductility, 26, 33, 35, 106
Duomo (Florence), 185, 187
Dynamic loads, 23–26

Earthquakes, 25–26
   domestic construction, 42
   energy dissipation system, 149
   horizontal loads, 139
   roof trusses, 115
Eiffel, Gustave, 17
Elastic buckling, 75
Elastic deformation, 32–33
Elastic design/analysis, 147
Elastic method, 43
Elevator, 138, 143, 145, 152, 232, 233, 239
Empire State Building (N.Y.C.), 143, 152, 153
Energy conservation, 42, 150
Energy crisis, 234, 239
Energy dissipation system, 149
Engineering, 233
Environment, 231–240
   energy crisis, 234
   history, 232
   structure and, 239
Equilibrium conditions
   loads and, 28–30
   spans, 56
Expansion joints, 27

Fire
   arches, 125
   Chicago, 51, 138
   commercial/industrial buildings, 42
   composite construction, 147
   domestic construction, 40
   history, 49
   loads, 21, 26–27
   service core, 145, 239
   spans, 43
   steel, 36
   tall buildings, 138
Fire endurance, 27

Fire loads, 21, 26–27
Fire stairs, 27
Flat plate concrete floor, 79, 80
Flat slabs, 80
Floor structures, 73, 79, 80, 147
Folded-plate roofs, 202, 206
Footings, 167–168, 169
Foundations, 164–180
   basement walls, 172–173
   combined column footings, 169
   footings, 167–168
   long-span buildings, 175
   materials for, 165–166
   piled, 169–172
   problems for, 176–180
   raft foundations, 169
   retaining walls, 172–173
   soil pressure and, 166–167
   tall buildings, 173–175
   See also settlement
Furniture, 21

Galvanization, 36
Garisenda Tower, La (Bologna), 174
Gaudi, Antonio, 127
Geodesic domes, 192–196
Gitterkuppel domes, 192
Glass, 233
Glass walls, 234
Gothic arch, 126
Great Pyramid (Gizeh, Egypt), 12, 49
Gregory, David, 126
Gropius, Walter, 233

Habitat apartments (Montreal), 151
Hagia Sophia (Istanbul), 186
Hammurabic Code, 15, 16
Heating, 232, 233
Height, 12–13, 137–163
   elevators, 49, 51 (*See also* Elevator)
   energy dissipation system, 149
   factors limiting, 152–153
   floor structures, 147
   foundations, 173–175
   history, 49, 138
   horizontal loads, 139–141
   loadbearing walls, 149–151
   low-rise reinforced concrete frames, 139
   low-rise steel frames, 138
   problems for, 154–163
   rigid frame theory, 147–149
   structural materials, 42–43
   structures, 16, 17
   vertical spans, 69

vertical structure, 142–146
Hilberseimer, Ludwig, 233
Honnecourt, Villard de, 232
Hooke, Robert, 126
Hoop forces and tensions, 182, 184–185, 186
Horizontal loads, 139–141
Hypars, 198–199

Illinois Institute of Technology, 233
Industrial buildings, 42
Insulation
   sound, 238–239
   thermal, 237
Iron, 138
Isidoros, 186
Isostatic lines, 105

Joints and jointing
   portal frames, 118
   structural assemblies, 111
   structural members, 86–87

Lamella roofs, 192–196
Laminated timber, 36
Lanchester, F. W., 214
Lattice vaults, 192–196
Leaning Tower of Pisa, 174
Lever principle, 53
Liberty Park Towers (Pittsburgh), 149–150
Lighting, 232, 233
   heat, 238
   natural, 237–238
   roof structures, 207–208
Lintels, 76–79
Live loads, 21, 59
Loadbearing walls
   environment, 232
   tall buildings, 149–151
Loads, 21–27
   equilibrium conditions and, 28–30
Louisiana Superdome (New Orleans), 17
Low-rise reinforced concrete frames, 139
Low-rise steel frames, 138

Masonry, 13, 79
   arches, 127
   brittleness, 35
   cooling, 232
   domestic construction, 40
   energy conservation, 42
   failure of, 30
   insulation, 150
   spans, 68, 125

   vault/dome construction, 183–186
Materials. *See* Structural materials; *entries under names of materials*
Matrix-force/displacement methods, 149
Maximum bending moment, 63, 64
Maximum permissible stresses, 43
Menai Suspension Bridge (North Wales), 211
Meridional forces, 182
Modern Movement, 233
Modulus of elasticity, 33
Moisture, 27–28
Moment concept, 51–54
Moment distribution method, 147, 149
Monadnock Building (Chicago), 149, 150, 165
Monolithic casting (concrete), 79, 87, 139

Navier, L. M. H., 77
Nervi, Pier Luigi, 17

Old Colony Building (Chicago), 143
Osborne, John, 137
Otis, Elisha Graves, 138
Otto, Frei, 213

Palladio, Andrea, 114
Pantheon (Rome), 12, 14, 15, 49, 187
Parallel-chord trusses, 116–117
Partitions, 21
Photoelasticity, 107
Piled foundations, 169–172
Pin joints, 111
   parallel-chord trusses with, 116–117
   portal frames, 118–123
   trusses, 114
Planar space frames, 216–217
Planetarium (Jena, E. Germany), 14
Plane trusses, 218
Pneumatic structures, 214–215
Portal frames, 118–123, 131–136, 143
Pratt truss, 114
Precast concrete, 87
Prestressed concrete, 83–85
   height, 42
Prestressed shells, 209
Principle of superposition, 72

Raft foundations, 169
Reinforced concrete, 35–36
   cracking, 46
   domes, 187–191

domestic construction, 41
failure of, 30, 83
fire, 42
flexural strength, 79–83
height, 42–43
jointing, 87, 111
prestressed concrete, 84
problems on, 95–97
vaults of, 187–191
Reinforced concrete frames, 139
Resisting moment, 54
Retaining walls, 172–173
Rigid frames, 118–123, 147–149
Rigid joints, 111, 117
Riveting, 87
Rockefeller Center (N.Y.C.), 153
Roofs, 117
folded-plate, 202, 206
side-lit/top-lit structures, 207–208
winds, 24
Roof trusses, 114–115
Rouen Cathedral, 17
Rust, 36

Safdie, Moshe, 151
Safety, 43
See also Fire
St. Isaac's Cathedral (Leningrad), 127
St. Pancreas Station (London), 17
St. Paul's Cathedral (London), 14, 15, 126, 218
S. Marco, 174
S. Maria del Fiore (Florence), 14, 15
Saracen arch, 126
Schwarzwaldhalle (Karlsruhe, W. Germany), 14
Schwedler, J. W. A., 192
Schwedler domes, 192–196
Sears Tower (Chicago), 17
Semicircular vault, 182
Serviceability, 46–47
Service core, 143, 145
Service floors, 145
Service load, 43
Services, 239
Settlement, 27–28, 165
Shear, 35
beams, 102–103
Shear forces, 57
Shear resistance, 102
Shear walls, 117
Shells
general theory of, 196–198
prestressed, 209

Simply supported beams, 63
Skeleton frame, 138, 149
Slabs
foundations, 175
maximum permissible load, 88
problems on, 95–97
Snow, 23
Soils, 165, 166–167, 169, 172, 175
Solar energy, 234–235
Solar radiation, 23
Sound, 238–239
Space frames, 126–127
Span(s), 12–13, 48–69
beams, 58–59
bending moments, 57
curved structures/trusses, 65–69
equilibrium conditions, 56
forces composition/resolution, 55
foundations, 175
history of, 49–51
masonry, 125
moment concept, 51–54
shear forces, 57
structure, 16, 17, 43
vertical, 69
Spandrel beams, 104
Statically determinate/indeterminate structures, 112–114
Static loads, 21–23
Steel, 33, 35, 36
beams, 78
buckling failure, 76
domes, 187
failures of, 30
fire, 42
height, 42–43
jointing, 86–87
reinforced concrete, 35–36, 83, 84
spans, 43
stress concentrations, 106
vaults, 187
Steel frames, 138
Steel plate, 103
Stone. *See* Masonry
Strasbourg Cathedral (France), 12
Strength, 33
Stress analysis, 106–107
Stress and strain
beams, 77
calculations of, 72–73
concrete, 83–84
Stress concentrations, 106
STRESS program, 112
Stress trajectories, 105

Structural assemblies, 72–73, 110–136
arches, 123–127
computerized design, 112–114
parallel-chord trusses, 116–117
pinned joints, 111
portal frames, 118–123
rigid joints, 111
roof trusses, 114–115
Structural materials
brittleness in, 33–36
choice of, 36–38
commercial/industrial buildings, 42
domestic buildings, 40–42
ductility, 33
elastic deformation, 32–33
reinforced concrete, 35–36
safety, 43
serviceability, 46–47
spans, 43
table of, 38–39
tall buildings, 42–43
Structural members, 72–73
handbooks for, 100
jointings of, 86–87
Structural steel. *See* Steel
Structure
design and, 16–17
environment and, 239
insulation and, 238–239
services and, 239
Suction, 24, 25
See also Wind
Sunshading, 237–238
Superdome (New Orleans), 17
Suspension cables, 43
Suspension structures, 211–214
Sydney Opera House (Australia), 199

Tall buildings. *See* Height
Telford, Thomas, 17
Temperature, 27–28
domes, 189
See also Climate
Tensile stress, 35
Tension members, 73
Thermal inertia, 235
Thermal insulation, 237
Three-dimensional trusses, 216–217
Ties, 88, 98–100
Timber, 36
beams, 78
domestic construction, 40
jointing, 30, 86
Torroja, Eduardo, 17

Torsion, 104–105
Triangle of forces, 55
Triangulated trusses, 114
Truss(es)
forces/moments, 72–73
plane, 218
problems on, 129–131
roof, 114–115
spans, 65–69
three-dimensional, 216–217

Ultimate load, 43
Ultimate strength, 43
Utzon, Joern, 199

Van der Rohe, Ludwig Mies, 233
Vaults
lattice, 192–196
masonry construction of, 183–186
reinforced concrete, 187–191
theory of, 182–183
Ventilation, 232–233
Vertical loads
low-rise reinforced concrete frames for, 139
low-rise steel frames for, 138
Vertical spans, 69
Vertical structure, 142–146
Vierendeel truss, 117
Vitruvius, 125

Walls
energy conservation, 42
footings for, 167–168
See also Load-bearing walls
Water table, 165
Welding, 86–87
Wind
domestic construction, 40, 41
energy dissipation system, 149
load-bearing walls, 150
loads and, 23–5, 139
roof trusses, 115
vertical spans, 69
World Trade Center (N.Y.C.), 25, 152, 153
Wotton, Sir Henry, 11, 234
Wren, Sir Christopher, 126, 218
Wrought iron, 33

Zinc, 36